Tracing Our Past

A Heritage Guide to Boundary Bay

Tracing Our Past
A Heritage Guide to Boundary Bay

Anne Murray
with photographs by
David Blevins

Nature Guides B.C.
Delta

Nature Guides BC

PO Box 18170, 1215c - 56 Street
Delta, British Columbia
Canada V4L 2M4

www.natureguidesbc.com

Library and Archives Canada Cataloguing in Publication

Murray, Anne, 1952 Feb. 16-
Tracing Our Past : A Heritage Guide to Boundary Bay / Anne Murray; with photographs by David
Blevins.
Includes Bibliographical references and Index.
ISBN 978-0-9780088-2-6

1. Boundary Bay Region (B.C. and Wash.)--History.
2. Human ecology--Boundary Bay Region (B.C. and Wash.)--History.
I. Blevins, David Preston, 1967-
II. Title.
FC3845.B66N87 2008   971.1'33   C2008-901357-3

Printed by Friesens in Canada, on chlorine-free paper made with 30% post-consumer
waste; EcoAudit information available.

*Disclaimer: While every effort has been made to make the information in this guide accurate
and up to date, the author and publisher take no responsibility for problems arising as a result of
readers' heritage-viewing activities or use of this guide.*

*The author has compiled the information in this guide from many sources including those of
Indigenous people. Absolutely no intellectual property rights are claimed over any stories or cultural
information belonging to those sources, who should be cited directly.*

Cover: Designs on a rock at Lily Point are believed to have been created in the 1970s, testifying to
the ongoing fascination with this historic and spiritually-important location in Boundary Bay.
Photograph by David Blevins

Page i. View of Mud Bay ca.1965; photograph courtesy of Surrey Archives.

This page: Fishing fleet at the mouth of the Fraser River, ca.1890-1910; photograph courtesy of
Delta Museum & Archives.

Page viii-ix. A foggy dawn in Burns Bog. Photograph by David Blevins

CONTENTS

MAP OF THE BOUNDARY BAY AREA   *VI*

A SENSE OF HISTORY   1

THE DAWN OF TIME   3

THE HIDDEN STORY OF MIDDENS   23

HALLOWED GROUND   43

SAILS IN THE MIST   61

PIONEERS IN A NEW LAND   79

METAMORPHOSIS OF LANDSCAPE   99

HARVESTING RIVER & SEA   119

THE RICHES OF THE LAND   143

CONSERVATION OF BOUNDARY BAY   163

HERITAGE DESTINATIONS   185

TRACES OF THE PAST   197

NOTES & SOURCES   *198*

BIBLIOGRAPHY   *212*

PHOTO CREDITS   *222*

INDEX   *224*

MAP OF THE BOUNDARY BAY AREA

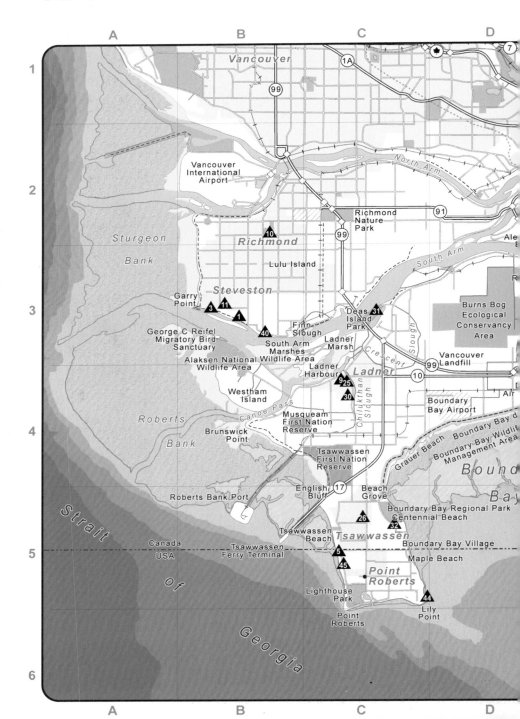

Heritage Destinations: numbers refer to locations listed on pages 186-196.

miles
0 1 2 3 4 5

kilometers
0 2 4 6 8

First Nation Reserve
Urban Areas
Parks
Marsh
Intertidal Mud Flats
Eelgrass Beds
Ocean Depth

Map Provided by
Backroad Mapbooks
... THE START OF EVERY ADVENTURE!!

For more information see Backroad
Mapbook: Vancouver, Coast & Mountains

Fraser River

stminster

Derby Reach
Regional Park

Tynehead
Regional Park
& Fish Hatchery

35

1A **Surrey**

Fort
Langley
2
13 14

99A

Surrey
Lake

Bear Creek

Sullivan

33

Cloverdale

168 street

River

15

Glover Road

Salmon

10

38

Nicomekl
Hatchery

River

e Brown
ark

Panorama
Ridge

10

12

Portage
Park 36

Langley

37

1A

Colebrook

99

Serpentine

Serpentine
Fen

Nicomekl

River

Kensington
Prairie

Brookswood

Anderson Creek

Murray Creek

Bay
rk

Mud
Bay

15

kie
t

34 **Elgin**

Chantrell Creek

**South
Surrey**

Campbell Creek

29

Sunnyside
Acres Urban
Forest

Redwood
Park

Campbell
Valley
Regional
Park

**Ocean
Park**

**White
Rock**

Hall's
Prairie

39

River

omais
Point

15
Pier 28

Semiahmoo
Park

Semiahmoo
First Nation Reserve

British Columbia

Canada

ada

8

Washington

USA

SA

Blaine Marine Park

543

Semiahmoo

Bay

42 27 41
Semiahmoo 6 **Blaine**
Spit Park

27

Drayton

Harbor

548

5

California Creek

Dakota Creek

43

N

Birch
Point

Birch Bay

Custer Prairie

Acknowledgements

A book like this could never be completed without the assistance and support of many people from diverse areas of expertise. I am very grateful for the patience and courtesy shown to me by the Coast Salish people, the staff of all the museums and archives that I have consulted, and the experts on different topics that I pestered with endless questions or asked to review drafts. Needless to say, opinions, errors and misconceptions are entirely mine.

My sincere thanks go particularly to Andrew Bak, Kathy Boissort, Bert Brink, Jack Brown, Jill Campbell and the Musqueam Language Committee, Roy Carlson, John Clague, Susan Crockford, Pauline De Haan, Michael Duncan, Hugh Ellenwood, John Fehr, Ryan Gallagher, Derek Hayes, Barb Joe, Sharon Kinley, Tanya Luszcz, R.G. Matson, Don McPhail, Kate McPherson, Don Meikle, Don Munro, Catherine Murray, Margaret North, Patricia Ormerod, the Point Roberts Historical Society, Sue Rowley, June Ryder, Dave Schaepe, Terry Slack, Barry Smith, Leona Sparrow, Hilary Stewart, Brian Thom, Nancy Turner, Don Welsh and Judy Williams. Thank you, too, to all my family and friends who have been enthusiastic about the project, and to my parents, W.J. and Margaret O'Hara, for fostering my lifelong interests in nature and history.

Special thanks go to my two editors, Eleanor Murray and Sarah Murray, without whom this book would have been far less readable and sprinkled with too many commas. David Blevins has been wonderful to work with, as always, and his photographs are a great asset to the book. I should also like to thank the folks at Backroads Mapbooks for letting us use their base map on pages vi-vii. Finally, this book could never have been written or published without the unfailing support of my husband, Len Murray.

Anne Murray

Creating the photographs for this book has been a wonderful experience. I am very grateful to Anne for asking me to help. I enjoy making photographs, but I enjoy it even more when making them has a purpose.

Some of the images in this book were quite challenging to make, and I was fortunate to have the help of some very knowledgeable and supportive people. Geoff Clayton, Brian and Rose Klinkenburg, Dave Melnychuk, Susan Leach, Eliza Olson, Doug Ransome, Russ Weisner, Lisa Zabek, and the UBC Botanical Garden each helped make some of the images possible.

I would also like to acknowledge the photographers who created the historic images reproduced in this book. Although most of their names have been forgotten, their photographs help take us back into the past in a way words can not.

Finally, I would like to thank my wife, Leandra, who accompanied me on many walks in and around Boundary Bay, helped me create some of the images, and without whose love and support this would not have been possible.

David Blevins

Boundary Bay Timeline

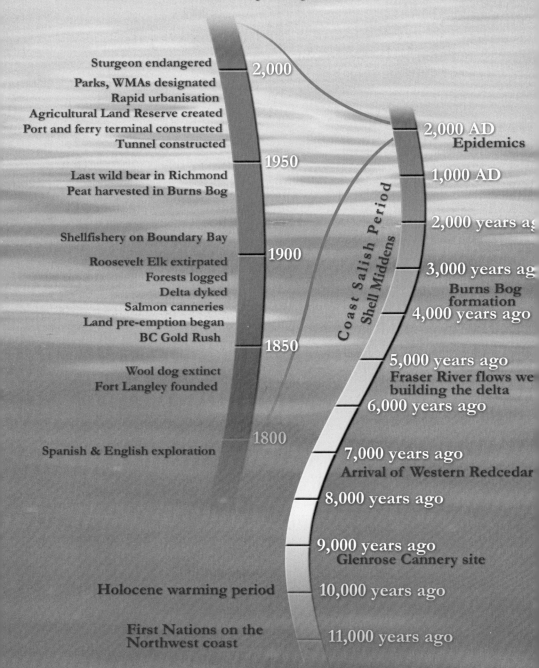

Sturgeon endangered

2,000

Parks, WMAs designated
Rapid urbanisation
Agricultural Land Reserve created
Port and ferry terminal constructed
Tunnel constructed

1950

Last wild bear in Richmond
Peat harvested in Burns Bog

Shellfishery on Boundary Bay

1900

Roosevelt Elk extirpated
Forests logged
Delta dyked
Salmon canneries
Land pre-emption began
BC Gold Rush

1850

Wool dog extinct
Fort Langley founded

1800

Spanish & English exploration

Holocene warming period

First Nations on the
Northwest coast

Pleistocene Ice Ages

Coast Salish Period
Shell Middens

2,000 AD
Epidemics

1,000 AD

2,000 years ago

3,000 years ago
Burns Bog
formation

4,000 years ago

5,000 years ago
Fraser River flows west
building the delta

6,000 years ago

7,000 years ago
Arrival of Western Redcedar

8,000 years ago

9,000 years ago
Glenrose Cannery site

10,000 years ago

11,000 years ago

A SENSE OF HISTORY

"...the dying of a generation is the vanishing of what was known."
~ Dr. Barry Leach, writer, conservationist

Tracing Our Past ~ A Heritage Guide to Boundary Bay is the story of a landscape and the people who transformed it. It is an exploration of local history from an ecological perspective: how human activities have shaped past and present landscapes, and how we are influencing our future on this coast. I look at the natural resources used by people for thousands of years and at the ecosystem changes that have occurred. This guide is intended to be a companion to my earlier book, *A Nature Guide to Boundary Bay,* which celebrated the wealth of wildlife to be found in and around Boundary Bay, from whales and eagles to sandpipers and salmon. In that volume, we saw how this exceptional area is a migration stopover and wintering area for millions of birds, as well as being rich in many other forms of life, including world-renowned Fraser River salmon.

In choosing topics for this current book, I could only skim the surface of the subject; the history of Boundary Bay is an ancient one. Many farming families are justly proud of having lived here for five generations, yet Indigenous coastal people have been here for 500 generations! Tantalising clues to their lives are found in archaeological remains, information which is supplemented by a rich oral history. European exploration came late and opened the door to an eclectic mix of settlers: their stories provide us with a glimpse of the landscape of the time. For years, agriculture and fishing were mainstays of local communities, reaping the harvest of land, river and sea. The twentieth century saw rapid settlement and land conversion, resulting in sweeping changes to the environment. In recent decades, conservation has come to the fore, as concern for ecological sustainability increasingly interests us.

I hope to provide here an enjoyable guide to this natural heritage. Signs of past activities create varied and interesting heritage destinations, such as Coast Salish traditional sites, old canneries, period houses, and historic trails. *Tracing Our Past ~ A Heritage Guide to Boundary Bay* covers the whole watershed of Boundary Bay, including both sides of the international border from Delta, Surrey, White Rock, Langley and Richmond, in British Columbia, to Point Roberts, Blaine and Birch Bay in Washington State.

The dawn redwood (Metasequoia) existed many eons before the Pleistocene Ice Ages and formation of Boundary Bay. Fossil leaves, dating back 70 - 80 million years, have been discovered in rock formations in B.C. and Washington State. It was not until 1946 that a living specimen of the dawn redwood was first discovered - in China. Dawn redwoods are unusual in being deciduous conifers, losing foliage in the fall. This tree was photographed in Redwood Park, Surrey (Map grid F4).

THE DAWN OF TIME

"We have been here since time immemorial." ~ *Chief Kim Baird,*
Tsawwassen First Nation [1]

It is difficult to imagine now, but the entire Strait of Georgia and Puget Sound were once completely buried beneath a massive layer of ice, which covered all but the tips of the highest mountains. Huge ice sheets, like those found today in Greenland and Antarctica, stretched across much of the northern continent during periodic cycles beginning about 2.6 million years ago.[2] This Pleistocene Ice Age, interspersed with non-glacial intervals, lasted until geologically recent times. Now extinct animals, such as mammoth, lived in the lower Fraser Valley during one such inter-glacial interval, about 25,000 years ago.

Following that milder period, the climate became colder. Mountain glaciers crept lower and massive ice sheets built up in British Columbia and started to move down the Georgia Basin. At its maximum extent the ice reached as far south as Olympia, Washington, until another change in climate stopped its advance. Over the next few millennia the ice retreated, melting from the hillsides and exposing vast quantities of pulverized rock, silt and sand in the outwash; the raw materials of a new landscape. As glacial meltwater flowed away to the sea, plants and animals moved in to colonize the emerging coastline, followed in time by people fishing, hunting and picking berries.

The human history of Boundary Bay is as old as the landscape. People lived at the former mouth of the Fraser River at least 9,000 years ago, when Point Roberts was an island in a much larger bay, and ocean waters lapped at the foot of Panorama Ridge. Douglas-fir and shore pines grew on sunny bluffs as forests began to invade the uplands. The early people set up camp on the beach close to where the Fraser River then flowed into the sea, and feasted on elk, black-tailed deer, salmon and starry flounder. They must have gazed out at a beautiful bay, ringed by steep-sided mountains, dotted with distant islands and full of porpoises, whales and fish. Their descendants include the Coast Salish people whose oral histories speak of living here since time immemorial, in a landscape transformed by spiritual forces of great power and presence.

---

1. References for quotes and information in this book are all in Notes and Sources, page 198 onwards.
2. See page 24 for information on calendar and radiocarbon dates; calendar dates are used in this chapter.

THE LAST ICE AGE

During the frigid conditions of the last Ice Age, the Georgia Basin was covered by the southwestern part of the huge Cordilleran ice sheet that extended from Alaska to Washington State. Ice flowed south down the Strait of Georgia and Puget Sound and westward along the Juan de Fuca Strait, and was about 2 km (1.2 mi) thick in the Boundary Bay area. Only the snow-encrusted tips of the North Shore, Olympic and Cascade mountains were exposed above the ice sheets. The scenery must have been much like parts of present day Greenland, a vast landscape of black and white, interspersed with jagged

American Border Peak was one of the many nearby peaks that remained above the ice sheets.

emergent peaks, the glaciers striated with giant crevasses and above all, the cold clear air.

The heavy weight of the Cordilleran ice pressed down on the earth's crust. This should have raised the sea level, yet so much water was locked up in the ice sheets that world-wide sea level at the glacial maximum was actually about 125 m (400 ft) lower than at present. The balance between lower sea levels and depressed land levels varied across the Pacific Northwest, because the ice sheet was thick inland yet thinner on parts of the coast. When the ice melted, sea level quickly rebounded; with time, more slowly, so did the land, resulting in a complex redistribution of submerged and emerging land forms.

The climate changed about 16,000 years ago and a rapid warming trend began, causing the ice to recede. The southern Strait of Georgia and the Lower Mainland were mostly ice free by 13,500 years ago, although glaciers still flowed down the Fraser Valley. As the ice melted, the sea rushed in, encroaching all the way to Pitt Lake, and completely covering the low-lying area of what is now the Fraser delta. The ancient shells of clams and cockles from this cold sea have been found in sediments in East Delta, Point Roberts, South Surrey and other locations near the bay. At this time, the relative sea level was 200 m (656 ft) higher than it is today. On the bare hillsides above the ocean, trees

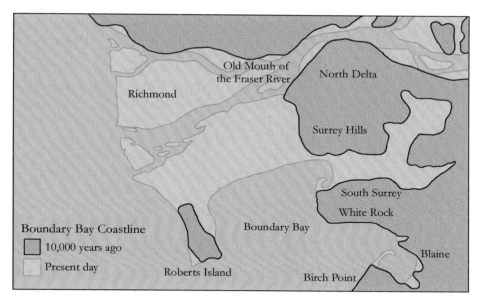

Formation of the Fraser delta: 10,000 years ago the mouth of the Fraser River lay near North Delta - based on a map by J.J. Clague et al. 1991.

and shrubs soon took root, their seeds brought on the southerly breezes. A wide variety of wildlife lived in this emerging landscape, including birds, rodents, and carnivores.

Another cold period, 15,000 years ago, caused ice to briefly advance as far as Aldergrove in the Fraser Valley. Meltwater from this glacier flowed along the upper reaches of the Nicomekl and Little Campbell rivers, the latter constructing a large delta near the present town of Langley. When this delta was formed, sea level was 40 m (131 ft) higher (relative to the land) than it is at present. The final cold phase did not last long in geological terms, and by 12,000 years ago the ice had finally gone from the valley bottoms.

In its wake was a new landscape of forests, rivers and streams, and giant boulders carved from mountains and carried many miles by the glaciers. These huge erratics can still be seen today, protruding above the soil in forests or lying on the shore in many places around Boundary Bay. The underlying bedrock rebounded once the huge weight of ice had gone, and relative sea level fell very rapidly, reaching its present level shortly after 11,000 years ago. Sea level continued to fall to about 12 m (39 ft) below current values by 8,000 years ago, and then rose to its present position about 5,000 years ago.

Meanwhile, the Fraser River was gradually establishing its course through the lowlands. At first, it

LOOKING AT ANCIENT LANDS

Signs of the Ice Ages

The sharp peaks of the Coast Mountains show that they stuck out above the ice sheets. In contrast, the rounded lower hills lay below the ice and were smoothed and polished by the movement of the glaciers.

Boulders brought down by the glaciers, called "erratics" are common around Boundary Bay and include the famous rock of White Rock beach. They can be seen on Drayton Harbor's intertidal, near the Semiahmoo Trail in Surrey, and many other places.

Beaches yesterday and today

Beaches from earlier times are found in unusual places. Redwood Park is ringed by a beautiful series of raised beaches, and there are thick sand deposits high on the hill in Joe Brown Park, Surrey.

Point Roberts uplands were an island 2,400 years ago; Beach Grove and Boundary Bay village lie on old beaches. A series of sandy spits at the north entrance to Tsawwassen helped link the island with the growing delta. Sloughs running between these spits were once used as canoe routes between the bay and the Strait of Georgia.

Modern spit formation by longshore drift can be observed on the west side of Boundary Bay, at Blackie Spit, Crescent Beach and Semiahmoo Spit, Blaine (for all locations see map page vi-vii).

probably flowed into large lakes or marine bays near Fort Langley, rapidly filling them with sediments and building a floodplain. About 10,000 years ago, the river entered the sea at New Westminster and began to extend the coast with its delta. By about 5,000 years ago, the delta reached half way to Roberts Island, eventually joining it to the mainland by about 2,400 years ago.

Vegetation sprang up quickly on the floodplain, and wildlife flocked to the wetland marshes and bogs of the newly-emerged coastline. In the still cool climate 11,000 years ago, lodgepole pines grew on drier hills and alder, shrubs and rushes invaded the soggy valley bottoms. Herds of caribou may have attracted hunters to the coast during this period. As marshland spread along the shore, migratory waterfowl descended in season, and elk roamed in to graze the meadows. Pollen grain studies show that there were no western redcedar growing at this very early time, but people were already beginning to fish salmon from the churning waters of the Fraser Canyon.

Earthquakes must have occurred periodically in this seismically active region. More recent signs of liquefaction and rapid submergence, such as liquified sand, buried tree stumps and intertidal mud layered with peat, were found in the bank of the Serpentine River, in South Surrey, dating to 1,700 and 3,500 years ago.

This old growth Douglas-fir in the UBC Malcolm Knapp Research Forest is similar to the forests that were once common on the uplands around Boundary Bay.

The weather became warmer and drier around 10,000 years ago, and Douglas-fir began to grow in great profusion. Wildfires swept through the forest, thinning out young trees and eliminating dense undergrowth, but encouraging the survival of fire-resistant forest giants, massive trees many hundreds of years old. Deer, elk, bear and wolves sheltered in the forests that lined the shore. Pollen remains of understorey spikemoss, bracken and grasses provide us with a picture of this ancient landscape.

The Fraser River was bringing a lot of silt and sand downstream and the delta was building, yet the mouth of the river still lay at the foot of the North Delta bluffs, near the present day Alex Fraser bridge, when some early people set up camp. Evidence of their presence, in the form of 9,000 year old pebble tools, was found at Glenrose Cannery, known in the Halkomelem language as *qwu-qwu-7ap-ulhp*[3] (see pages 10, 56). The same distinct style of technology occurs at coastal sites from northern Vancouver Island to Washington, and at Hatzic Rock, called *Xa:ytem,*[4] and Yale on the Fraser River.

3. This is an anglicized form of the spelling; the "7" represents a glottal stop. See notes page 198
4. Approximate pronunciation Hay-tum

Further information on the last Ice Age

After the Ice Age by E.C. Pielou
Vancouver, City on the Edge by John Clague and Bob Turner

STONE TOOLS

Stone tools were used for thousands of years by people all around the world.

Prehistoric Georgia Strait stone tools are characteristically large, round pebbles with edges chipped on one side (unifacial pebbles) or big, leaf-shaped stones with chipped edges on both sides (bifacial).

They differ markedly from the obsidian microblades of the outer Pacific Northwest coast and the fluted points of Clovis and Folsom hunting cultures which are found elsewhere across the Americas.

The exact origin of the people who made them is a mystery.

Chipped stone tools were in use for thousands of years, and had different, specialised designs. They were gradually replaced by ground stone tools and artifacts made from other materials.

Chipped basalt core from Tsawwassen site (UBC Laboratory of Archaeology).

PREHISTORY

Somehow the earliest people found this newly-formed landscape, which had emerged from under the ice sheets. It is not known where these first explorers came from or how they got here; however, genetic studies suggest that many thousands of years ago several waves of migration into North America originated from Northeast Asia.

Large scale human migrations are known to have occurred in other parts of the world. Around 40,000 years ago, for example, Australia and Indonesia were reached and settled from Asia, possibly in response to a population boom. In the same time period, there was a spectacular increase in cultural manifestations around the world. Tools, ceramics and carvings rapidly diversified and ritual burials began. When the great ice sheets advanced again over much of the northern hemisphere around 29,000 years ago, animals and humans sought refuge in ice free areas. Communities became isolated from each other for hundreds or thousands of years, and language and culture diversified.

Humans began to move up to the edge of the northern ice sheets in Europe and Russia from around 25,000 years ago. Across Siberia, the Dyukhtai people hunted woolly rhinoceros, mammoth and bison, using stone tools similar to those found in

British Columbia. It is thought that population expansions in Asia may have caused people to adapt their technology to Arctic and marine environments and move into new land, both across the Bering Strait tundra and along the shorelines of the Pacific Ocean.

The Bering Strait land bridge linked Russia with Alaska at a time when sea levels were considerably lower than at present, around 25,000 to 11,000 years ago. Bison, mammoths and musk ox roamed the emergent tundra plain and may have enticed hunters eastward; however, the earlier idea of a corridor south through the ice sheets is now generally discounted as an explanation for settlement of the continent.

There is evidence that some people moved northward within America and settled just south of the ice. Many ancient archaeological sites found in the Americas date from 12,000 to 11,000 years ago, a period corresponding to the use of sharp, fluted, stone points, known as "Clovis points." Attached to spears, these distinctive tools were used to hunt mammoth and mastodon, with deadly effect. Clovis hunters are thought to have caused or accelerated the extinction of most large mammal species across North America. However, it was not Clovis hunters who came across the Bering Strait land bridge, because no Clovis points have been found in Asia or in coastal British Columbia. The nearest Clovis site to the Strait of Georgia is 11,500 year old East Wenatchee in Washington State, about 150 km (93 mi) from Puget Sound, an area that lay south of the ice sheets.

There is growing interest in the idea of a coastal route from Siberia

One theory is that the northwest coast was settled in the far distant past by a coastal route via Siberia and Alaska. If so, the people who arrived on the coast as the landscape emerged from under its icy shroud were perhaps originally coastal dwellers from Asia and Russia.

Small nomadic groups could have spread east along the Alaska shoreline and then south down into British Columbia and Washington, taking advantage of the lower sea level of the time and ice-free coastal refuges. Boat travel would have made it easier and quicker to bring children, older people and possessions from camp to camp, while driftwood could have provided fuel for warming fires, a commodity that was lacking on the bleak, treeless tundra plains. Living off the rich resources of ocean and shore, any such migrants would have left few, if any, traces. In time, the sea would have risen and covered their tracks. Archaeological records are sparse for this period, and potential sites now underwater.

GLENROSE CANNERY
qwu-qwu-7ap-ulhp

Glenrose is one of the oldest known prehistoric camps on the south B.C. coast. The lowest layers are 9,500 old - the only known site in the Fraser delta with an Old Cordilleran Culture component (page 28). Upper layers are about 2,000 old. Archaeologist R.G. Matson reported on excavations in 1972-73. Over 600 pebble, bone and antler tools were unearthed and remains of elk, deer, beaver, dog, seal, salmon, eulachon, sturgeon, clams and bay mussels were found.

Today, Glenrose is hidden and not open to the public. Artifacts uncovered during excavations are at the University of British Columbia.

Anthropomorphic figure from St. Mungo layer at Glenrose. (UBC Laboratory of Archaeology)

It may remain a mystery how people first found the Fraser River. What is known is that people were in the fish-rich Fraser Canyon at least 10,000 years ago. The earliest camp found in the Fraser delta is the Glenrose Cannery site, lying at the foot of the North Delta escarpment (see sidebar). At the time the camp was first occupied it would have been near the mouth of the Fraser River, then well to the east of the present Strait of Georgia coastline.

Glenrose is one of only a handful of such ancient sites to have been discovered on the British Columbia coast, and no such early sites have been found along Puget Sound. Most shell middens discovered around the shores of Boundary Bay and in the San Juan and Gulf Islands cover the period from about 4,500 years ago to historic times. Local archaeological sites are described in the next chapter, page 23 onwards.

Further information on prehistory

Early Human Occupation in British Columbia edited by Roy L. Carlson and L. Dalla Bona

The Prehistory of the Northwest Coast by R.G. Matson and Gary Coupland

British Columbia Prehistory by Knut Fladmark

Bering land bridge animation: INSTAAR website at http://instaar. colorado.edu/QGISL/bering_land_bridge/

Boundary Bay is the meeting place of different Coast Salish peoples.

COAST SALISH LAND

The Indigenous people of Boundary Bay are often known as the Central Coast Salish, a collective name for a diverse people, traditional speakers of the Salishan group of languages. They are the probable descendants of the early people whose village sites and artifacts are located around the bay, showing a continuity of occupation for thousands of years.

Boundary Bay is the meeting place of two different Coast Salish peoples. To the south are the Northern Straits Salish speaking, reef-net fishers of the Strait of Georgia, living along the mainland coast, through the San Juan and Gulf Islands and in southern Vancouver Island. A second group, the Halkomelem speakers, live and fish along the Fraser River, and across its estuary to the east coast of Vancouver Island (see page 56 for more about Salishan languages, and pages 43-59 for information on Coast Salish bands and tribes).

It is difficult to learn about the long ago inhabitants who lived eons before the time encompassed by human memories. Oral histories of the Coast Salish emphasize their people's presence here since creation times, and the existence of close cultural and spiritual relationships with natural features in the landscape.

The shoreline near Ocean Park (Kwomais Point) appears today much as it did thousands of years ago.

First Ancestors & Transformers

Halkomelem tradition tells us that the Point Roberts peninsula was a particularly important place in Coast Salish history. This is "the place where the first man of this race was created" as Tsawwassen Chief Harry Joe told the McKenna McBride Commission in 1914, speaking of the high land of "scale-up" (English Bluff, *stl'elup* or *stl'alep*), that faces west across the Strait of Georgia.[5] The time when it was an island, anchored to the mainland with a cedar rope, is still remembered in the stories.

Many oral histories are known only to the Coast Salish themselves, particularly to those families who have ownership of songs and narratives. Written versions are limited in number. This knowledge is private, powerful and to be explained in the proper longhouse setting, not treated casually.

Fortunately for those of us interested in this history, a number of elders from the 1930s to 1960s shared their memories and stories with ethnologists, such as Diamond Jenness and Wayne Suttles, who then published them. Jenness spoke extensively in 1936 to Peter Pierre (known as Old Pierre), described as a Halkomelem-speaking, Katzie medicine man of high repute. Suttles conducted detailed studies of the Straits Salish in the late 1940s and early 50s, and continued to study and write about the Coast Salish people throughout his long and distinguished career. Older members of the Tsawwassen First Nation shared their memories with Arcas Consulting Archaeologists for a 1991 study, and the Stò:lo[6] people published a history and atlas of their nation that explains their origin in the valley.

In the local traditions, First Ancestors came down from the sky to the Fraser estuary. Point Roberts features in several stories. According to Chief Harry Joe, one of the First Ancestors was "*Stetson*" who gave Tsawwassen its name and also described it as "*Steilup*" or *stl'elup*, meaning "I want from now and everlasting." This is the name of the old village on the Tsawwassen First Nation land in Delta. Another First Ancestor, in the Katzie tradition, was called *Swaneset*.

The First Ancestor age was a spiritual age, where humans and animals were more closely related than now. Stories from this time tell of salmon, raven and dog people, among others. Following this age, one or more Transformers walked the earth. Transformers were mystical creators of the present world (some traditions describe them as leaders rather than creators). They had the power to change the Ancestors into stones, trees, animals or other

---

5. All sources: page 198 onward

6. Stò:lo is pronounced and formerly written as Stalo.

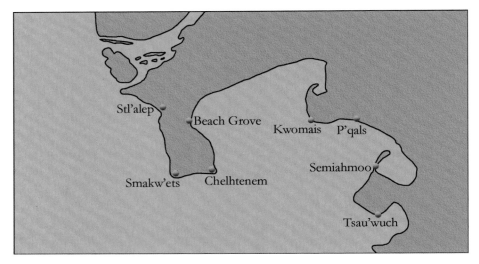

Some historic First Nations locations mentioned in the text. Spellings differ depending on the sources, as the Salishan languages were originally oral, not written.

natural objects, left as signs and instructions for the second people. Where the Transformers' work was done are spiritually potent sites, the massive rocks sacred and immovable. Even today, these sites are places of lingering power and beauty.

One such Transformer was the great and powerful *Xa:ls*.[7] *Xa:ls*, with his two brothers and one sister, arrived suddenly from the west in Boundary Bay. His first encounter took place at *Chelhtenem*,[8] (*sc'ultunum, tselhtenem*, Lily Point) in Point Roberts. Here below the cliffs, lies a rocky reef over which schools of migrating salmon are swept by strong currents around the point, up towards the mouth of the Fraser River. In the version of the story told by Old Pierre, "three brothers accompanied by 12 servants.... appeared suddenly at *Chelhtenem*... in front marched the eldest of the three brothers, a being of marvelous power named Khaals [*Xa:ls*] who could transport them wherever he wished by his mere thought. Khaals approached an Indian and his wife who were sitting on the beach … changing their bodies to stone. To the woman he said: 'You shall help the people who come hereafter. If they speak fair words to you, you shall grant them fine weather.' What he said to the man, who sank into the ground deeper than the woman, we no longer remember."

A woman later saw the stone man, but when her friend searched the same area he had sunk back into the ground.

7. Approximate pronunciation Haals, where the "h" is breathy, like the ch in the Scottish word "loch."
8. Approximate pronunciation Slethenum

Xa:ls then confronted a woman near Boundary Bay who never shared clams, but cooked them only for herself. He transformed her to stone declaring "you shall dwell among the clam beds for ever." Old Pierre described how a Tsawwassen man much later came across the woman and sold the transformed stone to a white man, but shortly after that he and his family all died. *Xa:ls* then moved on up the valley following the path of the First Ancestor, *Swaneset.*

Chelhtenem (*sc'ultunum, tselhtenem*) is also described by the Lummi, whose ancestors are buried there, as the "centre from which the underground passages radiate." The First Ancestor of the site, *Smekwats*[9] was given "power over all the underground channels that lead from *Chelhtenem* to Sechelt, Pitt Lake, Orcas and Cultus Lake." The pools leading to these power channels are located above the bluff and "people drowned at distant places would be found floating off the shore at *Chelhtenem.*" The ancestral name was also used as a place name: *Smakw'ets*[9] refers to the grassy lowland in the southwestern part of Point Roberts (Lighthouse Park area), or sometimes the Point in general.

P'qals,[10] meaning "white rock," is a legendary site lying close to the international border on Semiahmoo Bay. This huge glacial erratic gives the modern town its name. According to

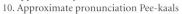

9. Approximate pronunciation Smak-wich
10. Approximate pronunciation Pee-kaals

Carvings on a rock at Lily Point testify to the lingering power and beauty of this coast.

Old Pierre, it was *Smekwats'* son who was responsible for this huge boulder, one of four rocks thrown by the young man, desperate to find his power after catching his mother sleeping with a stranger. After proving his strength, he was able to throw the guilty pair far away, the stranger to the east and his mother to the south, where she was transformed into the south wind, "her tears turned into raindrops." *Smekwats* himself was destined to become the north wind, the warrior spirit, but "his vitality went into the deep water off Point Roberts where it gave rise to the innumerable monsters that haunt that place." The waters were long a place of sanctity and purification rituals.

Old Pierre's story is quite a different acccount from the popularised one, expressed in more westernised language, of the great Cowichan "sea god" casting the stone before bringing his "princess" to Semiahmoo Bay and founding the local tribe.

Petroglyph at Kwomais Point. (Photograph by Don Welsh.)

Sqwema:yes,[11] or Kwomais as it is generally written, is another Transformer site, which lies below the high cliffs of Ocean Park. The shoreline is strewn with boulders and gravel and the water is deep and clear. Every April, plainfin midshipman, strange little fish with wide heads and suckers on their stomachs, make nests under the rocks. The males guard the eggs and young, and the abundance of fish attracts noisy groups of bald eagles, gulls and crows. The meaning of "*Sqwema:yes*" is "dog-face," an apt description of the curious little fish, and perhaps the source of the name. Kwomais Point beach was an important location for petroglyphs, or carvings on stone, some of which remain (see page 38).

Ancestral Coast Salish lived off the abundant, seasonal resources of land, river and sea. The climate 9,000 years ago was warm and dry. Douglas-fir and shore pines grew on the uplands, while wet prairie grasses and shrubs flourished on the delta. The presence of bay mussels at Glenrose Cannery site, an early instance of shellfish consumption, suggests the shoreline was rocky below the cliffs, much like it is at Ocean Park today. Coastal conditions were adequate to support people year round, although each type of food was seasonal. Eulachon, sticklebacks and starry flounder could be netted in spring; salmon and sturgeon were caught in summer.

11. Alternative spelling *k'wumayus*; approximate pronunciation Kwo-may-is

Millions of shorebirds flew in from Arctic nesting grounds; these are dunlin at the mouth of the Fraser River, near Swishwash Island.

There was plentiful game to hunt, in the forest or at sea. In late summer and fall, millions of shorebirds and waterfowl flew in from Arctic nesting grounds and descended in hordes on the sand and sedges of the delta. Geese, cranes and ducks fed in the marshes; porpoises, whales and seals swam in the fish-rich waters. Black-tailed deer and black bears were common, sharing the forests with now unfamiliar sights: marten, wolves and Roosevelt elk, which were much prized by early hunters.

There were wolverines at Beach Grove, and even the huge Californian condor soared above the coast. This impressive bird, far bigger than a bald eagle and much heavier, scavenged on sea mammal carcasses washed ashore. Perhaps it inspired stories of Thunderbird, reputed to carry off whales in its talons, with wings that made thunder, and flashing eyes that struck lightning.

Further information on Coast Salish traditional history

Coast Salish websites: pages 54 - 59

The Faith of a Coast Salish Indian by Diamond Jenness

A Stò:lo Coast Salish Historical Atlas, edited by Keith Carlson

Handbook of North American Indians of the Northwest Coast Vol. 7, edited by Wayne Suttles

Coast Salish Essays by Wayne Suttles

Western redcedar.

THE DAWN OF THE CEDAR AGE

A cataclysmic geological event occurred 7,700 years ago with the eruption of Mount Mazama, Oregon. This created Crater Lake and showered volcanic ash over a huge area, including Boundary Bay and the Fraser estuary. Remnants of the thin layer of white ash deposited by the eruption are preserved in bogs and lake floors in the area, where they help in dating sediments.

During the next thousand years, the climate changed again. It became wetter and cooler, favouring the spread of western hemlock, western redcedar and yellow cedar from the south. Seeds dispersed by wind and water grew quickly sheltered beneath conifers, and formed dense forests with grand fir, salal and ferns. By 4,000 years ago, these trees had achieved a position of dominance in the landscape and a remarkable synergy between cedar and the coastal people had developed. It was a relationship that uniquely distinguishes the Pacific Northwest coastal culture, based on resources from one remarkable tree. The versatility of the cedar was fully recognised and every part of the wood, bark, needles and pith was used, made into boats, tools, houses, clothing, artwork, storage boxes, medicine and fishing nets.

Cedar boards were split with stone wedges and adzes and fastened using cedar branches and cedar rope. Even in drier southern areas, like Boundary Bay, western redcedar was used extensively. Canoes were marvels of seaworthiness, speed and design, built from hollowed-out cedar logs and powered with cedar paddles. Salmon were caught from weirs built of western redcedar timber, and scooped from the water with dip nets woven from the tree bark. So much use obtained from a single tree species must surely be unique.

The life of the cedar forests is intimately connected with that of Pacific salmon. The life cycle of the salmon has been well studied. Emerging from eggs laid in the gravel of coastal and inland streams, the juvenile salmon head downstream to the ocean, spend several years of adult life in the ocean and then return to their natal stream for a once in a lifetime chance to reproduce. As they move upstream, their bodies greatly change in colour and form, and worn out by their long migration, they spawn and die. The carcasses litter the river and provide a huge feast for bears and mink, raccoons and eagles. These predators transport the dead salmon up the river banks and into the forest and leave droppings and fish in the undergrowth, effectively fertilizing the forest. The trees benefit from the fish and the fish survive best in rivers sheltered by the forest. Salmon found more and more places to spawn as glacial meltwaters carved valleys in which cedar and hemlock trees took root.

The First Salmon Ceremony greets the incoming migration......

The coastal tribes caught salmon where they surged up the Strait of Georgia and into the rivers, streams and tributaries. Knowledge of life cycles, migrations and seasonal abundance became of paramount importance, and correct behaviour was essential for sustaining the life-giving fish, revered as transformed humans and supernatural beings. Even today, the First Salmon Ceremony, used to greet the incoming migration and celebrate the start of the fishing season, occurs all along the coast from northern California to Alaska. Many etiquettes are observed by the Coast Salish. The Tsawwassen, for example, cook the first salmon in a special way and return their bones to the water in a sacred ritual. The Lummi traditionally address the salmon as elder brother or elder sister. Children are taught to show respect for the fish and not to treat it as a plaything.

Sockeye and chum salmon dry and store well and are rich in nutrients. The salmon fishing people prospered, particularly if they had sufficient family members to do the processing. The fish had to be gutted and sliced and hung on racks to dry in the sun and wind, for winter use. Storing food, trading it for distant goods, and exchanging it in times of celebration and in times of need were all features of an increasingly complex society. Cedar and salmon provided the staples of life in abundance.

Besides forest, river and ocean, the local people had access to yet another resource-rich habitat: a vast sphagnum bog lying in the heart of the delta. Here they hunted game and gathered berries, herbs, moss, lichens and other essential plants. It was this diversity of food sources that enabled people to thrive for many generations in the Boundary Bay area.

A LANDSCAPE IS BORN

Every year, the Fraser River brings down millions of tons of silt, sand and clay from upstream erosion of river banks, glacial deposits and land-slides, together with piles of tree trunks, branches and stumps. For 4,000 years after the glaciers retreated, the Fraser flowed south from North Delta towards Boundary Bay and the river gradually built a vast delta of sand and mud.

Around 6,000 years ago, the river diverted from its southward route and began to flow west into the Strait of Georgia. This created the new, active front on the western shore of the delta. Boundary Bay became defined by the shores of the Fraser delta floodplain, and by the deltas of five much smaller rivers flowing into the bay: the Serpentine, Nicomekl and Little Campbell Rivers, and Dakota and California Creeks, near Blaine. The bay was buttressed by the high land of South Surrey, White Rock and Birch Point to the east, and by Lily Point on the new peninsula of Point Roberts to the west. Where the tide flowed in and out of Boundary Bay, thick sands eroded from nearby cliffs were deposited over the old river mouth's mud and silt, and gradually the sea shore moved south.

Burns Bog, lying between the Fraser River and Boundary Bay, originally extended to the shores of Mud Bay, an area about 40 sq.km (16 sq.mi). The origin of the bog dates back 6,000 years, to the wetland plant communities of the low-lying delta.

All of Boundary Bay as seen from the dyke at 72 Street, Delta.

As the active front of the delta moved west, sedges and grasses replaced salt marsh. Shrubs like sweet gale and hardhack were eventually able to grow on slightly drier ground, and contributed leaves and woody debris to the soil. According to botanist Richard Hebda, these "plant communities and their peaty leavings were crucial in raising the surface of the fledgling bog."

Once the flat, poorly-drained surface of the delta was slightly elevated in this way, it was no longer regularly inundated by the river. Instead it became dependent for its moisture on rainfall, particularly the heavy winter rains. This is the typical situation for bog formation, and by about 3,500 years ago, bog plants appeared, including Labrador tea and

Bog-rosemary in Burns Bog.

sphagnum moss. Spongy sphagnum mosses draw up groundwater as they grow, and also increase the water's acidity. In Burns Bog, sphagnum moss accumulated in ever deepening layers, gradually rising in the centre to create a domed bog, as high as 5 m (16.3 ft) above the surrounding delta. The nutrient-poor, acidic bog has a specialized plant assemblage that is very uncommon on the southwestern B.C. coast.

As it approached the sea, to the west of Burns Bog, the Fraser River flowed through marshes and sand flats, that extended every year as the river brought down sediments. These wetlands, and the salt marshes of Boundary Bay, were habitat for myriad of insects, fish, and birds; the wildlife-rich landscape of the estuary and bay had been born.

For more information

Discover Burns Bog by Bill Burns

A Teachers' Guide to Burns Bog by Burns Bog Conservation Society

Midden excavations at Whalen Farm in 1949; photograph by archaeologist Carl Borden (UBC Laboratory of Archaeology).

THE HIDDEN STORY OF MIDDENS

"In the neighbourhood of Boundary Bay [middens] stretch almost continually for miles along the sound." ~ Charles Hill-Tout, 1895

When the sea level stabilized, clams, cockles and mussels soon made their home on the shore. These molluscs were collected and eaten in vast quantities by early inhabitants of the area. Huge piles of discarded shells, known as "middens," can be found along the shores of Boundary Bay, silent witness to thousands of years of occupation by family groups enjoying the fruits of the sea. The word "midden" is derived from a Danish word for domestic refuse; archaeologists use it to mean the mounds of shells and other materials that indicate the presence of prehistoric settlements.

Many clues to the lives of early people come from shell middens. Most archaeological sites in the Boundary Bay area are shell middens on ancient beaches, from about 4,500 years old[1] to relatively recent times. Rainwater percolating through the layers and layers of shells produces an alkaline solution that when combined with acid soils results in a neutral environment, ideal for preserving bones, seeds and other items. The large volume of these shell deposits and the relatively low density of artifacts is always a challenge to the archaeologist. Weathering by tides and wind, disruption from changing sea levels and discontinuous, intertwined layers make interpretation difficult. Nonetheless, shell middens are the chief source of prehistoric relics on the Pacific Northwest coast, including Boundary Bay.

Thousands of years are spanned by the midden deposits, with many of the sites showing signs of continuous occupancy and a gradual specialization and sophistication of technology. Ancient artifacts found around Boundary Bay include a 3,000 year old stone bowl with a carved human face, heavy, smooth stone mauls, chisel blades, arrow heads and beautifully crafted harpoons for hunting sea mammals. Items found in middens can be viewed in museums around the region (see page 188 onwards).

---

1. Dates are in calendar years, see page 24 for information on radiocarbon dates.

ARTIFACT SURVIVAL

Archaeological discoveries provide insight into the long human history of the area, as well as the changing landscape and ecology. Tools and implements found in middens add to our knowledge of what was eaten and how it was prepared. Animal bones can tell us about the prehistory of the bay's wildlife, the diving and dabbling ducks that migrated through, and the gulls and crows which scavenged the fishing camps, as they do today.

Particular characteristics of artifacts, together with radiocarbon dating, help decide on the cultural affinity of sites. As tools and materials show changes through time, archaeologists assign them to local and regional "cultural types" or "phases," and attempt to envision the living conditions and customs of the long ago communities. (Cultural types around Boundary Bay are listed on page 28.)

Interpretation of local archaeological sites is challenging, as everything made of wool, wood, cattail or other plants has long ago decayed. The formerly high sea levels mean that some settlement sites are now on heavily timbered or developed uplands. Bones seldom survive more than a hundred years if laid directly in soil, accessible to oxygen and rainwater. However, once shellfish became an important part of the indigenous diet, the alkalinity of the shell middens shielded some artifacts and bones from the acidic soils.

"Wet" sites are those where artifacts are completely submerged and no oxygen can reach allowing organic materials to be preserved. Baskets, twine or nets, for example, are sometimes protected by the mud of the Fraser River. Once exposed to the air, they soon decay. Such fragile materials need to be handled by experienced archaeologists if they are to survive.

RADIOCARBON DATING (C-14 years)

Radiocarbon dating is based on the fact that carbon-14, a radioactive isotope of carbon, slowly decays into stable nitrogen-14. Measurement of the remaining carbon-14 in an organic sample can be used to date certain prehistoric items, in conjunction with calibration from other known natural calendars, such as tree rings.

Radiocarbon dates are given as years BP, "before present," with 1950 taken as year zero. Older sites have calendar dates somewhat greater than the uncalibrated radiocarbon age, because of variation in the carbon-14 from the sun.

An estimate of the precision of the date is given as a $\pm$ value. In this book, most prehistoric dates are given as calendar years (calibrated radiocarbon dates), rounded to the nearest five hundred years.

ARCHAEOLOGY OF ANCIENT TIMES

Older, lower archaeological layers contain thousands of chipped stone tools, similar to those used across the world in ancient times. It is believed that local pebble tools were used for wood-working, hunting and preparing a wide variety of meat, fish and vegetable foods. Being ideally suited to these tasks and readily made from available resources, they remained in use for many generations. If they blunted, a few quick strikes could make a new blade.

Ground stone tools gradually came into use around 4,000 years ago. Later, stone mauls, adzes and wedges were increasingly used for cutting and carving cedar, the raw material for boats, paddles and housing. By 3,000 years ago, weavers had long needles, or awls, made from animal bones, just like those used in historic times. Salmon and other fish were harvested, and fishing nets from this period have been found at Beach Grove (for archaeological sites see page 30). Five hundred years later, barbed antler points were used for harpooning sea mammals, spear fishing and waterfowling. At Whalen Farm a different style of hafted toggling harpoon was found. There were dozens of different tools involved in collecting, hunting, storing, processing and preparing the wide variety of fish, shellfish, plants, birds and mammals that were

Barbed antler point from Beach Grove. (UBC Laboratory of Archaeology)

consumed and used. Woven cattail, root and bark bags held the produce.

These practical tools continued in use for centuries, with refinements being made along the way. By 2,500 years ago, copper items appeared in middens, traded in from distant locations, and decorative objects became more frequent than in earlier times. There are signs of many more people year-round and more evidence of their cultural activities, including carvings, weavings and ornaments. Upper class women wore facial decorations, such as stone or wood "labrets," ornaments that were inserted under the lip or in the chin, as well as nose rings, earrings, necklaces and bracelets.

1,500 years ago, specialised wood and bone instruments were used for harvesting plants, fishing and hunting, and far fewer chipped stone tools appear in the middens. Very thin, sharp knives, with ground slate points, were used for butchering game, gutting fish and working with plant materials. Bone whistles were played for musical enjoyment and perhaps when hunting birds.

A very curious practice of head deformation began. As babies grew, their heads were slowly flattened by binding them to a board, a custom that continued until the 1850s. The strange high slope this gave to the forehead can be seen in pictures of Coast Salish during the period of European contact. Slaves taken in battle were not permitted to bind the heads of their children. Beads, necklaces, pendants and earrings

Coast Salish Woman, Caw Wacham, 1847, sketched by Paul Kane (1810 -1871).

remained popular, just as they do with many of us today, but southern upper class women stopped wearing labrets around the time that head flattening became fashionable.

Many kinds of plants must have been employed for woven clothing, mats, baskets, twine, dyes, food and medicine. There are, however, very few plants of any kind catalogued for the Boundary Bay middens. Usually it is only charred remnants that are preserved. A rare find of skunk cabbage was made at Glenrose Cannery and rush baskets were uncovered at Beach Grove golf course. Plank house remains do not appear in the earliest archaeological sites, although many wood-working tools have been found. It is assumed, however, that long, post and beam, shed-roofed plank houses were the

Clam basket with a bird cage weave, a traditional style that spanned millennia. (White Rock Museum & Archives)

normal type of dwelling on this part of the coast by 2,400 years ago. Unlike on the northern coast, totem poles are not part of the Central Coast Salish tradition, but house posts were carved with symbolic animal and human designs.

Women were responsible for spinning wool from Salish dogs and mountain goats, using long wooden spindles and whorls, of the kind found into modern times. The round spindle whorls were often beautifully carved, a Coast Salish speciality.

The Coast Salish have serious concerns about inappropriate contact with the dead

Many midden excavations have unearthed skeletons. The Coast Salish have serious concerns about inappropriate contact with the dead, which they believe may cause harm to the living. Archaeologists are interested in burial practices and anatomical details that can give clues about social culture, diet and the health of the population. Since 1970, local excavations have generally been undertaken with Aboriginal approval, representation and involvement.

The earliest graves discovered date from 4,200 to 2,400 years ago. At this time people were buried simply, perhaps just wrapped in cedar mats, with some red ochre but with few additional grave items. It seems that most people were of equal status in the community. However, ascribed status may have arisen earlier than has been supposed: a Tsawwassen adult and teenager, for example, who died about 3,800 years ago, were found buried with 75,000 carefully crafted stone and shell beads. Perhaps these beads had been sewn onto a decorative blanket wrapped around them, but now decayed. In later centuries this type of recognition for high status individuals became more common, and some members of the community were buried with many extra objects and greater ceremonialism.

About 700 years ago, in ground interment seems to have ceased. Tree burials were used into the historic period, as described by many early European explorers. The deceased was laid to rest in a canoe or box, and placed in the branches of a tree. Middens sometimes give clues to how the person died; for example, three young women found buried at the foot of Tsawwassen bluff had been scalped. They appear to have been non-local slaves, perhaps the victims of an inter-tribal raid.

The health of the people around Boundary Bay, discerned from the condition of bones and teeth in the skeletal remains, appears to have been generally good. The diet was varied, with most protein obtained from marine sources.

ARCHAEOLOGICAL SEQUENCES IN THE LOWER FRASER

Oral history and human memories can reach back a long way, yet become lost with time. We owe most of our knowledge of ancient history to the work of archaeologists. Around Boundary Bay, recent excavations have been conducted by teams from local universities and museums, as well as by specialist consultants.

Professor Roy Carlson, in *Early Human Occupation of British Columbia*, describes three archaeological time periods for the province:

Late Period: *2000 BP to European contact time*
Middle Period: *5000 - 2000 BP*
Early Period: *Before 5000 BP*   (BP = before present, measured from 1950)

Within the Early Period, Carlson lists five cultural traditions found in British Columbia, one of which is the "Pebble Tool Tradition" typical of the Georgia Strait. In the Fraser delta/Boundary Bay area, Glenrose Cannery on the Fraser River is the only known site dating to this time period. However, many sites have been found dating to the Middle and Late Periods.

Regional Culture Type	Local Phase	Time Range
Gulf of Georgia	Stselax	600 BP to historic
	Marpole/Stselax Transition	1,000 - 600 BP
Marpole Culture Type	Marpole	2,500 - 1,000 BP
Locarno Beach Culture	Mainland Locarno Beach	3,500 - 2,500 BP
Charles Culture	St Mungo	4,500 - 3,500 BP
Old Cordilleran/Early	Glenrose	Before 4,500 or 5,000 BP

This archaeological sequence for the Strait of Georgia has regional cultural type names, and local phase names assigned for the sequence within the Fraser Delta.[1] Approximate time ranges for the local phases are given.

1. Based on a variety of sources, see page 198.

Stone head found at Beach Grove.
(Vancouver Public Library)

LOOKING AT BOUNDARY BAY MIDDENS

Charles Hill-Tout, an English settler and anthropologist, wrote about the Boundary Bay middens in 1895: "the shores of the [Fraser] estuary and of Puget Sound ... are literally covered with them. In the neighbourhood of Boundary Bay they stretch almost continually for miles along the sound. Between Ladner at the mouth of the Fraser and Point Roberts in Washington State I found them in scores, sometimes situated several miles back from the water in the midst of thick bush and timber. These latter were generally specimens of the older kind, and like those on the Lower Fraser were composed before the forest grew there, when the delta was less extensive than at present and the salt water reached farther inland."

Today, almost all the middens have been lost to suburban development,

with little to show for thousands of years of history. This section gives an overview of past archaeological work at selected local middens.

On the west side of Boundary Bay the shoreline has been building steadily, and sites that were once beside the sea are now over 500 m (1/4 mi) inland. Midden ridges on these relict beaches can be seen in parts of Tsawwassen and in Boundary Bay Regional Park, but are mostly overgrown with vegetation. Surface shells seem to be mostly clams, but cockles, mussels and barnacles occur. The shellfish beds from which they were harvested may have been carefully tended, as they were by later Coast Salish, who handed the rights to ownership and use down through families.

Crescent Beach, *k'wumayus*[2]

Crescent Beach has a large and important shell midden, lying below the cliffs near the village (see map page 30). Lower levels of the midden are 4,500 years old, while upper layers date from 700 AD to a historically recent 1800 AD. Excavations for the railway and sewer lines uncovered many human skeletons and resulted in emergency excavation of the site.

In 1989/90 excavations, thousands of fish bones and abundant shellfish were unearthed, together with a few mammal and bird bones. Bay mussels were common in the lower layers, and clams in the recent strata.

2. See notes page 200 for orthography details

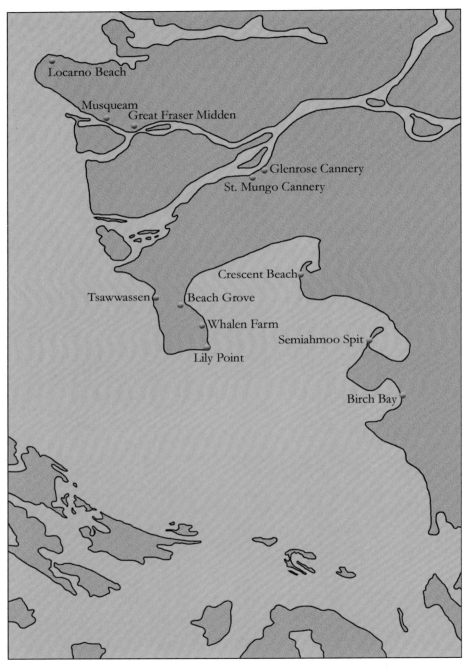

Map of selected archaeological sites and locations around Boundary Bay. See the accompanying text for Coast Salish names.

Beach Grove, *ttunuxun*

People were regularly living and fishing on Beach Grove sand spits 4,000 years ago, in a location now occupied by the South Delta Recreation Centre, a golf gourse and housing developments. The large Beach Grove midden covers two different eras; the oldest portions are about 4,000 to 3,000 years old, and the more recent deposits date between AD 40 and AD 700.

Unusual horseshoe-shaped mounds were the site of a Coast Salish village

Eleven large, horseshoe-shaped shell mounds, dated AD 300 - 600, were located in the southern portion of the Beach Grove site. They lay in a row, opening to the east along the old shoreline, now some distance from Boundary Bay. Only a few have survived urban development. The depressions, many metres across, are surrounded by thick ridges of shell middens. These unusual depressions would probably have contained shed-roofed plank houses, characteristic of the Coast Salish cultural area. To archaeologist Julie Stein, however, they appeared similar to an unusual pit house found at English Camp on the San Juan Islands. There is a small woodlot within the midden area where hollows and ridges hint at two of the original house structures.

Mallard - the wild duck of choice.

The Beach Grove site has a long history of archaeological work, often under salvage conditions. Dozens of bird bones were unearthed, showing that people relied on the huge flocks of migratory waterfowl that arrive in the estuary every fall. In fact, the name "*ttunuxun*" refers to ducks.

Most bones were from mallard, the wild duck of choice for modern hunters, with some gulls, cormorant and even great horned owl. Probably the carcasses had been scavenged by bald eagles and crows, which were also identified. The presence of owl bones is curious, since in most Pacific Northwest cultures owls were spiritually significant and never eaten. Dog bones were also found, as well as elk, deer, harbour seal and other animals. Herring, sculpin, four species of salmon and four species of flatfish must have been caught and processed at the beach. At a later salvage dig, fish nets and wooden artifacts were hurriedly uncovered.

Other Beach Grove finds include a carved stone head (page 29) and a stone sculpture of a human figure, found by Harlan Smith in 1937, and now in the Canadian Museum of Civilization in Ottawa. This figure is linked with the Transformer story on page 14-15.

Tsawwassen, *stl'elup*

Herring fishing camps, cemeteries and the village site of *stl'elup* have been uncovered at the foot of the north-facing part of English Bluff, on Tsawwassen First Nation land. The site dates back 4,200 years and was originally beside a freshwater spring, on a sandy beach beside the sea.

The old village at Tsawwassen is recorded in the Northwest Boundary Survey Report of 1857. At this time the village had seven longhouses.

Hand maul from Tsawwassen site.
(UBC Laboratory of Archaeology)

Prior to realignment of Highway 17, that leads to the B.C. Ferry terminal, an extensive site study was made by Arcas Consulting Archaeologists, 1989-1991. It was determined that at least eight separate prehistoric settlements existed there.

The first camp, belonging to the St. Mungo phase (see page 28), was used for harvesting bay mussels, showing that the shore was at that time rocky, with strong surf. It was followed by a herring fishing camp but later the site was used for a cemetery, and many people were interred there. 109 bodies were recovered during the excavations.

The next series of camps date to the Marpole phase, 1,500 to 2,000 years ago, and include clam gathering camps and more cemeteries. For hundreds of years, the site once more became a herring camp, and finally the old village was built. Evidence of house platforms, built before the recent historic period, date to about 900 years ago. The archaeologists described finding decorative objects, wedges, barbed points and awls made from deer and elk bones. A study of burial practices revealed both in-ground and above-ground inhumation, depending on the time period. Besides bay mussels and herring bones, a variety of other faunal remains were unearthed, including elk, deer, seal, ducks, salmon, flatfish, littleneck clams and butter clams. The large number of dogwinkles, a type of marine snail, is a bit of a mystery. It has been speculated that this shell may have been used as a dye.

Other middens were located further south along the bluffs: English Bluff, and Tsawwassen Beach. A traditional camp at *c'a:yum* just south of *stl'elup*, is also known to the Tsawwassen.

Whalen Farm

More than 2,000 years of continuous use were revealed during a series of excavations at the Whalen Farm site in Maple Beach, WA. The lowest level dates to about 3,500 years ago (1,500 BC). Midden ridges, stretching north-south for 800 m (0.5 mi) lie beneath what were once pastures in Michael Whalen's farm and are now mostly houses and cabins in a seaside community. The midden crosses the international border, north into Boundary Bay village, and adjoins several other even older ones to the west. Together they form the largest midden complex in the Point Roberts peninsula.

When the shells were first discarded, Point Roberts was still an island at high tide, although there may have been a sand and salt marsh causeway to the mainland at low tide. The beach site appears to have been a popular area for shellfish gathering and processing. Heart cockles, horse clams, butter clams and bay mussels were roasted and steamed with red

COAST SALISH DOGS

Coast Salish wool dogs (Drawn from forensic reconstruction by Cameron J. Pye, Susan Crockford).

The "wool" dog is known from Coast Salish oral history and from records of the first European visitors. This small dog had long hair and a curled tail, and was bred for its soft wool.

It was often kept on islands, the isolation preventing interbreeding with the larger, short-haired, "village" dogs, a necessary precaution since the woolly gene was recessive. The wool dog died out some time after European contact, around 1858.

Village dogs were used for hunting and belonged to a breed common across western North America. It was rangier, and about the size of a dalmatian. It is also now extinct.

Dog bones have been found in middens around Boundary Bay and on islands in the Strait of Georgia. Judging by the careful placement of dog skulls within graves, dogs held a special spiritual significance for the early coastal people, as well as being valued companions.

elderberries in pits, probably wrapped in the giant leaves of skunk cabbage. Over 400 artifacts were collected during a series of excavations by Carl Borden in 1949 and 1950. Thirteen human skeletons were exhumed, three of which were 2,000 years old. Another skeleton was exposed in 1955 when a bulldozer began preparing foundations. Wilson Duff, an archaeologist who was visiting the site by chance, did a hurried salvage. Further salvage excavations took place in 1972.

Faunal remains found at Whalen Farm include numerous birds, as well as dog, deer, elk, harbour seal, porpoise and marten. Archaeologist Dimity Hammon excavated in 1985 on the Canadian portion of the site, and some of the artifacts are on display in the Delta Museum. They include bone harpoon points, ground stone beads, pendants and bone awls, hard stone chisels, adzes and axes.

Net anchor or sinker stone (Tsawwassen). (UBC Laboratory of Archaeology)

The Canadian half of the site is now flattened, although extensive middens lie further west on fossil sand spits in the Tsawwassen Forest and under privately-owned fields, east of 56th Street. Mounds also remain on the American side of the site.

Semiahmoo Spit, *nuwnuwuluch* and *tsi'lich (s7iluch)*

This narrow sandy spit, which bars the entrance to Drayton Harbor, is an important site for the closely related Semiahmoo and Lummi people. These people speak the same dialect of Northern Straits Salish and have marriage and kinship ties. Two winter villages lay on the east and west sides of the spit. A burial ground is also located on the spit. This was the subject of lawsuits a few years ago, when human remains of over a hundred people were seriously disturbed during construction of a new sewage treatment plant.

The midden on Semiahmoo Spit is 4,200 years old. Traditional activities of this area included duck hunting with nets, fishing for smelt and salmon, digging horse and butter clams, and hunting Steller sea lions, bear, elk and black-tailed deer.

Although there are many other important archaeological sites in the Blaine - Semiahmoo - Birch Bay area, including *Si'ke* village, Lummi policy is to generally keep information on cultural resource properties private.

The shoreline at Lily Point is very rich in marine resources.

Point Roberts, *Chelhtenem, sc'ultunum,* **or** *tselhtenem*

This important village and reef net fishery site at Lily Point lies on the southeast corner of Point Roberts (see page 53 for more about reef netting). This stretch of shoreline is very rich in marine resources of all kinds and the location was in use well into historic times. Ever since its first recording by archaeologist Harlan Smith in 1898, the site has been a fertile source of information on the past. Archaeological remains discovered here include extensive shell middens, burial cairns, petroglyphs, anchor stones associated with the reef fishery and evidence of tree burials on the bluff. The pools above and below the bluff, as well as the beautifully coloured cliff itself and the water seep across it, are associated with spiritual powers residing at the site in the Northern Straits Salish tradition.

Birch Bay, *Tsau'wuch*

European settlers arriving in the Blaine - Birch Bay area in the 1800s were puzzled to find ruined villages along the coast, deserted following smallpox epidemics (see page 48). *Tsau'wuch*, described as "*Strav-a-wa*," a place of clams, was home to many Northern Straits Salish people, who harvested huge numbers of shellfish. Pebble tools for scraping mussels from rocks were among the artifacts found at Birch Point.

St. Mungo Cannery,
suw'q'weqsun

On the Fraser River, south of Annacis Island, two closely adjacent sites take their names from later canneries: St. Mungo and Glenrose (map page 30). Both sites are of great antiquity, but the lower layers of the Glenrose Cannery site are exceptional. This site is briefly described on page 10.

The St. Mungo Cannery site dates from about 5,300 to 2,300 years ago, and gives its name to the local phase of the Charles Culture Type (see page 28). It once lay beside tidal flats and marshes below the west edge of the North Delta escarpment. The midden was originally over 300 m long and 100 m wide (975 ft x 325 ft), but was

Chipped basalt point from St Mungo site.
(UBC Laboratory of Archaeology)

subsequently traversed by the railway and by River Road. Much of it was then destroyed when the Alex Fraser bridge was constructed in 1986.

Excavations at the site yielded hundreds of artifacts of stone, shell, and antler, and a rich variety of bird and animal bones, testifying to the varied diet of the inhabitants. Among these, it is interesting to note the remains of tundra swan and greater white-fronted geese, which are relatively uncommon birds today, as well as gulls, loons, herons and crows.

The Great Fraser Midden,
c'usna7um

The Great Fraser Midden, Eburne Mound, was found at Marpole on the North Arm of the Fraser River, and once covered 1.8 ha (4.5 ac) of Musqueam territory. Artifacts characteristic of the Locarno Beach and Marpole eras were excavated from this site (see page 28). A nearby wet site, Musqueam, has yielded fine examples of basketwork, for which the riverside people were renowned.

C'usna7um (the "7" is a glottal stop) has been built over in recent decades and like so many of the local sites, there is no longer any sign of this long established settlement. Many artifacts from the excavations are in the Vancouver Museum collection (see page 190).

ART AND CULTURE IN PREHISTORIC TIMES

The flat delta landscape around Boundary Bay has few places for lasting works of art comparable with the cave paintings found in parts of Europe. This is a land of pebbles not rock, and of wood not metal. The earliest inhabitants had to exercise their artistic and cultural flair through other means.

Red ochre is the world's oldest paint

Red ochre, the world's oldest known paint, was found at every site. It was held in large mud clam shells for easy application, and was probably used for painting wood and as body paint, as it was by later generations here on the coast and by many people around the world. An Upriver Halkomelem story tells how Swainson's Thrush, *xwut*, instructed her tribe in the preparation of red ochre paint.

Red ochre is a form of iron oxide. While the compound is non-magnetic, some tiny grains of the magnetically-sensitive mineral can be contained within it. This has led to the fascinating supposition that coating the body with red ochre not only served ritual, decorative and possibly hygienic roles, but also gave the wearer a way of sensing the earth's magnetic field. Australian Aboriginal tradition tells of feeling their way along traditional trails in this way. Red ochre was also used in many cultures, including those of the Coast Salish, to decorate the bodies of the dead, perhaps to help them find their way to a safe resting place. Five people buried at Glenrose Cannery over 4,000 years ago had been placed in graves with ochre nodules and powder, and one skeleton was smeared red over the ribs, shoulders and hands. Ochre was used at the 5,000 year old St. Mungo site to demarcate an area of the floor, perhaps for a house shrine or ritual.

Besides ochre, other materials used for body paint and tattoos could have included carbon black from soot or devil's club charcoal, mixed with bear grease, and white or grey ash. Face and body painting continued into contact times and was remarked upon frequently in the writings of Spanish and English explorers.

Numerous plant species were used for colouring and decorating cloth, baskets and mats and for dying fishing nets to make them invisible to fish. One of the most renowned dyes was bark of the red alder tree that grows commonly in wet, marshy areas and yields beautiful shades ranging from very dark brown to bright orange red. Other dyes, such as Indian paint fungus (red), oregon grape (yellow), western hemlock bark (red), grand fir bark (brown) and lichen (yellow), were easily obtained from the surrounding forest.

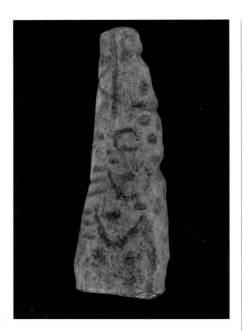

Zoomorphic incised antler haft,
Tsawwassen. (UBC Lab. of Archaeology)

A few enigmatic carvings have been excavated of animals, human faces and seated human figures, whittled from wood or antler. These might have been decorations for favourite tools or weapons, or perhaps served as totems and charms. One of the oldest carved objects, and the earliest anthropomorphic figure in British Columbia, is an antler tool handle, dated between 2,600 and 1,300 BC, found at the Glenrose site (see page 10). Such objects become increasingly common within the last three thousand years. Stone bowls in the form of seated human figures are a more frequent feature of later periods, around 500 BC, and are characteristic of the Coast Salish region.

Several petroglyphs (stone carvings), have been found at Kwomais Point, near Ocean Park, Surrey (page 16). One stone now stands in Heron Park, Crescent Beach (page 192), and one is in the Surrey Museum, Cloverdale. Some are still *in situ*, but are very difficult to see. The White Rock Museum has casts from other petroglyphs in Semiahmoo Bay. (For museums see pages 185 onwards.)

Crafts take up hours of time when all the raw materials must be laboriously collected and prepared prior to making the required clothes, baskets or housing. No doubt some families and individuals were more talented than others at particular activities, handing down their knowledge from grandparent to grandchild, with subtle shapes and colours distinguishing the work. There was scope for individuality in the design of baskets and blankets, which took months of work to complete, and in the way food was prepared and served, the clothes and ornaments worn, and the painting of bodies and faces. The craft tradition is still carried on by some families, and Coast Salish baskets, for example, command high prices.

Singing has always been used around the world to pass on oral histories, since it is one of the easiest ways to memorise words. Tribal songs were sung from generation to generation, carrying the stories through time. Judging by the importance of song, dance, oratory and story-telling to the people of the Pacific Northwest, these would have also been a vital part of the Coast Salish ancestors' culture, integrated with celebrations and with spiritual, healing and ceremonial occasions. The Spanish explorers in the Strait of Georgia recorded how quickly a song they sang was picked up and repeated. This was a trait also noted by the Russians at Sitka, Alaska in 1860. When a competitive singing match ended in ignominy, the Russians were witness to an outbreak of violence among the contestants!

Winter dances were spiritual occasions

The Coast Salish winter dances were spiritual occasions, in which dancers were moved by their guardian spirits, most often, but not always, in the shape of animals. Such traditional activities continued long into the historical period and gradually adapted to the modern era.

The Coast Salish have always been great story-tellers. Anthropologists, listening to older band members relating their memories, described how mere words could not do justice to the power of the story. The written language always seems quite inadequate, even baffling, when it comes to reading traditional legends. Dramatic costumes, vivid atmosphere and powerful oratory are all necessary components for the full blossoming of this art form, which must have riveted audiences in the longhouses on dark winter nights.

Just how much of the remembered historical cultures of the Coast Salish people existed in these very ancient prehistoric times is not known, since few societies remain unchanged for millennia. The few clues we have look back on a vibrant world of art and culture, now dissolved into the landscape. A life beside the sea, with wood and shell and antler, has disappeared beneath us, leaving only the faint echoes of early stories.

Culturally modified western redcedar: the ancient skill of cedar bark stripping is still known to some elders today.

CARE OF ANCIENT PLACES

At least 39 archaeological sites have been discovered near Boundary Bay. Each site has been given a Borden number by the B.C. Heritage Branch, or a USA registry number. Some middens are on private or reserve lands and are not accessible to the public. Others are in parks or are visible from nearby roads. Visiting these locations, and recognizing the extraordinary length of history that they represent, can give one a greater appreciation for the Aboriginal inheritance and a deeper understanding of our local landscape.

A proper respect must be shown for the antiquity of the middens

Naturally, a proper respect must be shown for the antiquity of the middens, most of which reveal little on the surface. Artifacts are in museums or university collections and nothing is left to see. In some cases, the vegetation is overgrown and only unusual mounds, or layers of white shell glistening through the black soil, betray the presence of thousands of years of history. Elsewhere, developments have rolled right over sites, building new homes where hundreds of years ago people also sang and laughed, ate and slept.

Wondering about prehistoric artifacts on your land?

In British Columbia contact:

B.C. Heritage Branch
250 952 5021
www.tsa.gov.bc.ca/heritage/

In Washington State contact:
Department of Archaeology and Historic Preservation
360 586 3065
www.dahp.wa.gov

It is **illegal** to collect archaeological artifacts or disturb middens anywhere in the Province or State without a permit.

It is a great loss that so many ancient sites were destroyed. Even recently, archaeological work was often done in a rush as bulldozers stood by and developers hovered, waiting to build. In 1973, the B.C. Heritage Conservation Act was created to conserve archaeological sites and artifacts as heritage resources. In Washington State, the Archaeological Resource Protection Act serves the same purpose. Archaeological sites must not be disturbed. If artifacts are discovered, the applicable authorities must be contacted (see inset). Local First Nations/Tribes should be approached for advice on cultural matters. For example, the Stò:lo have chosen to share information on

Ground slate biface from Crescent Beach.
(UBC Laboratory of Archaeology)

An understanding of past cultures cannot necessarily be derived either from the historical record or from archaeological traces, yet excavations have revealed many exciting clues to supplement oral histories.

As yet, no coordinated collection exists to reflect this fascinating history of the Boundary Bay area. Artifacts are scattered in a number of locations, both public and private. It would be a wonderful addition to the region to have a local First Nations' Centre that could tell the full 9,000 year story.

culturally-important places through their historical atlas. In contrast, Lummi Nation and Nooksack Indian Tribe policies may restrict the sharing of information on cultural resources, so as to protect the integrity of sacred, private or otherwise special locations and knowledge.

Human burial sites need very special care. These ancestors are family members of living Coast Salish, and naturally their exhumation can provoke emotional responses. Often, Aboriginal people will prefer to re-inter the bodies rather than subject them to scientific investigation. This can lead to misunderstandings and sad situations, such as at the Semiahmoo Spit site in 1999, where dozens of human remains were insensitively and roughly disturbed during construction work.

Further information on Coast Salish history and culture

First Nation websites: pages 54 -59

The Prehistory of the Northwest Coast by R.G. Matson and Gary Coupland

Stò:lo Coast Salish Historical Atlas, edited by Keith Carlson.

Coast Salish Essays by W. Suttles

Plant Technology of First Peoples in British Columbia by Nancy J. Turner

A Field Guide to WA State Archaeology by M. Leland Stilson

Cedar by Hilary Stewart

Stone, bone, antler and shell: Artifacts of the Northwest Coast by Hilary Stewart

Indian Art Traditions of the NW Coast, edited by Roy Carlson

Museum of Anthropology, UBC (see museums, page 188)

The Boundary Bay coast south of Ocean Park.

HALLOWED GROUND

"The smoke from their morning fires covered the country."
~ Old Pierre, Katzie Elder, 1936

Boundary Bay is the meeting place for several Indigenous peoples, speakers of the Salishan languages who are collectively known as Coast Salish, a name of convenience that can obscure the diversity of its members. Known in British Columbia as the First Nations and in Washington as American Indian Tribes, the original inhabitants of the coast have a long and interesting history.

The Semiahmoo, Saanich, Lummi, Samish and Songhees are all traditional speakers of the Northern Straits Salish language (see page 54). They are salt water people, living around salmon routes up the Strait of Georgia. They characteristically used reef nets to catch fat, inward-migrating salmon. The Straits Salish speakers' territory lies mostly on the south and east side of Boundary Bay, through the Gulf and San Juan Islands, down to Deception Pass and southern Vancouver Island.

Tsawwassen, Musqueam, Stò:lo and Cowichan are among traditional speakers of the Halkomelem group of languages (page 57). Their homeland lies along the lower Fraser River, around its mouth at Roberts and Sturgeon Banks, and over the water in eastern Vancouver Island and parts of the Gulf Islands. The Katzie, another Halkomelem-speaking First Nation, define their traditional territory as the entire Pitt River watershed south across Boundary Bay. For many generations, Halkomelem families built and owned specialised fish weirs and traps on tributaries of the Fraser River and in the Fraser Canyon. They fished for salmon using nets, spears and rakes.

Another tribe, the Nooksack, had traditional lands high up in the Little Campbell River - Drayton Harbor watersheds and along the Nooksack River, and spoke yet another distinct language. They had different traditions and relied more on game meat than the coastal people (page 59).

Only the Coast Salish themselves can truly speak for their history, so this chapter is based entirely on published literature and conversations. To learn more, visit with local members or consult their websites.

TRADITIONAL LIFE

The tradition of the Coast Salish is to live in family groups and small settlements, and this seems to have been the practice for many generations. Up to historic times, the typical living accommodations in winter were shed-roofed cedar longhouses close to the sea. Summer time meant camping trips to take advantage of seasonal fish, game and berries.

The salt water people had elegant, high-prowed cedar canoes and were at home on the water. Seasonal movements, trading forays and marriage alliances were all aided by the ability to move along the two great waterways of the region: the Georgia Strait and the Fraser River. Extended family and kin groups met together at fishing grounds and berry fields, where traders from other locations could also mingle.

Geographic isolation did not prevent resources making their way to coastal communities, whether it was goat wool from the mountains, sharp blades of obsidian from Oregon, or nephrite from Sumas Mountain. Nephrite, a type of jade, was used for sharp knives and adzes. California mussels, larger than the local bay mussels, were obtained from the Nuu-Chah-Nulth people on the west coast of Vancouver Island. From Boundary Bay, many items were sent inland, including tule mats and baskets, dried

salmon, shellfish, and seashell rattles and ornaments. It is not known how exactly this trade was carried on in earlier centuries; it could have been a series of exchanges from village to village or a more organized trading trip. Some clans may have been more itinerant than others, and some goods were probably taken during raids. In later years, traders from the inland tribes came down the valleys during certain seasons.

Copper has been found in shell middens dating to at least 500 BC

Copper was an important item of trade from very early times: it has been found in shell middens dating to at least 500 BC, and was therefore in use long before European contact. It was perhaps traded from the Copper River in Alaska or the east coast of Vancouver Island. Much later, it was the subject of a lucrative exchange with Europeans. Iron was also valuable. Spearheads may have been traded from Russia via northern tribes, or recycled from Asian shipwrecks.

Other relatively recent sources of metals are more mysterious. A Point Roberts' boy told Jose Narvaez and his crew in 1791 that "there is some flat country through which many people come to trade for fish and stay for two moons, bringing iron, copper and blue beads, wearing distinct dress and having different bows and arrows." They also had horses, which the boy identified from a picture. This story suggests that contact with far interior tribes was a regular event at this time. (Horses were used on the Great Plains after about 1730.) The local people also described much larger ships sailing up the Strait of Georgia, from which they had obtained fine quality brass bracelets. Joseph Whidbey, in Puget Sound the following year, described meeting "another band of migrant Indians, who carried all their possessions on their backs, including many trade items of European origin."

The average coastal family, however, may have lived quite isolated from other groups, judging by the diversity of dress, decoration and facial characteristics encountered by early European explorers. In later historic times, a series of inter-tribal raids, wars and slavery took place. Northern tribes came on forays to the Fraser delta in the early nineteenth century, using guns they had bought from the fur traders.

By no means only hunter gatherers, the Coast Salish actively maintained certain crops, digging and nurturing wapato marshes and camas prairies, clearing clam beds of stones and broken shells, and setting regular fires to limit forest growth and foster berry plants (see page 116). Ownership of resource areas was passed down through families who cared for them

COAST SALISH CROPS

Wapato leaves.

Wapato, or swamp potato, is a wetland plant with large heart-shaped leaves and edible tubers. It is abundant in freshwater marshes at locations such as Ladner Marsh and the Pitt Valley, north of the Fraser.

Blue camas is an attractive flowering plant, with an edible bulb. It grows on warm, sunny bluffs and islands, in the coastal Douglas-fir - arbutus (madrone) ecosystem.

and controlled their use. Most of all, then as now, the Coast Salish relied on fishing, watching and waiting for the coming of the first salmon and harvesting enormous quantities of this remarkable fish, which was consumed fresh or dried for use through the winter.

Blooming of the bigleaf maple signalled the time to strip cedar bark.

COAST SALISH SEASONS

Boundary Bay has abundant wild food but its occurrence is strongly seasonal. The Coast Salish adapted their lives to this natural cycle and each month was associated with specific activities.

In spring, fresh green shoots of salmonberry and nettle provided welcome salads. The blooming of the bigleaf maple in April signalled the time to strip cedar bark for processing. Oily little eulachon fish returned from the ocean to spawn in late April or May and were fished with wooden rakes. Hooker's willow bark was ripe for peeling in May and June, when the pith could be removed and twisted into ropes. Camas bulbs were harvested in May. Trips were made to the prairie patches scattered among the upland forests or out to the Gulf Islands. Clams were dug in spring and sockeye salmon arrived in the river during July and August, both resources requiring intensive activity, drying them for winter use, trade and potlatch sharing.

Mosquitoes were present in dense clouds in the wetlands, and often the only place for relief in the summer was out on the water or where sea breezes graced the shores. As the summer faded, native cranberries and blueberries ripened, followed by wapato tubers. In the fall, traders arrived in the valley and families gathered together in longhouses where religious ceremonies, winter dances, potlatches, oratory, story-telling and craft work took place. Women caught up with their weaving and basketry, using bulrushes, cattails and cedar that they had gathered and dried for the purpose. They made mats, wall hangings, leak-proof roofs and baskets of many kinds, tightly woven waterproof ones for cooking, and more open, cedar bark ones for fruit picking. Dog wool and mountain goat hair were woven into blankets and cloaks.

Late fall was the time for deer and elk hunting and the arrival of hundreds of thousands of ducks and geese at the start of the waterfowl season. In midwinter, sturgeon were harpooned and bears hunted in their dens.

For generations, the Coast Salish people lived within their landscape and used its resources to the full. They possessed detailed knowledge of plants, trees, fish, fowl and animals. The ethnographer Wayne Suttles interviewed older people about their memories, learning the words and uses for different plants and the way in which practical information, appropriate etiquette, moral principles, genealogy and family stories were passed down through the generations. He learned that families who were in possession of this "advice" were considered upper class and those who no longer remembered their history and family traditions became lower class. Coast Salish priorities today continue to be focused around family, community, longhouse culture and a sense of place, despite over a hundred and fifty years of very fundamental changes.

This, then, is the bare bones of what we know about Coast Salish life before the modern era. The individual stories, feats of bravery, invention or daring, joys and sadness, beliefs and understanding are held only through the memory of their descendants, the First Nations. For thousands of people, those memories were obliterated in the late eighteenth century, when the first of the Eurasian plagues descended on the Pacific Northwest.

COAST SALISH SEASONS

Spring: fiddlehead.

Summer: salmon.

Fall: blueberry.

Winter: bear.

SHADOW ON THE LAND

Smallpox is a horrific disease. It starts with a raging fever, followed by a rash all over the body that rapidly transforms to oozing blisters. These lead to intense skin and tissue damage and the pain and itching are unbearable. *Variola major*, the most serious form, causes mortality in 30% to 100% of cases. A less severe form, *Variola minor*, with a one percent fatality rate, can last for a month. Survivors are immune from future attacks, but their bodies and faces are deeply scarred. Some become blind from loss of eyes to the disease, and many suffer depression, deep grief and sometimes mental breakdown.

DISEASE CONTROL

A form of inoculation against smallpox, called variolation, was practised in China as early as the tenth century. In European towns, gradually improving sanitation and nutrition in the early 1800s controlled many infectious diseases, but smallpox still killed tens of thousands.

Not until the widespread distribution of Edward Jenner's 1798 vaccine, the first in the world, was smallpox finally conquered in Europe.

In North America, the worst effects of imported diseases occurred between 1800 and 1900, when up to 90% of the Indigenous population are thought to have died.

The disease is highly infectious, even up to three weeks after death. Items that have come in contact with the sufferer can remain infectious for over a year, because the microscopically tiny virus is extremely stable in the environment.

The last case of smallpox occurred in 1978. The mild form was once a childhood disease in Europe, China and India. In North America, where smallpox had been unknown before European contact and people had no natural immunity, its arrival caused devastation. The worst form of the disease hit the Pacific Northwest coast from an unknown source and with disastrous consequences. As historian Robert Boyd describes, the epidemic arrived in the Strait of Georgia and Puget Sound sometime in the mid 1770s, ten to twenty years before Europeans are known to have sailed the Juan de Fuca Strait.

Regardless, the disease must have had an Old World origin and been spread by travelers or trade goods. European and Russian fur traders were sailing the northeast Pacific and western Vancouver Island and may have come into the Strait of Georgia. A Russian sea otter hunter, on a ship bound for the Aleutians, unwittingly transmitted smallpox to the Kamchatka Peninsula, eastern Siberia, starting an outbreak in 1767 in which over 5,000 people died. Perhaps the disease moved east along the islands and then down the coast. Alternatively, the virus could have travelled with trade goods overland from the south or east, where communities had already been exposed. One way or another, it was a dread plague that came to the coast.

Entire families died of the disease

Smallpox reached virgin soil in the Strait of Georgia. Villages around Boundary Bay suffered immediately, and many people died. Survivors were taken in by neighbours, and communities relocated in response to the situation. Another wave of epidemics hit the area in 1782. In just two months during the late summer, it is believed that two thirds of the Stò:lo people died. People travelling away from their villages would return to find their entire families lying dead of the disease. A story circulated of finding a single baby at its mother's breast, the lone survivor of an annihalated community. Several Lummi villages were also nearly wiped out by smallpox and measles, forcing relocations from the San Juan Islands to the mainland.

George Vancouver and his crew, sailing into Boundary Bay in 1792, noted the aftermath of the epidemic. Smallpox was "very fatal among them" wrote Vancouver, "its indelible marks were seen on many and several had lost the sight of one eye." This observation was echoed by Lt. Peter Puget in his journal: "the smallpox most have had and most terribly pitted they are; indeed many have lost their eyes."

The expedition botanist, Archibald Menzies, recorded tree burial sites and deserted settlements along the coast, and thought there were far too few people for such fine territory. On the south side of Birch Bay, he saw "the site of a very large village, now grown over with a thick crop of nettles and bushes….. in one place in the verge of the wood, I saw an old canoe suspended five or six feet from the ground between two trees and containing some decayed human bones, wrapped up in mats and carefully covered over with boards."

Halkomelem stories relate what happened; Joe Splockton described the pestilence as *stalacom,* a word indicating a supernatural, powerfully negative quality, that he encountered on a canoe trip up the Fraser. "The people became frightened and began to shoot." Afterwards it wasn't long before all the people died "and Tsawwassen, where many people had lived, was almost deserted for many years." The pestilence in this story might be smallpox or another fatally infectious disease of the time. Oral history, and the archaeological record at *Stl'elup,* the old Tsawwassen village, suggest that it was temporarily deserted in the early 1800s.

Infectious diseases took a steady toll on the population throughout the nineteenth century. By now, the deadly nature of smallpox was known, and villages, bodies and belongings were cleansed in all-consuming fires. Nettles grew quickly on the burnt lands, hiding the land of the lost settlements. Often only old people, the survivors of earlier epidemics, were left to mourn their children and grandchildren. The population plummeted. It is thought that the vast majority of the original Aboriginal population of the Pacific Northwest died out as a result of Old World diseases, such as smallpox, measles, influenza and scarlet fever, to which people had no immunity. The spread of syphilis caused sterility, reducing birth rates and further contributing to rapid depopulation.

The exact dates of the different epidemics is the subject of research. Botanist Dr John Scouler noted in his August 1825 journal that one of a group of natives with Cowichan Chief Chapea, at Point Roberts, was deeply marked by smallpox, yet he does not recount any local outbreak. The 1836-37 epidemic passed by the Lower Fraser, although it reached every Coast Salish village along Puget Sound. Measles, raging up the coast, arrived at Fort Langley on the Fraser River in April 1848, perhaps aboard the Hudson Bay Company ship, the *Beaver*, or via a trading canoe. Everyone was affected, as at that time

Rocky Mountain juniper.

there was neither cure nor vaccine. Several hundred Coast Salish died and many settlers' children perished as well. In midwinter 1852-53 yet another epidemic affected the south coast, rapidly spreading to settlements on the Skagit and Nooksack Rivers and up into Boundary Bay and the Fraser estuary.

Many Semiahmoo died: "When smallpox came to Birch Bay, they buried people in one house after another. Except for those who were away at the time, the only people who were saved were one family who used the wood of a little tree called *pcinelp* for firewood. That killed the germs." "*Pcinelp*" or *put'thune7ilhp* is the strong-smelling Rocky Mountain juniper, *Juniperus scopulorum*, which grows on dry, rocky shores. It was widely used by First Nations as a house disinfectant and fumigant, and the Saanich hung branches of it around house walls to ward off disease.

Tens of thousands died in the 1862 epidemic that hit the northern B.C.

communities of Haida, Nisga'a and Gitsan. When the disease reached the Fraser estuary, a major vaccination program helped control it locally, but nonetheless, many people died. That year a huge salmon run came up the foggy Strait of Georgia to a people and countryside in mourning.

With so many people having lost family members through a century of disease, the traditional Coast Salish culture was nearly lost. Family history, skills, ecological and medical knowledge, all the social "advice" that had distinguished the upper classes, used to be passed on by word of mouth. This was impossible where families and communities had died or disintegrated. The impact of the pandemics on the First Nations, coinciding with a period of rapid settlement by outsiders and technological changes, lasted for many long years. It is only in recent times that the effects are finally being overcome, with increasing birth rates and revitalization of culture and community.

Further information on the time of the epidemics

The Coming of the Spirit of Pestilence by Robert Boyd

Social Power and Cultural Change in Pre-colonial B.C. by Cole Harris

A Stò:lo Coast Salish Historical Atlas, edited by Keith Carlson.

EXPLORING COAST SALISH HERITAGE

The Boundary Bay area is home to many different First Nations bands and American Tribes, belonging to the larger cultural group known as the Coast Salish. This section gives an introduction to the Indigenous people of the Boundary Bay area and some of the special places associated with their history.

Family Networks

Fishing, hunting and gathering at central locations and travelling with the seasons meant that neighbouring bands or tribes met and intermarried. Building extensive marital networks diversified the family's access to resources. Marriage was therefore a strategic approach to territory enlargement, and society was bilateral, with both male and female lineages holding the rights to certain resource areas.

Family networks stretched beyond language and dialect groups. Semiahmoo people, for example, have family connections with such neighbours as the Saanich, Nicomen, Lummi, Matsqui, and Nooksack. Even in the early days, some people welcomed marriage with Europeans.

A renowned Saanich hunter, known as Chanique,[1] lived with, and later became, Semiahmoo. Chanique's

---

1. pronounced Cha-hay-nack

Canoes at Goodfellow's Camp, north of Lily Point on the shores of Boundary Bay; APA cannery in background (Delta Museum & Archives)

daughter, Cecelia, married C.B.R. Kennerly, the head surgeon and naturalist on the 1858 Boundary Commission, as it was thought by the family to be a good move to ally with the newcomers. Sadly, she was soon abandoned by her husband as he headed east with the survey team, leaving her alone to bring up their son, George Kinley. She later moved onto the Lummi Reservation, marrying a local man.

Point Roberts peninsula ~ a shared-use fishing hot spot

For many thousands of years, Point Roberts has been a favoured salmon fishing location, washed by the salty water of Boundary Bay on one side and the Fraser River plume on the other. Historically, the peninsula was

also a source of halibut, sturgeon, shellfish, seal, porpoises and water birds from Boundary Bay, and deer, elk and hares from the woodlands. Tsawwassen and other Halkomelem bands utilised the southern and western shores, particularly, for such activities as fishing and clam digging. The Tsawwassen village, *Stl'elup*, was on the Canadian end of the peninsula, facing the Strait (see page 32). The wealth of resources attracted potlatching, feasting and trading.

In the summer, from July through October, the Semiahmoo, Saanich, Lummi, and others gathered for reef net fishing at Lily Point, on the southeast tip of Point Roberts. The largest sockeye run up the Strait of Georgia sweeps across Boundary

REEF NET FISHING

Reef nets, like those used by Lummi and Saanich at Point Roberts, were made from hand-twisted rope and strung between canoes, being kept vertical by cedar wood floats and stone anchors. When the salmon shoals headed on migration towards the river mouth, they were swept over the reefs and into the strategically placed nets.

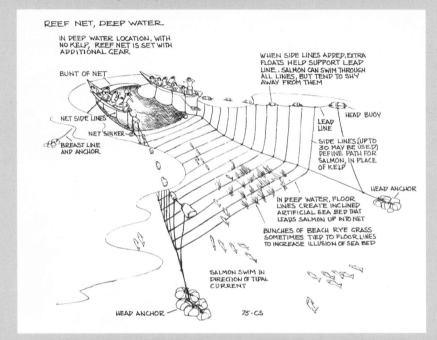

REEF NET, DEEP WATER

IN DEEP WATER LOCATION, WITH NO KELP, REEF NET IS SET WITH ADDITIONAL GEAR

BUNT OF NET

NET SIDE LINES

NET SINKER

BREAST LINE AND ANCHOR

WHEN SIDE LINES ADDED, EXTRA FLOATS HELP SUPPORT LEAD LINE. SALMON CAN SWIM THROUGH ALL LINES, BUT TEND TO SHY AWAY FROM THEM

HEAD BUOY

LEAD LINE

SIDE LINES (UP TO 30 MAY BE USED) DEFINE PATH FOR SALMON, IN PLACE OF KELP

HEAD ANCHOR

IN DEEP WATER, FLOOR LINES CREATE INCLINED ARTIFICIAL SEA BED THAT LEADS SALMON UP INTO NET

BUNCHES OF BEACH RYE GRASS SOMETIMES TIED TO FLOOR LINES TO INCREASE ILLUSION OF SEA BED

SALMON SWIM IN DIRECTION OF TIDAL CURRENT

HEAD ANCHOR   75·CS

Reef net, deep water: drawing by Hilary Stewart, from "Indian Fishing."

Bay and past Lily Point, as it makes for the Fraser River. The tall cliffs overlooking the bay were the site of the annual First Salmon Ceremony, when a ritualist looked for signs of a good season, such as a wet streak appearing across the north bluff. *Chelhtenem (sc'ultunum, tselhtenem)* was also a seasonal village: the name means "to hang salmon for drying."

The beautiful shoreline, meadow, sandy 60 m (200 ft) bluff and forested slopes combine to create an idyllic landscape. Juan Pantoja, in 1791, describes an "incredible quantity of rich salmon and numerous Indians" at the *Isla de Zepeda* (the Spanish explorers' name for Point Roberts). A year later, Peter Puget saw a large, deserted village "on the White Bluff" that he estimated would house about 400 people (see page 74-75). As it was June and too early for the sockeye migration, the inhabitants were absent. *Cadboro*, a Hudson's Bay Company ship, anchored off Point Roberts on July 14 1827 where Sir George Simpson noted "a number of Indians in groups on the beach and in canoes round the vessel." At Point Roberts was "a large camp of about one thousand.....inhabitants of Vancouver's Island, who periodically cross the gulf to Fraser's River for the purpose of fishing." Similarly, on August 20 1829, at least 200 canoes were assembled off the Point.

Little now remains of the past at Lily Point except for the posts of later canneries scattered on the shore, and post holes thick with rockweed.

Semiahmoo Bay and Birch Bay ~ Straits Salish speakers

The shell middens around Birch Bay show a long history of habitation: Lummi and Semiahmoo ancestors lived there until smallpox and tribal warfare killed them or they were forced to move. Later generations were limited to Indian Reserves or Reservations in the vicinity of their traditional lands.

The Semiahmoo First Nation has reserve land at the mouth of the Little Campbell River on Semiahmoo Bay. The village is just east of White Rock, within steps of the tide. In earlier days there were winter villages at Birch Bay and on Semiahmoo Spit (Tongue Spit - Map F5). At the time of the Spanish explorers there was a village close to where it is now; houses are marked on Jose Narvaez' 1791 map (see page 67). Today, the Semiahmoo population is quite small, with many living away from the reserve.

Fishing has always been the way of life

Fishing has always been the way of life for Semiahmoo. Reef net fishers caught salmon at Point Roberts and at a flat-topped rock near Birch Point, where salmon would swim north before swirling across Boundary Bay.

In the old days Semiahmoo people would travel as far as Waldron Island, in the southern Gulf Islands, to tend and gather blue camas bulbs. Camas also grew on prairies near the mainland villages. Game was hunted in the California and Dakota Creek watersheds, and waterfowl on Semiahmoo Spit. Large nets were hung between poles to catch ducks and geese as they flew into the sheltered waters of Drayton Harbor. The beaches here were always excellent for clams, crabs, mussels and cockles.

Semiahmoo
www.semiahmoofirstnation.org

Semiahmoo Spit (Tongue Spit) today.

In the 1800s, Lekwiltok people from Johnstone Strait came raiding down the coast, killing, looting, and taking slaves. The Semiahmoo protected themselves in two defensive forts: an old one on the bluff between Crescent Beach and Kwomais Point (Ocean Park), looking out over Boundary Bay, and a historic construction where Blaine now stands (see page 86). No sign of them now remains.

Lummi

Ancestral Lummi land, *Ske'lot'ses*, includes mainland south of Birch Bay. The Lummi also used reef net fishing locations along the coast, including Point Roberts. Their language is a Northern Straits Salish dialect (see page 56). The tribe's reservation, established in 1855 by the Point Elliott Treaty, is located at Lummi Island, overshadowed by the still-active volcano, *Kul-shan* or "bleeding wound" (Mount Baker). Salmon has always been a major source of food and wealth for the Lummi. As with other Coast Salish, the traditional round of seasonal activities included gathering plant materials and shellfish, waterfowling, fishing, and hunting of sea and land mammals. Birch Bay was a favourite gathering place for clams and Lake Whatcom provided beaver, unusual land-locked coho salmon and freshwater mussels. Roosevelt elk were hunted inland, up river valleys and around Lake Terrell.

Recent generations of Lummi have a history of serving in the American military and are still strongly involved in the fishing industry (page 121).

Lummi
www.lummi-nsn.org

SALISHAN LANGUAGES

Coast Salish languages belong to the Salishan language family. They are in danger of dying out as older people who conversed in them have passed on. Efforts are being made to revive their use.

Halkomelem group

Halkomelem is an anglicised word for the dialects spoken along the Lower Fraser River:

Hul'qumi'num

Island dialect; spoken from Malahat to Nanoose Bay, Vancouver Island; some differences between Nanaimo, Chemainus and Cowichan.

Hun'qum'i'num

Downriver dialect (Tsawwassen, Musqueam, some Islanders).

Halq'emeylem (Halkomelem)

Upriver dialect spoken along the lower Fraser River.

Northern Straits Salish group

This Central Coast Salish language is spoken from south Vancouver Island to the islands of Haro Strait, and around Bellingham, WA. It includes *Sencot'en* (meaning "our language" and pronounced sen-chaw-then) spoken by the Semiahmoo and Saanich, and dialects spoken by Lummi, Sooke, Songhees and Samish.

Nooksack

Te'celesem was spoken in the Nooksack Valley and Chilliwack area.

Chinook Jargon

This was a trade language used in historical times along the coast and on the Fraser River, mixing Coast Salish, French and English words.

Saanich

The Saanich, including the Tsawout of the east Saanich peninsula, the Tseycum, Tsartlip, Tseycum, Malahat, and Pauquachin, have land on Vancouver Island and the Gulf Islands. Historically they travelled by canoe each summer to the reef-net fishing locations at Point Roberts and in the San Juan Islands. They are mentioned in the 1827-30 Hudson's Bay Company journals as having come from Point Roberts with salmon and beaver pelts to trade at Fort Langley. Some also stayed near the fort for the winter. Their name is written in the journals as "Sanch," though this appellation may have loosely included others, such as Lummi or Semiahmoo, who fished in the Boundary Bay area. The Saanich also went regularly to the Gulf Islands to hunt deer, fish and gather camas. A winter home was in Brentwood Bay, north of Victoria. In 1852, the Saanich signed the Douglas Treaty, which transferred some lands to the Crown in exchange for the undisturbed continuation of traditional hunting, fishing and other resource use.

In 2001, the Tsawout, Tsartlip, Pauquachin and Semiahmoo formed the Sencot'en Alliance, to further their cultural and political goals.

Sencot'en (Saanich) speakers
www.tsawout.ca
Saanich language website:
www.cas.unt.edu/~montler/Saanich

Coast Salish canoes on the river at New Westminster, 1887. (Vancouver Public Library)

The Fraser ~ River of Salmon

The Halkomelem-speaking people of the Georgia Strait range from the Fraser Canyon to the eastern shores of Vancouver Island. Traditionally, there were different dialects for different stretches of the river and estuary (page 56). The rich resources of the river encouraged seasonal mobility. There were camps for fishing sturgeon and salmon, game hunting, tule and berry gathering and other such activities. At the end of the year, people headed back to their winter villages.

Stò:lo and Kwantlen

A vibrant presence in the Fraser Valley, the Stò:lo are people of the river. A collective name for up to 24 lower Fraser River bands, about 5,000 people, the Halkomelem word literally means "river." The internal politics are a complex web of relationships. Eleven Aboriginal communities in the region belong to the Stò:lo Nation Society, and a group of eight bands has recently allied as the Stò:lo Tribal Council, among them the Cheam, Scowlitz and Kwantlen. Historically the Kwantlen was a large group with a village, *Skaiametl*, where New Westminster now stands. Later they moved up river to Langley, nearer the Hudson's Bay Company fort. Some downriver bands, such as the Tsawwassen, Musqueam, Katzie, Coquitlam and others, remain independent of these two Stò:lo organizations.

The Stò:lo have a heritage centre and Transformer site at *Xa:ytem* or Hatzic Rock near Mission (see page 187).

Stò:lo

www.stolonation.bc.ca
www.gov.bc.ca/arr/firstnation/stolo_tribal
www.xaytem.museum.bc.ca

Tsawwassen

On July 25 2007, the Tsawwassen First Nation became the first urban B.C. band to vote in favour of a treaty. The treaty expanded reserve lands, transfered governance authority and confirmed fishery allocations.

The Tsawwassen traditionally fished for herring and salmon, hunted deer in the upland forest, and harvested clams. In spring and early summer, they lived at *Tl'ektines*, a large prairie camp on Lulu Island, in the Fraser River, where they fished for sturgeon and salmon, and picked berries. In July, Tsawwassen travelled to *Kikayt*, across the Fraser River from the Kwantlen village of *Skaiametl*, where sockeye were fished and dried. The Tsawwassen have close historical and familial ties with the Kwantlen.

Tsawwassen village once had seven longhouses, but only three were still standing in the late 1920s. The very last longhouse was demolished in the early 1950s to make way for the ferry causeway. Within the last fifty years, the environment on Roberts Bank has been greatly changed (page 165). The Tsawwassen, like so many bands, has struggled with the loss of traditional resources as a consequence. Their community of about 300 people is nonetheless working on economic and cultural rejuvenation activities.

Tsawwassen
www.tsawwassenfirstnation.com

American wigeon flying over abundant waterfowl in Boundary Bay.

"Tsawwassen" means "land facing the sea," a name now shared with the community in South Delta and the ferry departure point for the Gulf Islands and Vancouver Island.

Katzie

The Katzie have many cultural ties with the Boundary Bay area, and many families around the Fraser estuary have relatives among the Katzie people. First Ancestor and Transformer stories of Old Pierre, a Katzie elder, were recorded in 1936 (see page 13). Katzie relationship with the land is complex: "it would likely be as true to say the land owned the Katzie people as it is to say the Katzie people owned the land."

Katzie
www.katzie.ca

Musqueam

The Musqueam have land around the North Arm of the Fraser, and a small property at Brunswick Point. About 900 Musqueam registered with the B.C. treaty process. Their traditional territory includes both Vancouver and Richmond, and many generations fished and hunted in the Fraser estuary marshes. Sturgeon traps were owned and operated along the river. The archaeological site of Eburne Mound (see page 36) is on Musqueam land.

"Musqueam" means the place of the *muthkwuy*, a type of tall grass. The Musqueam have a special affinity with the wetland grass, which flourished or dwindled just as the population of the Musqueam people themselves rose and fell over the years.

Musqueam
www.musqueam.bc.ca

Cowichan

Cowichan winter village is near Duncan, on Vancouver Island, where their band office is now located. The Cowichan are Hul'qumi'num speaking people, who lived, travelled and fished in the Strait of Georgia and the lower Fraser River. In summer, the Cowichan and Nanaimo (*Snunéymuxw*) crossed the Strait to fish, collect berries and harvest other plants, based at the large summer village of *Tl'ektines* in Richmond. The Cowichan are part of the Hul'qumi'num Treaty Group, which comprises six Vancouver Island First Nations, with over 6,200 members.

Hul'qumi'num Treaty Group
www.hulquminum.bc.ca

Nooksack

Nooksack land is in the watershed of the Nooksack River, just south of the Boundary Bay watershed. Nooksack spoke the language *Te'celesem*. The name "Nooksack" probably refers to the ferns that grew so profusely in the valley. The land was overlooked by the awe-inspiring heights of Mount Baker, described either as *Pekows* or *Quck-sman-ik*, the White Mountain.

This small peaceful tribe was badly affected by epidemics in the last two centuries. A further tragedy occurred in 1855, when they were unable to make it to the Point Elliott Treaty meeting due to heavy snowfalls. Missing the short process, they were denied a reservation of their own and were told to join with the Lummi. This was not something they wanted to do, so they opted for homesteading.

Nooksack
www.nooksack-tribe.org

Further information on local Coast Salish people

Brian Thom's Coast Salish page:
http://home.istar.ca/~bthom

Coast Salish Artist Joe Jack's website:
http://joejack.com/homepage.html

Halkomelem Ethnobiology (SFU)
www.sfu.ca/halk-ethnobiology

Native Peoples of the Northwest ~ A Travelers Guide to land, art and culture by J. Halliday, G.Chehak

A portion of George Vancouver's map of the "Gulph of Georgia" and Juan de Fuca Strait from the voyage of 1792.

SAILS IN THE MIST

"Two places in that direction had much the appearance of large rivers, but the shoals ....have prevented our having any communication with them."
~ *Peter Puget, 1792*

It was inevitable that, sooner or later, the outside world would discover the misty coasts and calm waters of the Strait of Georgia, but a mystery remains as to who made it here first. Was it the Spanish explorers of the late eighteenth century or that most notorious of English seamen, Sir Francis Drake, more than two hundred years earlier? Who were the mysterious traders that met with the Coast Salish but left no written record of their voyages? The stories lie hidden in elusive maps and journals or buried along the coast. The search for answers has engaged the interest of historians and geographers through the ages.

The inland seas of Puget Sound and the Strait of Georgia were the last part of temperate North America to be accurately drawn on world maps. Not until the 1790s did the combination of better weather, improved technology and medicinal advances allow these inland seas to be regularly reached from abroad. The climate was colder in the Pacific Northwest in earlier centuries, which proved an obstacle to sea voyages. In the 1500s, Spanish galleons used mid-Pacific currents to speed trading voyages between California and Asia, but showed little inclination to brave cold and foggy northern latitudes. Only a very few mariners ventured further, such as Bartoleme Ferrelo, who sailed from Mexico in spring 1543 and reached 44 deg. north before abandoning due to cold.

Although the Pacific Northwest coast is an oceanic neighbour to northern Asia, various constraints prevented the connection being made. The Chinese, under Emperor Zhu Di, had a magnificent fleet of ships in the Ming Dynasty of the 1420s, but the cost of its upkeep became prohibitive. The fleet was destroyed, and despite a high demand for sea otter furs, Chinese boats did not cross the Pacific for four hundred years, except when the occasional fishing junk was blown astray and wrecked on American shores. The Russians too were familiar with their home stretch of shoreline but took years to venture from Siberia east across the Bering Strait. They only developed a presence on the Northwest Coast in the mid-1700s, after Vitus Bering's ill-fated voyage that precipitated the fur trade.

SIR FRANCIS DRAKE'S MYSTERY VOYAGE

The English mariner, explorer and privateer, Sir Francis Drake, made a historic voyage around the world in 1579. Local writer, Samuel Bawlf, has hypothesized that Drake reached latitude 57 deg. north and returned south through the Strait of Georgia, becoming the first European in Boundary Bay. According to Bawlf, details of the voyage were kept secret and maps changed due to the sensitive political climate during Queen Elizabeth I's reign.

Maps of the voyage show that Drake actually reached only 42 deg. north, (now the California- Oregon border), although it seems to have been unusually bad weather that year, with "vile, thicke and stincking fogges" and "extreme and nipping cold."

Bawlf's book describes Drake's purported voyage south through the Strait of Georgia, landing at Comox, his proposed *Nova Albion* (meaning New England), past the mouth of the Bay of Islands (Fraser River) and into Beautiful Bay (Boundary Bay). From here, he sailed down to the Juan de Fuca Strait, and after stopping to careen his ship in Whale Cove, Oregon, set sail across the Pacific Ocean.

While an intriguing idea, this theoretical extension to Drake's voyage has been largely discredited. However, this notorious seaman will no doubt continue to be the subject of speculation among enthusiasts for many years to come!

EXPEDITIONS TO THE PACIFIC NORTHWEST

Generations of maritime explorers were deterred by stormy seas and foul weather through the Little Ice Age, a cold period which seems to have spanned several hundred years between the 15th and 19th centuries. Raging storms and high winds prevented Juan Hernandez from landing at Nootka on the west coast of Vancouver Island in 1774 and gale force winds similarly assaulted the ships of Spanish explorers Bruno de Hezeta and Bodega y Quadra the following year, blowing them far offshore.

The famous explorer, James Cook, sailed a very erratic track up the northwest coast because of the weather, being repeatedly driven south and offshore by severe gales. Cook's crew member James Burney recorded in his journal on March 7 1778 that "bad weather again obliged us to keep to sea." Even Cook, a very seasoned mariner, recorded a month later "exceeding tempestuous weather" at latitude 50 deg. north. A Spanish naturalist at Nootka wrote: "the north wind in winter is extremely strong and its duration almost continual. It roots out trees and puts any vessels which may be anchored in the port in great danger." Winds from the south and southeast during the rest of year he says were accompanied by "thick fogs and continuous rains."

Storms hindered navigation for the early European expeditions to the Boundary Bay area.

In 1789, American trader Robert Gray sailed some distance into the Juan de Fuca Strait, but could not continue because of bad weather. Many others entirely missed the mouth of the Strait due to poor visibility.

The 1790s were somewhat warmer than previous years, and calmer weather must have prevailed to allow so many traders and explorers into the Strait of Georgia. Once inside the strait, sailing was easier in the mild climate, but rain continued to make life difficult for the seamen. James Douglas, writing in 1842, summed up the experience of many when he described "the dreary wilderness of the northwest coast."

Finding the Way

Until accurate timekeepers were invented by John Harrison in the form of portable, sea-worthy chronometers, the only method of establishing longitude was by a complicated calculation involving planetary motions. To do this, one had to be able to observe the position of the sun or planets, so heavy clouds and rain made voyages in these uncharted northern waters extremely challenging. Furthermore, quoted latitudes for the Pacific Northwest voyages varied widely, and inaccuracies of one or two degrees were quite regular in historical sources. In 1792, the Spanish explorer Dionisio Galiano had to correct many mistakes in longitude and latitude on Narvaez' maps drawn just a year earlier. Galiano had the benefit of more advanced instruments, such as chronometers.

The Urge to Explore

It was not only bad weather and poor technology that kept the outside world away. For many years there had been no economic motive to explore these distant lands. This changed in the mid-1700s when sea otters, fur seals and foxes became scarce on Russian shores, causing the Czar to initiate an Alaskan expedition to find new sources of fur. The 1741 voyage ended tragically for its commander,

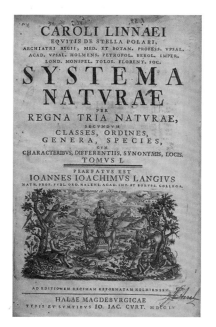

The cover of a 1760 edition of "Systema Naturae," Carl Linnaeus' book on nature classification.

Vitus Bering, yet together with the well-publicized discoveries of Captain Cook, it began the lucrative Pacific Northwest fur trade. The trade attracted English, French and American adventurers, and soon decimated many marine mammal populations. Northern coastal people traded furs for iron, copper and guns, which led to intensified intertribal conflict and some aggressive raids by northerners into the Strait of Georgia.

a time of renewed interest in the natural world...

In contrast to these commercial interests, the mid-eighteenth century was also the "Age of Enlightenment" in Europe, a time of renewed interest in the natural world, the collection and cataloguing of wildlife, the cultivation of new and strange varieties of plants, and an interest in human cultural diversity. It was the time of Carl Linnaeus, Voltaire and Descartes, the royal societies and botanical gardens of Madrid and London, and the founding of the American Philosophical Society. European royalty and governments of the time were supportive of scientific discovery.

It was during this period that a number of important explorations set out, including global voyages such as those of James Cook and Alejandro Malaspina. They took with them naturalists and scientists who were instructed to collect every kind of nature specimen and cultural artifact. Much of our earliest natural history knowledge of the Pacific Northwest is based on the writings of Archibald Menzies, botanist with Vancouver's expedition, Georg Wilhelm Steller of the Bering expedition, and José Mariano Mozino, naturalist with the Spanish expedition of Don Bodega y Quadra.

Increased scientific knowledge also led to better health for the travellers. For many centuries before bacteria were discovered, it was believed that infectious diseases were caused by bad air, humidity or rotten smells. Improved hygiene in the eighteenth century slowly began to reduce the prevalence of many such diseases, although this was tragically not the case for North American Aboriginal populations. The convergence of calmer weather, better technology and improved health set the stage for the European exploration and mapping of the Strait of Georgia.

Further information on European exploration

The World Encompassed by Sir Francis Drake

Voyages of Delusion by Glyn Williams

Mr. Menzies Garden Legacy by Clive Justice

Longitude by Dava Sobel

FIGHTING SCURVY

Western hemlock - a cure for scurvy.

Scurvy, a painful disease caused by lack of vitamin C, killed many seamen on long voyages in cold climates. The disease had afflicted people since ancient times, yet the role of diet was not recognised by Europeans until about 1600, and it took another three hundred years to discover vitamins.

Even once it was known that citrus fruits prevented scurvy, the Spanish navy seldom carried them and the British were equally lax, until James Lind in the mid-1700s published two books on the health of seamen. This persuaded the British navy to adopt the use of limes (and so the seamen were called "limeys").

James Cook and George Vancouver, who had sailed with Cook as a midshipman, both knew of Lind's work and took precautions against scurvy. They fed their crews lemons, brewed "spruce" beer from western hemlock, and served up native plants such as orache, all good sources of vitamin C. The sailors survived the voyages in good health.

THE EXPLORATION OF BOUNDARY BAY

On a summer evening in June 1791, a Spanish schooner sailed north through the San Juan and Gulf Islands. Juan Pantoja y Arriaga, pilot of the *Santa Saturnina*, under orders from Don Francisco Eliza, the Commandant at Nootka, was awed by what he saw as he cleared the reefs on the east end of Saturna Island. A vast unbroken stretch of water, rimmed with distant snow-capped mountains, reached to the far horizon. Pantoja named the strait *El Gran Canal de Nuestra Senora del Rosario la Marinera*, after a famous shrine in Seville. A month later, José Narvaez, another of the expedition's pilots, returned to these waters to complete the exploration, sailing into Boundary Bay and making the first detailed chart of its coastline.

Pantoja and Narvaez were probably not the first outsiders in the Strait of Georgia that year. At least five foreign trading ships were in the Juan de Fuca Strait in 1790, and many more were working the west coast of Vancouver Island, north to Alaska. Two years later the Scottish botanist, Archibald Menzies, counted "no less than 30 vessels on the NW coast." The Lummi told stories of three white men coming into Bellingham Bay before the arrival of any known vessel, who were killed by the now extinct Mar-ma-sece tribe and buried near Sehome. Settlers to the area in 1853 exhumed some bones and found brass buttons, a knife and an old flintlock pistol. The Russians had well established settlements on the northern coasts, whaling ships were arriving from Boston and Nantucket, and adventurers from France and Britain were afloat, sometimes using Portuguese and American ships to evade national trading agreements. Although his voyage was later overshadowed by those of other explorers, José Maria Narvaez Gervete was credited as "discoverer" of the Vancouver area.

They anchored off a sandy point .... Semiahmoo Spit

Narvaez was born in Cadiz, Spain and joined the navy as a teenager. He sailed to San Blas, California, in 1788, spent two years at Nootka on Vancouver Island, and was the first Spaniard to confirm the existence of Juan de Fuca Strait in 1789. Two years later, he was sent to explore the inland waters. Sailing the tiny, unseaworthy *Santa Saturnina*, he and his crew headed up the mainland coast, past Bellingham Bay and into Birch Bay, before anchoring off a sandy point that they named *Punta de San Jose*, now Semiahmoo Spit. Someone must have gone ashore and met with the local Semiahmoo, because Narvaez drew the position of their village on his chart, as well as the elk hunting grounds of Lake Terrell.

Continuing north they mapped the cliffs at *Punta de San Rafael* (Kwomais Point), then sailed on a full tide into Boundary Bay. The shallow depth of water over the sand flats was not at first apparent, but soundings proved that progress with the schooner was impossible. Anchoring in mid bay, Narvaez took to the longboat, a shallow-drafted vessel fitted with oars and sails, in an attempt to reach the river mouth.

Land and water merged on the undyked delta floodplain, where marshland and bogs stretched north as far as the Fraser River. Rain and cloud obscured the view and the flooded ground was choked with a tangle of logs, stumps and willows. After attempting to reach the river, they marked its approximate site as *Boca de Bodega*. The expedition's commander, Francisco Eliza, in a politically correct moment, later changed the name to *Boca de Floridablanca*, after the Spanish Prime Minister of the time. Later still, this name was moved to Burrard Inlet, when it was realized that Boundary Bay was not the mouth of the major river in the valley.

Returning to the *Santa Saturnina*, Narvaez approached what seemed

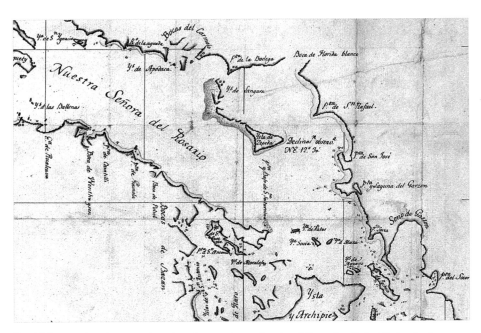

José Narvaez' 1791 map showing the southern Strait of Georgia; Point Roberts (in centre) appears as an island.

to be an island on the northwest side of Boundary Bay, naming it *Isla de Zepeda* (Point Roberts). Numerous Coast Salish families had gathered for salmon fishing and immediately set out in their canoes. They were watched anxiously by the Spaniards, who feared an attack; however, the visitors just wanted to trade freshly caught fish, including the most delectable salmon. Despite the language barrier a lively conversation was exchanged.

Although Narvaez' personal log book has never been found, he reported his experiences to Juan Pantoja, who wrote: "At the *Isla de Zepeda* there is an incredible quantity of rich salmon and numerous Indians....They speak an entirely different language." The Spaniards learnt that larger vessels than the *Santa Saturnina* had been seen farther up the Strait, from which the fishermen had obtained brass bracelets of exceptional quality, quite unlike any of the "trifles which the foreign vessels have brought." He also heard of overland traders on horseback: "An Indian boy obtained by the store-keeper told of people coming to trade copper, iron and blue beads" (see page 44).

This evidence that the Spaniards were not the first visitors to the area explains why many Coast Salish were friendly, relaxed and confident in their communication and trading, unusual behaviour for absolutely

Chief of Puerto del Descanso drawn by José Cardero, 1792.

first contact between peoples. The story is also illustrative of how little we know about the individual impact of explorers, traders and adventurers on people's lives, given that a boy could be "obtained." The Spanish were under the clear instructions of Conde de Revillagigedo, the Viceroy of New Spain (Mexico), that "nothing will be acquired from the Indians against their will; but rather by barter or by them giving it out of friendship; all must be treated with affability and gentleness, which are the most powerful means to attract them and to firmly establish esteem, so that for those who return to those places with the intention of settling, if such is determined, they will be so treated." The newcomers were anxious to establish trade.

The *Santa Saturnina* sailed on round the point and up the strait. Continued bad weather and shoals of sand kept them offshore as they sailed northwards. However, Narvaez and his crew knew they were close to a mighty river when "the schooner being anchored two miles out they collected and drank sweet water" and they saw a "line of white water more sweet than salt." The mouth of the Fraser River eluded them still. The presence of a great number of "gulls, tunny fish and immense whales" made them suspect the channel opened to the sea, and they followed it probably as far as Texada Island before heading back to the Juan de Fuca Strait.

Sadly, only a chart and secondhand reports have survived to record Narvaez and his crew's historic three week voyage.

THE GALIANO AND VALDES EXPEDITION

The following year, two young naval officers trained with the latest skills in hydrography, cartography and astronomy were sent by Alejandro Malaspina, Commander of the Spanish Royal Navy's scientific expedition to the Pacific, to complete the mapping initiated by Narvaez. They were to chart the full coastline of the *Gran Canal,* and explore in particular the *Boca de Floridablanca* for any possible routes through to the

northwest. Dionisio Alcala Galiano in the *Sutil* and Cayetano Valdes y Flores Bazan in the *Mexicana*, together with seconds in command Secundino Salamanca and Juan Vernaci, expedition artist José Cardero and other crew members, accordingly sailed from Mexico in March 1792, up the Pacific coast to Nootka and from there in early June into the Juan de Fuca Strait.

Galiano died fighting at the Battle of Trafalgar

Galiano had studied astronomy in Cadiz and sailed as a midshipman to South America, later serving as a naval astronomer on the Pacific Coast and in Spain. He was to end his days fighting the British, and like their hero Horatio Nelson, died at the Battle of Trafalgar. Valdes, the younger of the two commanders, was born in Seville and had sailed with Narvaez on the previous year's voyage, learning some of the Salishan dialects. He too eventually returned to Spain and took part in the battles of La Coruna and Trafalgar, surviving to become Governor of Cadiz. He was later exiled for many years to England for political intrigue but finally rose to the position of Captain General of the Spanish Fleet.

José Cardero was originally a cabin boy from southern Spain, who had sailed in earlier voyages with Valdes;

The Sutil and Mexicana approached by Coast Salish canoes, 1792, drawing by José Cardero.

his natural talent led him to become the ship's artist and his beautiful drawings are one of the few visual records of the time. He may also have compiled the account of the voyage, published by Espinosa y Tella.

Sailing up the Juan de Fuca Strait, the explorers heard from a local chief that foreign vessels were ahead of them. By a curious convergence of exploration, their detailed mapping of the Strait of Georgia coincided with the arrival of Captain George Vancouver for the same purpose.

Galiano and Valdes made landfall in Bellingham Bay one evening in early June, and spent an anxious night in shallow waters, their anchors dragging as the tide ran out. The distant volcano, *Montana del Carmelo* (Mt. Baker), rumbled and glowed in the dark. A wind came up strongly in the morning, both ships fouled their anchors, and the Mexicana ran aground. Finally they set sail, eventually sighting two small coastal vessels (Joseph Whidbey charting Bellingham Bay) and later on the lights of the English ships at anchorage in Garzon Creek (Birch Bay). The Spaniards planned to enter Boundary Bay at dawn, to allow plenty of time to explore an area they felt "would be full of interest." A fresh wind blew through the night forcing them into shallow water. They cast anchor. Early on June 13 they found themselves in a "closed bay with trees all around," between Kwomais Point and Point Roberts. The longboat led the way into what they believed was the *Floridablanca*

River, with the schooners following under light canvas. It soon became obvious that the channel ended in "low land, marshy and full of trees" and barely a fathom of water. Naming the marshy bay *Ensenado del Engano* (Deceptive Bay), Galiano and his crew tacked towards the *Isla de Zepeda*, soon realizing on that clear summer morning that what had looked like an island was in fact a peninsula. They hastily renamed it *Punta de Zepeda*.

They named the marshy bay "Ensenado del Engano" - Deceptive Bay

It was still early in the morning when a British brigantine, *Chatham*, sailed across the bay, coming alongside the *Sutil* off Maple Beach, Point Roberts. The brig's commander, Lt. William Broughton, was invited aboard, and there followed an exchange in which both parties were frank about the condition of their vessels and their achievements in mapping the coast.

After this meeting, the boats sailed around *Punta de Zepeda*, the *Sutil* and *Mexicana* trailing the larger, faster British ship by some considerable distance. The Spaniards took hours to work their way out of a back eddy before rounding the Point. They were still straining with sail and oars, on a cloudy evening with rain looming, when they were overtaken by the outgoing tide off the Fraser River. Too exhausted to fight the freshet, the sailors headed for the islands and eventual safe anchorage on Gabriola Island. Here they rested and took on water and wood before resuming their exploration of the Fraser estuary.

They met with a number of Natives of various tribes, both on land and in canoes. Journal descriptions give us a fascinating glimpse of these encounters, which were generally friendly. The people of each village or canoe flotilla varied in appearance, behaviour and often in language too. Galiano was "astonished at the difference in face, build and character among the natives of this strait in the distance of a few leagues."

Crossing back to the mainland, the Spaniards noted brackish water and almost collided in the dark with a great floating tree at the mouth of the Fraser. On June 20 they anchored off *Punta de Langara* (Point Grey) where they were approached by Vancouver and Puget in the English longboat.

On a cloudy evening with rain looming ...

GEORGE VANCOUVER'S EXPEDITION

George Vancouver had been a midshipman with Captain James Cook on two of his historic voyages. Now the Commander of the British Naval Expedition to the Northwest Coast, Vancouver was entrusted with mapping the coastline from Juan de Fuca Strait to 60 deg. north. The English had still not entirely dismissed the belief and hope of a Northwest Passage that would speed their merchant navy to the Orient, although Vancouver himself was less than optimistic about the prospect of finding such a route. The *Discovery,*, commanded by Vancouver, and the *Chatham,* under Broughton, left England on 1 April 1791. They took a full year to reach the Oregon coast. Crew members included Lieutenant Peter Puget, Lieutenant Joseph Baker and Sailing Master Joseph Whidbey, all of whom were destined to leave their names on features of the landscape.

To satisfy its scientific patron, Sir Joseph Banks, fellow of the Royal Society and Director of the Royal Gardens at Kew, the expedition even had an official botanist. Archibald Menzies was instructed to examine in detail all the plant and animal life encountered, as well as recording geology, soils and native cultures. An indomitable Scotsman, Menzies accepted this daunting task with relish. He spent a considerable time on the voyage arguing with George Vancouver over the importance of his plant collections, particularly the live specimens kept unsuccessfully in a box on the *Discovery's* quarterdeck.

Vancouver and his crew sailed and rowed up the Juan de Fuca Strait and into Puget Sound, methodically surveying the coast and naming geographic features after benefactors and colleagues. They believed that in this way it was claimed as British territory. They charted much of Puget Sound and then turned north towards the Gulf of Georgia, with which Vancouver honoured King George III.

Vancouver missed the initial meeting with Galiano and Valdes...

Totally unaware that the Spanish were ahead, Vancouver sailed through Rosario Strait and Bellingham Bay, eventually making anchorage off Birch Bay, where there was a supply of fresh water. He sent Joseph Whidbey back south with two boats to map the coast in detail, then left with Peter Puget on a week long trip in boats up the Strait of Georgia. He thus missed the initial meeting with Galiano and Valdes, whom Whidbey had spotted sailing up through the Gulf Islands and Broughton had seen from Birch Bay.

ARCHIBALD MENZIES

Botanist and surgeon,
1754 - 1842

Archibald Menzies was born in Perthshire, Scotland, and studied at the University of Edinburgh. He sailed around Cape Horn in 1786 and subsequently joined Captain Vancouver's expedition to the Pacific Northwest, as botanist and ship's surgeon.

In later years, he visited the Sandwich Islands (Hawaii) and with Joseph Baker, made the first recognized ascent of Mauna Loa. He served in the British Navy in the West Indies and ended his days as a London surgeon.

Menzies introduced Europeans to such northwestern plants as the western hemlock, salal, Pacific dogwood, cascara, Oregon grape, flowering currant, chocolate lily and ocean spray, many of which were planted in gardens around the world.

His name is recognized in plants he discovered, including arbutus (Pacific madrone), *Arbutus menziesii*, false azalea, *Menziesia ferruginea*, and coastal Douglas-fir, *Pseudotsuga menziesii* ssp.*menziesii*. (Botanist James Douglas introduced the Douglas-fir to England in 1827.)

Left: The scientific name for Douglas-fir, Pseudotsuga menziesii, commemorates botanist Archibald Menzies.

Peter Puget described seeing the high white cliffs of Lily Point in 1792.

The Village at Lily Point

While Vancouver went north, Peter Puget explored the Point Roberts peninsula, landing at Lily Point. It was June but the Coast Salish families had not yet returned to their salmon reef netting villages. Puget found "a habitation of near four hundred people, but was now in perfect ruins and overrun with nettles and some bushes."

He went on to describe three rows of houses, divided by narrow lanes and containing the frames of about six large houses, made of heavy, long timber logs, placed on top of 4.2m (14ft) tall standards. Each standard was notched to receive the giant rafters. Planks lay nearby, ready to be placed on the walls. Puget may not have known that the Coast Salish carried tule and cattail mats with them from village to village to complete the dwellings. He described the high white cliffs of Lily Point and the sandy area a mile or so to the southwest, dropping off quite suddenly into deeper water.

Meanwhile, the remainder of the expedition camped in Birch Bay to take astronomical observations, check chronometers and attend to repairs. This was an ideal opportunity for Menzies to botanize in the vicinity of Semiahmoo Bay where he observed "in full bloom diffusing its sweetness that beautiful shrub, the *Philadelphus*" (native mock orange). At Birch Bay,

Menzies found black cottonwood, quaking aspen and black birch, the latter inspiring the new name for the bay. Rocky Mountain juniper grew in gnarled and twisted forms on sunny shorelines, the same shrub that was said to have saved one Semiahmoo family from the smallpox (see page 50). The energetic botanist also collected specimens of willowherb, Hooker's onion, death camas, cattail and stinging nettle. There is no record of the English talking with any of the Semiahmoo, who perhaps were away sturgeon fishing at this time.

Vancouver described the exploration of Semiahmoo and Boundary Bay very prosaically: "I departed at five o'clock on Tuesday morning. The most northerly branch......caused little delay; it soon terminated in two open bays: the southernmost which is the smallest.....extends in a circular form to the eastward, with a shoal of land projecting some distance from its shores. This bay affords good anchorage from 7 to 10 fathoms of water: the other is much larger and extends to the northward; these by noon, we had passed around...."

He named the point after Henry Roberts, his friend and former Commander who had also served under James Cook. He gave neither Boundary Bay nor the headland at Ocean Park a name and did not remark on any possible river opening in the bay.

Sea asparagus (Salicornia virginica) is a common plant of the salt marshes.

Once around Point Roberts, the boats kept offshore to avoid the tidal sand flats on Roberts Bank. The result of this was that Vancouver completely failed to record the mouth of the Fraser River; however, Puget noted that "two places in that direction had much the appearance of large rivers, but shoals hitherto have prevented our having any communication with them."

The marshy terrain was fringed with trees and awash with the spring freshet, which made it hard to discern any features. As Puget described it, the view from the boats was of a "very low land, apparently a swampy flat, that retires several miles, before the country rises to meet the rugged snowy mountains…this low flat being very much inundated and extending behind Point Roberts to join the low land in the bay to eastward of that point, gives its high land, when seen at a distance, the appearance of an island.….the shoal continues along the coast to the distance of seven or eight miles from the shore, on which were lodged….logs of wood and stumps of trees innumerable."

Doggedly, Vancouver and his crew rowed and sailed nearly 400 km (200 mi) up to Jervis Inlet before turning back to rejoin the ships. It was in Burrard Inlet that Vancouver had his first sight of the Spanish vessels, met with Galiano and Valdes and learned of the scientific exploration they were conducting. It is an indication of the contrasting characters of the two English officers that Vancouver grudgingly states Galiano "spoke a little English" while Puget wrote "this gentleman spoke English with great ease and fluency!"

Vancouver was a difficult character, often unpopular with his crew and suffering from a debilitating illness that not only worsened his moods but caused his death a few years after the voyage. It is strange that he should have been the one to leave his name, not only to a city but to a whole island. It was also a strange coincidence in time and place that two such similar European voyages should converge after so many years of ignorance about the Pacific Northwest.

After his encounter with the Spanish explorers in Burrard Inlet, George Vancouver rowed back to his ships at Birch Bay, pausing on the 50 km (30 mi) journey to purchase some huge sturgeon from the Musqueam at Sturgeon Bank. In a light breeze, he and his crew rowed against two flood tides on the fifteen hour trip to Point Roberts, where they spent the night. In the morning they rowed east across Boundary Bay, to where their ships were waiting in Birch Bay.

The ships' crew were sorry to leave Birch Bay with its abundant roses, soft grass, copious fruit bushes and clean water, "by far the most pleasing place we have been at on the coast of America." On that midsummer morning they sailed away up the gulf, meeting with the Spanish schooners as arranged, so that they could coordinate their explorations. In clear weather and a fresh wind, a host of whales accompanied the four ships as they sailed north together.

The inner waters of the Georgia Strait had been measured, surveyed and charted, a slow and painstaking work. The English expedition's atlas and three volumes of journals were published a few months after George Vancouver died. They proved conclusively the absence of a northwest passage. The Spanish surveys were published as a book and atlas in 1802, but most of the other journals and charts were secretly stored away in the Madrid archives.

The names that Vancouver had given to the bays and headlands, the inlets, channels and islands, made their way onto geographers' world maps. Only a few places hold the Coast Salish names that had existed for thousands of years, or Spanish names given by the explorers who arrived ahead of the British. The names of Tsawwassen and Semiahmoo Bay, Nicomekl River and Chilukthan Slough are small reminders of the first millennia of human habitation, while those of Saturna, Galiano, Valdez and Whidbey Islands, Narvaez Point, Puget Sound, Vancouver Island and the city itself serve as memorials to the foreigners who braved wind, tides, rain and homesickness in the Pacific Northwest.

Further information on Boundary Bay exploration

A Spanish Voyage to Vancouver and the Northwest Coast of America, 1792, published by Espinosa y Tello

Malaspina and Galiano: Spanish Voyages to the Northwest Coast by Donald Cutter

The Voyage of George Vancouver 1791 - 1795, Vol. 11, by W. Kaye Lamb

Historical Atlas of Vancouver and the Lower Fraser Valley by Derek Hayes

Vancouver Maritime Museum
www.vancouvermaritimemuseum.com

B.C. Maritime Museum, Victoria
www.mmbc.bc.ca

Flooding on Delta Street, Ladner, in 1895. (Delta Museum & Archives)

PIONEERS IN A NEW LAND

"The pioneer spirit lived strong in those days and we do well to call to mind the much so few did, with so little to do with."
~ *Rebecca Jeffcott, settler*

The pioneers hold a special place in western history because they epitomize so strongly the North American ideal of adventurous, self-reliant people coping with hardship and surviving by their own efforts. The frontier, fount of dreams, limitless possibilities and unending resources, kept moving further and further west and north, driving forward all those who dreamed of a better life.

The arrival of early settlers, alone or in small groups, without apparent bonds to ancestors or history, was in contrast to the lives of the Coast Salish people, traditionally supported by numerous kinfolk and with deeply rooted ties to the landscape. At first, the homesteaders were few in number, living a tenuous existence close to the land. Their new world was often harsh and lonely. Children died, food was scarce, floods, wild animals and disease were constant hazards and the task of "taming" the marshes, bogs and forests almost insurmountable. Those who arrived in search of gold and riches were often disappointed and many left the Northwest, never to return. The pioneers that stayed to build homesteads and clear the land were spurred on by thoughts of the future, yet could have had no concept of the changes ahead. Within thirty years railroads arrived, industrialization and mechanization spread to every walk of life and the landscape was transformed. Within a hundred years of the immigrants setting foot on the Fraser delta, the marshes were drained and dyked, the surrounding forests were logged, and the institutions of Western European government, law and trade were entrenched in society.

The coastal landscape was awe-inspiring for the newcomers, surpassing anything that they had ever seen. It was full of superlatives: the largest trees, biggest salmon runs, longest river and highest mountains. Thousands upon thousands of birds migrated through the estuary and game was everywhere and free for the taking. Cougars, wolves and bears roamed the forests and storms brought the trees crashing down. In the bright morning light, wide open skies and glorious views renewed the pioneers' courage. Life was both a challenge and an adventure and nature's bounty was considered everlasting.

BY LAND & SEA

After the voyages of 1791-2, no other Europeans led expeditions to the area for many years with the exception of Simon Fraser, who descended the Fraser River in 1808, and for whom the river is named. He reached as far as the large Musqueam village near the mouth of the North Arm.

McMillan's expeditions

There was then another long lull in exploration until December 1824, when a group from the Hudson's Bay Company fort on the Columbia River arrived in Semiahmoo Bay. James McMillan and his multicultural team of forty-two men, including Hawaiians, Canadians, Iroquois and an American, were sent in the middle of winter to check out fur trading possibilities on the lower Fraser. Led by Coast Salish guides, they travelled by boat up the narrow, winding Nicomekl River. When this got too shallow, they made a portage across Langley Prairie to the Salmon River and then rowed downstream to the Fraser.

The back country at that time was deeply forested and was abounding in elk and beaver. Willows choked the streams, making travel along them almost impossible, and mud clogged the 7 km (4.5 mi) portage. Having ascertained the feasibility of building a fort on the main river, now confirmed as that described

Portage Park where James McMillan's expedition left the Nicomekl River on their route to the Fraser.

by Simon Fraser, the weary, rain-soaked men travelled downstream through the marshes and south past Point Roberts. McMillan noted that "the Indians are very numerous and collected in villages along its banks, they were overjoyed to see us." Many "*Cowitchens*," as they were known at the time (Halkomelem speakers, such as the Musqueam, Stò:lo, Tsawwassen, Katzie and Cowichan), had heard about the foreigners, but had not yet encountered them. Still, they had obtained Hudson Bay blankets, knives and other goods from tribes in the Interior. We might suspect the "joy" attributed to them by McMillan was just interest in new visitors!

The following August the Hudson's Bay Company brig, the *William and Ann* under Captain Hanwell, sailed on a trading voyage to the mouth of the Fraser. On board was the botanist Dr. John Scouler. The captain was too timid to spend much time on land and refused to enter the South Arm of the Fraser, which must have been frustrating for Scouler. Hanwell was wary of the local Coast Salish, even though they came peacefully in canoes to greet the ship off Point Roberts. He refused to return to the river.

James McMillan and his men had to subsequently sail on another ship, the *Cadboro*, to map the estuary shoals and find a site for the fort. Weather conditions were poor and progress was very slow for this voyage. On Friday July 13 1827 they anchored in the bay near Point Roberts, and McMillan went ashore to scout possible locations, without success. The wind was from the northwest and it took them until Monday to reach Sturgeon Bank. It was another week before they were into the Fraser River. Only at the end of the month could they land their horses and begin work on the long-awaited Fort Langley, in a location downstream of its present site. The Fort soon became an established trading post, with the Halkomelem-speakers of the river bringing in salmon for salting, and the Nooksack people trading furs from the lands around Mount Baker.

DR. JOHN SCOULER

Botanist and surgeon,
1804 - 1871

Like Menzies, John Scouler was a ship's doctor as well as a botanist. He was educated at the University of Glasgow, Scotland, and later became Professor of Geology.

Scouler and David Douglas, another famous Scots botanist, had previously sailed on the *William and Ann* to the Galapagos Islands, where they were the first European explorers to study the native plants. They were also among the first scientists to collect botanical specimens in the Columbia Valley and British Columbia, sending them to Sir William Hooker of the Royal Horticultural Society, England.

Scouler gave his name to many local plants, including northern wormwood, *Artemisia scouleriana*, Scouler's campion, *Silene scouleri*, and Scouler's willow, *Salix scouleriana*. He also recorded some wildlife: "the sea yields an abundant supply of fishes of the most delicious kinds ... every rivulet teems with myriads of salmon."

Coho salmon, recorded by Scouler.

Hudson's Bay Company farm, Fort Langley, ca.1895. (BC Archives)

The Hudson's Bay men settled in at the fort, married Coast Salish women and planted potatoes in gardens. Chief Factor John Work, who was with the first McMillan expedition, visited the fort as part of a September 1835 tour. He found the long haul up the Fraser time-consuming and difficult, writing in his journal how "the woods on both sides of the river are all on fire" following the dry summer and that they "met great numbers of Indians ….. going down the river from the fishing ground, their canoes all loaded with baggage and proceeding on to Vancouver's Island where they generally winter." Fire was a common occurrence and is often mentioned in writings of the time (see page 96).

After the Hudson's Bay Company, the next arrivals were mostly trappers, drifters or itinerants, moving up from the south along the borderless coast. In 1818, the United States and Britain had agreed to define their North American territories east of the Rockies along the 49th parallel of latitude. The western portion of the continent was still in dispute. The Spanish claimed California, the Russians occupied Alaska south to 54 degrees 40 minutes, and the land in between was known as Oregon Country, open to equal access by both American and British citizens. Fur traders from the Hudson's Bay Company roamed there and adventurers prospected for gold and silver.

"Fifty-four forty or fight"

Not until 1846, with settlers pouring into Oregon Country from the east and President James Polk declaiming they should push the border to "fifty-four forty or fight," did the American and British governments agree on a boundary line. The Oregon Treaty (or Treaty of Washington, depending on which side of the line you reside) declared that the 49th parallel should from henceforth be the border all the way to the mainland coast. West of the mainland, the boundary looped through the Georgia Strait to include the British colony of Vancouver Island. This division had the effect of slicing the Point Roberts peninsula in half. The southern, American portion was now separated from its closest national community at Blaine by the 20 km (16 mi) wide waters of Boundary Bay. The British, noticing this, tried unsuccessfully to have the Point annexed to British Columbia.

Point Roberts has retained a unique and charming character, but has to be expensively serviced from the mainland. The border line through the San Juan and Gulf Islands was to be disputed for several more years, until ownership of the San Juan Islands was ruled in favour of the USA.

Boundary Bay was put firmly on global maps in the 1850s, when the task of surveying the border was assigned to two Boundary Commissions: the Americans, led by Archibald Campbell; and the British Commission, under John Summerfield Hawkins. Both teams surveyed the line between 1857 and 1862, felling trees in a swathe along the parallel and placing marker posts and cairns. The most westerly cairn, Boundary Marker One, is on the high bluff at the western end of Roosevelt Road, Point Roberts (see page 187).

Derek Hayes tells the story in his Historical Atlas. Neither commission seems to have really had their hearts in the job. Both wrote reports, since the governments were uneasy about cooperation, but both reports were promptly lost, with the result that the boundary had to be re-surveyed between 1901 and 1907. Years later it was determined that the border had been inaccurately surveyed and should lie 300m (984ft) further south, an error that remained uncorrected when the Oregon Treaty was reconfirmed in 1908.

There was continuing confusion as to where the boundary lay even after the land was cleared. A map of Semiahmoo Bay, Washington Territory, printed in 1858 by the US Coast Survey, has absolutely no indication of the international border at all, suggesting they still supported President Polk!

For many years, settlers on the border roamed freely to and fro, bringing cattle up to graze in Canadian pastures or wandering down to join in 4th of July celebrations. In those more informal times, a transborder shooting, where the Peace Arch now stands, led to a hot dispute about jurisdiction. The offender had fired a shot from the St. Leonard's hotel in Canada and hit an old man in a shack across the road, on American soil. Where should the assailant be tried?

As time progressed customs houses were built and duties collected. The famous Peace Arch was built in 1920. It marked the completion of 100 years of amicable coexistence

Boundary Marker 1 at Point Roberts

between the two countries. Today, the international border through the Boundary Bay watershed is an active zone in the 6,500 km (4,062 mi) frontier. Lying between the quiet country roads of Zero Avenue in B.C. and Boundary Road in Washington, it has been the scene of night manoeuvres by drug and gun smugglers and police, trying to outwit each other in the darkness. Security towers, motion sensors and night vision goggles have become the norm in this security-conscious age.

Back in the nineteenth century, one challenge facing the colonists was getting the paperwork completed. As the land boundary was mapped and cleared, Captains James Prevost

and George Richards charted the ocean boundary through the Strait of Georgia and took soundings at the mouth of the Fraser River. Their maps show an intricate mass of depth records from Sand Heads to Derby Reach, the site of the first Fort Langley.

Before it was dyked, the estuary had to be constantly re-surveyed because the river kept shifting channel across the marshes. Meanwhile, Colonel Richard Moody led the Royal Engineers in a survey across the delta to Boundary Bay. His quick sketch map of February 1858 shows most of the lowland as marsh, with "moderately high ground" on Point Roberts and South Surrey-White

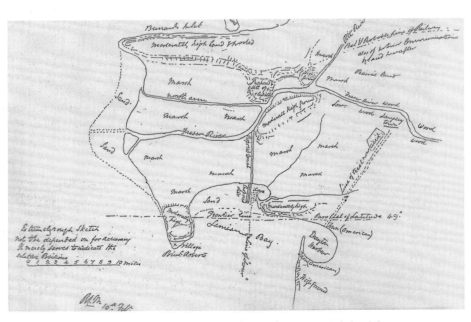

Colonel Richard Moody's 1858 sketch map of Boundary Bay and the delta.

Rock, a "proposed canal" between the Fraser River and the bay, along the route of present-day Highway 91, and a "proposed pier" jutting across Mud Bay. In 1979, the notebooks of Moody's Royal Engineers were used for a study of historic vegetation in the region (page 101).

The Royal Engineers set up camp in 1857 at the mouth of the Little Campbell River, where fresh water flowed into the bay. The estuary was much more extensive than it is today. Back then, the river curved north to flow over land that has since been filled for White Rock's 8 Avenue, a park and the railway. This whole area was then sedge grasses, bulrushes and tidal swamp. During their brief stay, the Engineers established a trail to the border (now Beach Road) and cut a trail along the river and across to Fort Langley. A year later, the British soldiers moved on to Sumas and their camp buildings became the base for Archibald Campbell and the American Boundary Commission.

The Point Elliott Treaty

The Semiahmoo people had at the beginning shown an interest in the newcomers, helping them find food and later building a mission church with the Oblate Fathers. However, the imposition of an international border, through traditional hunting and foraging grounds, presented a problem for the Semiahmoo, Lummi and Nooksack people. In 1855, the

JOHN KEAST LORD

Naturalist and veterinarian,
1818 - 1872

The English Boundary Commission hired John Keast Lord as a naturalist, veterinarian and transport manager from 1858 to 1862. During this time he was stationed at the Commission's depot at Chilliwack and travelled extensively west of the Cascades, from B.C. to California.

He wrote a two volume book, *The Naturalist in Vancouver Island and British Columbia,* describing in energetic style his observations of dozens of wildlife species. Carefully annotated lists include sightings of the Californian condor at the mouth of the Fraser river, flocks of swifts and goatsuckers (nighthawks) eating winged ants in June, and juncos - "the most abundant small bird in B.C."

He recognised the value of salmon to the native people, and described fishing with rakes, by moonlight, for glittering shoals of eulachon: "the labour goes on until the moon has set behind the mountain peaks and the fish disappear."

Dark-eyed junco, one of the bird species seen by Lord in 1858.

government of the new Oregon Territory met with tribes from all around Puget Sound, persuading them to accept treaties.

Along with more than twenty other tribes, the Lummi signed the Point Elliott Treaty on 22 January 1855. Negotiations for this pivotal treaty were conducted in Chinook Jargon, a trade language in use on the coast at that time. Despite the necessity of laborious translations into Northern Straits Salish and English legal language, the treaty was hastily concluded and reservations were designated.

Semiahmoo were left out of the Point Elliott Treaty negotiations of 1855

The Lummi were allotted land on Bellingham Bay, but the Semiahmoo, who lived both sides of the international border, were left out of the process. Although they never signed a treaty, those living on the American side were expected to move onto the Lummi reservation. The Semiahmoo First Nations Reserve on the Little Campbell River was not established until the 1880s.

Around this time, the Coast Salish on Boundary Bay and the lower Fraser River were suddenly faced with increased aggression from the Lekwiltok people (also known as Yukulta) from Johnstone Strait.

Stockaded fort at Fort Langley, 1862.
(BC Archives)

These northern tribes, armed with guns purchased from fur traders, made sweeping raids into the heart of the Fraser valley, taking slaves and food supplies.

On Boundary Bay, the Semiahmoo responded by erecting stockaded coastal forts, similar in style to Fort Langley and using the natural landscape for defence. One at an old defensive site between Kwomais Point and Crescent Beach was high on the bluff, with a good view of the bay, ravines down either side and a sheer cliff in front. The other fort was on the shore near the town of Blaine. These forts were pulled down some years later (see also page 55).

Raids continued to occur for a period of about ten years in the mid-1800s. Probably attacks or counter attacks took place along the Serpentine River because the Stò:lo atlas records two place names in the valley suggestive of this: *Kwo'tsesleq*, meaning "a look out for enemies," and a little further up river, *Th'emqellem*, meaning "battleground with ambush."

THE GOLD RUSH

In 1858 gold was discovered up the Fraser River and word quickly spread south to San Francisco, sparking a gold rush. In the space of a few months, over 30,000 miners, fired by the prospect of easy riches, arrived in Victoria and Whatcom (now Bellingham). These towns were the entry points for the trek up river. As they scrambled to get to the gold fields as quickly as possible, the crowds of newcomers overwhelmed local communities.

James Douglas, Governor of the British Colony of Vancouver Island, responded assertively to the influx of prospectors by claiming a two dollar head tax on everyone entering the Fraser River, as well as boat taxes ranging from six to twelve dollars and a ten per cent import tax on goods. Tariff collection was enforced by the brig *Recovery*, moored at the river mouth. Seemingly unconcerned about the questionable legality of this taxation, the British government rewarded Douglas in November 1858 with the position of Governor of the new Crown Colony of (mainland) British Columbia. To avoid the taxes at the mouth of the Fraser, miners bought passage across the Strait of Georgia to Point Roberts, or were towed in a small boat by steamer and released at Roberts Bank, where they could make their way up Canoe Pass, a short-cut to the Fraser River. The miners then rowed on an incoming tide up to Fort Langley, bypassing the gunboats. This was the route taken by the Ladner brothers, arriving from San Francisco via Victoria (page 91). A supply depot sprang up on the southwestern shore of Point Roberts. Known as Robert's Town, this cluster of six wooden buildings, including a store and saloon, barely lasted a year.

Alternatively, miners landed at Semiahmoo Spit on the shore of Drayton Harbor before taking a trail across Hall's Prairie. The Spit which had long been a clam-gathering camp and wildfowling location, now also became a supply depot, complete with a hotel and saloon. Attempts were made to push through an overland route, the Whatcom Trail, from Bellingham, bypassing the Fraser estuary entirely and joining the Anderson Brigade Trail that led from Yale up the Fraser Canyon. While initially successful, the track foundered on the high mountain peaks, snow bound for most of the year, and it never proved to be a viable route.

1904 Miners cabin. (BC Archives)

The gold rush was a disaster for thousands of people. Many miners died from starvation, sickness or accidents, and greed for gold led to murders and fights. For the Coast Salish whose territory was invaded it meant exposure to guns, diseases and yet more hungry strangers needing to share food. After only a year, disillusioned miners started to drift south again. Some tried their hand at other occupations or took up land, but most left the territory. Speculators who had bought land in Whatcom, Semiahmoo and Blaine in the expectation of it being a central staging area for the north country, were soon bankrupt. Cabins fell into disrepair, claimed by the wilderness.

Roadworks, carried out along White Rock's beach road many years later, uncovered the skeletons of two miners with bullet wounds in their skulls lying beside the ashes of a fire. The evidence suggests that their companion waited until nightfall to rob and kill them, then slip south across the border to a new life.

Boaters at White Rock in 1917, with the station in the background. (BC Archives)

TAKING UP LAND

One of the most noticeable aspects of western North America is the checkerboard appearance of the landscape, a geometrical network superimposed on natural features of the countryside. The Boundary Bay watershed is no exception, although some of the earlier highways lie at angles to the grid and newer subdivisions are trying to escape it.

Holdings were allotted on a regular grid alignment

The pattern is a product of the rectangular survey grid, commonly used throughout the west. It is quite different from the more haphazard, organic patterns of development familiar to Old World immigrants. Also known as "township and range," the rectangular survey method was advocated in the United States by Thomas Jefferson. He saw it as a way of maintaining and imposing order and authority in a vast uncharted landscape. After it was initiated, this development pattern permanently shaped the landscape.

The Boundary Bay watershed, with its curving coastline, marshes and rivers and irregular topography, would have been more attractively developed by using geographic features and landmarks to delineate the land parcels. Instead, in the hurry to bring the new colony into existence and provide land for settlers, holdings were allotted on a regular grid alignment. However, the process of pre-emption, by which choice areas went first, meant that sometimes the north-south grid was less than perfect. For example, reservation of land for railway companies created a stretch of the valley in Langley with lots aligned at an angle (see sidebar).

Each six mile square of the grid induced the need for boundary lines, either a road, fence, ditch or all three, and subsequent roads naturally followed the grid when land was subdivided. No one thought to include parks, common lands or walking trails in the picture, other than one or two existing paths. When neighbours knew each other well and shared resources, as they did in pioneer times, this was not a problem. As the population grew and property became a private domain, it meant that many people, including the original inhabitants, the Coast Salish, were excluded from the land.

With roads carving up the landscape and the common lands rapidly disappearing, hunting and gathering became increasingly difficult. The lowland areas were the first to be settled, yet these were also the most productive wildlife habitat. In this way, the ecological integrity of the land was compromised even before the full impact of the technological era began.

LOOKING AT THE SURVEY GRID

The rectangular survey grid formed today's landscape and molded many cultural aspects of life around Boundary Bay.

This survey method divided land into 6 mile by 6 mile squares (9.6 km), each covering 640 acres (256 ha). A half section was 320 acres and the quarter section, the regular pre-emption or homesteading parcel, was 160 acres (64 ha).

The basis of the land survey in the lower Fraser valley was the Coast Meridian, a line running due north from the intersection of Semiahmoo Bay and the international border. This survey line was cut by the Royal Engineers, and later became 168th Street, Surrey.

The typical grid pattern is obvious in most of the Lower Mainland. Early deviations from the grid can be seen in roads like Arthur Drive, Delta, Glover Road, Langley (the Railway Belt) and Old Yale Road (Highway 1A). Aerial photographs show the checkerboard pattern of the survey grid.

Surrey landscape from the air.

Land clearance in Steveston, 1890. (City of Vancouver Archives)

SETTLING & CLEARING

In 1860, the government allowed the pre-emption of land in the new colony of British Columbia. This opened the door for settlers to occupy and clear unsurveyed land, with the understanding that a fee would be due once surveys were completed. At that time there were only about 300 immigrants in the lower Fraser Valley, and a handful more around Semiahmoo Bay.

South of the border a somewhat different program applied. A quarter section was granted free to homesteaders in return for a commitment to settle and "improve" the land. This condition motivated the pioneers to clear, fell, fence and "tame" the landscape, an urge that is still strong in many people today. The homestead system was adopted in British Columbia in 1873, after complaints about the inequity of having to pay for land. By that time,

nearly all land around Boundary Bay, apart from the boggy heart of the delta, had been taken up by hopeful farmers or land speculators.

At first, people settled mostly at the shoreline. James and Caroline Kennedy and their young son arrived in 1859, and lived in a tent on the banks of the Fraser River across from New Westminster. A few Norwegian fishermen also settled on the river. In the 1860s, demand for the easily accessible riverbank land increased; sections from Crescent Slough to the river mouth were claimed first and remaining properties, including Deas Island, were taken up by 1880.

The first men to get their name on the registry in Boundary Bay were Samuel Hardy and Hugh McDougall in spring 1861, for land at the mouth of the Nicomekl River. Settlement on this sodden, marshy land was slow, and the 1874 voters' list for Elgin has only fourteen men's names. The

Mud Bay uplands were eventually homesteaded by families such as the Dinsmores, Collishaws and Loneys. A little school was set up in the marsh (later moved to Elgin), the children walking there each morning along the Nicomekl River. In the 1870s, Walter Blackie bought land at Crescent Beach, near the spit that now bears his name. Although several claims were recorded elsewhere around the bay, many were not taken up, and it was only in the 1870s and 1880s that families like the McKees, Bensons, Taskers and Booths pre-empted the Boundary Bay shoreline.

It must have been a daunting proposition to settle the land. Many pioneers had already tried their hand at different occupations in other places. The Ladner brothers, Thomas and William, arrived in the delta from San Francisco in 1858, hoping to make a fortune in the gold rush. After a spell as teamsters, they turned to raising cattle. They pre-empted land at Chilukthan Slough in 1868, soon called "Ladner's Landing," where the south arm of the Fraser ran full and wide past the shore. After marrying the Booth sisters, the Ladners made a sound living on the Delta, rising to local prominence in the community (page 145). They were joined by the Kirkland, Guichon, Burr and Arthur families, among others. Their stories and those of other Delta pioneers are engagingly related in Terrence Philips' book, *Harvesting the Fraser.*

Mrs M. Beveridge, her son Alex and two daughters, clearing land in Surrey, 1918.
(Surrey Archives)

In 1875, the Innes brothers came from Ontario intending to work in the Cariboo, but settled instead on Langley Prairie, at Innes Corner (now Langley City). The "prairie" was a relatively open area in the forest, one of several in Boundary Bay's watershed, including Clover Valley, Hall's Prairie, Kensington Prairie, and Custer Prairie. A few Surrey settlers also chose these areas, including George Boothroyd who pre-empted a quarter section in Surrey Centre, now the heart of a major metropolis, and Henry Thrift and family who settled on Kensington Prairie.

Land clearance was slow

Land clearance was slow, hampered by the timber monopoly in place at this time, which prevented the settlers selling any logs or shakes cut from felled trees. Even though land was being taken up, people actually living on their holdings were few and far between. The magic number of 30 residents, necessary for incorporation as a municipality, was only reached in Surrey in 1879.

One of the first families to homestead in Drayton Harbor was John Harris, his wife and four children, who lived in a one-room cabin near the mouth of Dakota Creek. They arrived by boat at Semiahmoo Spit and waded ashore, then rowed to the creek mouth at high tide. Their new home was inaccessible at low tide, when thick mud covered the harbour, and they had to clear log jams to keep the river navigable. They soon had neighbours. California Creek was settled by 1873, with about a dozen families setting up permanent residence in heavy timber. Two years later, Amos Dexter dammed a stream entering the harbour and set up a waterwheel-driven sawmill. It was not a very successful enterprise but was the first of the lumber businesses in the region. By that time, the Harris family had moved on to Point Roberts, where they took to stock raising, eventually owning 40 head of cattle. "Long-haired Harris" became renowned for running up debts and toting a gun, so it was not too surprising that his life ended tragically, killed by a neighbour at the age of 68.

Tongue Spit (Semiahmoo), where the steamboat from Puget Sound docked, developed a small town and later a cannery, all dependent on a freshwater well that tapped into the artesian strata far below sea level. As was typical for the time, the Chinese men who worked at the cannery were brought up from Oregon under a Chinese contractor and came without wives or family. They bunked at the workplace and did not acquire land, instead sending any saved money home to their families in China. Later when Blaine on the mainland became established, the town on the spit declined and soon only the cannery was left.

Government assistance encouraged a community of Icelanders to settle at Blaine, Birch Bay and Point Roberts in the 1880s. They worked in canneries and on fishing boats, raised cows, and grew vegetables. However, they could not pre-empt land at Point Roberts, as it was held by the US government as Army Reserve. Only two of the squatters, Kate Waller and Horace Brewster, who had settled prior to 1880, were allowed legal homesteads. The other families only gained the right to own land after petitioning President Roosevelt in 1905. The Icelanders' grateful present of a hand-crafted sheepskin rug was placed in a White House bedroom.

Sarah Olson, a widow with 7 children, was Point Roberts' lighthouse keeper in 1906. (Point Roberts Historical Society)

ROADS, RAILS & TELEGRAPH TRAILS

For most of human history around Boundary Bay, villages and camps were built close to shore and people traveled everywhere by canoe. The first Europeans came by sailing ship and boats were used by everyone for fishing and waterfowling and for ferrying people and goods. It was only when more people came and land was taken up away from the coast that the need for roads and rail lines was felt, spurred on by the growth of small towns.

The present day scenery around Boundary Bay is marred by overhead wires and pylons, which often stand in the middle of fields, forcing farmers to plow around them. Wires are fatal for flying birds that fail to see them in dim light. Hydro crews spend hours felling and pruning trees away from lines to prevent power outages. The wires are above ground because the water table is at, or very close to, the surface in the lowlands. This also presented a problem for road building. Many early roads were made of logs, laid side by side in a "corduroy" pattern, but first attempts had the logs sinking into a muddy morass.

Road-building was a labour intensive, summertime task and in the meantime people walked on footpaths through the forest. Few people had horses and most trails were too narrow for heavy ox carts. James Kennedy cleared a trail in 1861 along the foot of the North Delta escarpment, from his farm beside the Fraser River down to Mud Bay. Gold miners followed a path from Semiahmoo to Fort Langley, through Hall's Prairie, that was known as the Fort Langley Trail. Later, the Coast Meridian survey slash line (now 168th Street) became a shortcut through to the trail.

The Semiahmoo Trail linked New Westminster with Blaine

In 1865, the Collins Overland Telegraph Line marched through Bellingham Bay, Birch Bay and along Kennedy's track to New Westminster, in an attempt to link San Francisco through Russia to Europe. This ambitious project was destined for failure. No sooner had the line been laid through the British Columbia back country as far as the distant wilderness of Telegraph Creek on the Stikine River, than news came that the trans-Atlantic cable had already been laid. The line was discontinued. The old Whatcom Trail was revived as the route of another short telegraph line, later becoming a wagon road, known of course as Telegraph Road.

The 1873 Semiahmoo Wagon Road, also called the Semiahmoo Trail, was accessible to horse, pedestrian and ox cart traffic, and linked New Westminster with Blaine. Some of the trail in South Surrey is now protected as a public footpath.

By 1880, Delta's old Kirkland Road, (now Ladner Trunk Road), was built up with soil and cedar logs, lined on each side with a drainage ditch. All roads through the delta have to be elevated above the water table in this way, and even today dips in the highway can flood during a winter storm. Kirkland Road crossed the delta north of Boundary Bay, linking the new community of Ladner's Landing with Mud Bay. Another early route, the Old Yale Road, begins at New Westminster. Passing between the tall Douglas-fir of Green Timbers Urban Forest, this 1874 road (now the Fraser Highway) heads directly towards the snow-capped peak of Mount Baker, an outstanding landmark for the early settlers.

In 1885 the Canadian Pacific Railway reached Vancouver, launching the era of the powerful railway barons. The Blaine to Brownsville line was then constructed under the auspices of the Great Northern Railways, providing the north-south route so desired by some settlers, as well as a branch line to Ladner. Brownsville was where the road from Blaine and Elgin arrived at the Fraser River, and was just a ferry ride from New Westminster and the rail connection to Vancouver. The new rail line was a useful, if slow, link for communities south of the river. The Langley Railway Belt, federally administered land reserved for rail companies, was eventually surveyed and opened up for homesteading. The railways increased interest in settlement and the local population rapidly grew. The beginning of the twentieth century heralded a very new era of technology, mobility and growth, closing the chapter on the old way of life for Boundary Bay.

Ecological impacts

From an ecological perspective, roads and railways have many impacts. As they carved into the wilderness, they facilitated human settlement while accelerating the decline of wildlife habitat. Islands of habitat were left, with a resultant loss of biodiversity.

A few generalist species, such as crows and coyotes, thrived as the country opened up, but many forest and wetland species disappeared. Roads and railways made it easier for exotic, non-native species to invade. Dandelions and plantains spread along the roadsides, while earthworms and black slugs took over newly turned soils (see page 117). Railways continue today to introduce species: yellow-bellied marmots from the Interior of British Columbia have been found at Deltaport on Roberts Bank. Road traffic kills and injures hundreds of animals and birds every year, including rare owls and hawks. More settlers meant more pets, many of which are very ecologically destructive. Cats kill hundreds of

Mud Bay rail crossing.

millions of birds and small mammals every year in North America, and loose-running dogs disturb nesting and roosting birds. Other pets, such as rabbits, turtles and exotic fish, have been released into parks and ponds, displacing native species.

As dependence on roads and rail for transportation grew, the importance of rivers and sloughs waned. Many silted up or disappeared entirely. Chilukthan Slough in Ladner, once a navigable waterway, is now a narrow stream which disappears after a few miles. The Serpentine, Nicomekl and Little Campbell were once routes into the interior of the countryside. Control dams, built 1910 onwards, effectively closed them to anything larger than a kayak.

Ladner fire of July 6 1929. (Delta Museum & Archives)

SUMMER FIRES

Wild fires were a normal part of the seasonal landscape cycle before the last century, even on the damp west coast. They were started by lightning strikes or as regular burns set by the Coast Salish to keep berry areas productive.

John Work traveling up the Fraser in September 1835 noted how the woods were on fire the whole time, with a dense fog of smoke obscuring the hillsides. John Fannin, exploring the Fraser Valley for the government in 1873, mapped numerous areas as "burnt timber" or "burnt ground." From the Semiahmoo Trail eastwards he found fern land and alder belts,

places where fire had passed and the vegetation was regenerating. As the country opened up, fires became a constant problem for towns and structures that were built entirely of wood. A logging camp fire got out of hand in August 1912, sweeping along the top of the White Rock bluffs. Another big fire burned for two weeks across the hillsides in July 1916. In Blaine Harbor, the Jenkins Mill was destroyed by fire in 1908 and a lightning fire took the Morrison Mill in 1939. Blaine sent fire fighters to help with a big dock fire in White Rock in 1919.

Similar fires took place in all of Boundary Bay's early communities from Ladner in the west, where the old Chinatown burnt to the ground, to Cloverdale and Langley in the east. An 1898 fire consumed a third of New Westminster, and Steveston burnt disastrously in 1918. Fires throughout the 1920s and 30s caused hardship and sometimes tragedy. Three cabins and the Waters Pavilion store and dance hall at Boundary Bay, burned in 1935. This resulted in the death of Mrs Provency, the operator.

Layers of charcoal in the soil and charred stumps are evidence that Burns Bog has often been swept by fire, a hazard that continues today. Recent blazes occurred in 1990, 1994, 1996 and 2005, with smoke spreading across the lower mainland. Bog fires are difficult to extinguish, because the deep layers of peat keep the fire burning underground. There is an increasing risk of future fires if optimal water levels are not maintained in the bog.

Further information on pioneer history

A Historical Atlas of Vancouver and the Lower Mainland by Derek Hayes

The Fort Langley Journals edited by Morag MacLachlan

Nooksack Tales and Trails by Percival Jeffcott

Point Roberts - History of a Canadian Enclave by Richard Clark

Harvesting the Delta by T. Philips

The Surrey Story by G. Fern Treleaven

Along the Way by Margaret Hastings

The Surrey Pioneers by R.V. Whiteside

The Langley Story by Donald Waite

Gwen Szychter's series on Ladner and Tsawwassen-see bibliography for titles.

Museums and heritage buildings - see page 185 onwards.

Aftermath of the Steveston fire, 1918. (City of Richmond Archives)

Irrigation systems bring freshwater to Boundary Bay fields.

METAMORPHOSIS OF LANDSCAPE

"It was nothing to see forty or fifty grouse sitting in a cottonwood tree."
~ Margaret Stewart

The Boundary Bay landscape that European explorers first described has undergone an enormous transformation in the last two hundred years. It is difficult to imagine now how the original dense forests that covered the uplands could all have been logged. Yet that is what has happened. The few remaining tall Douglas-fir and western redcedar trees in Surrey and North Delta are regrowth over the last hundred years. The wetlands that presently surround Boundary Bay, so rich in life, are only remnants of what was once a great expanse of wetland, teeming with wildlife and covered with clouds of insects. Our mighty Fraser River once ran wild and free, finding its own way to the ocean and carving its own banks, but today it is bound to its course by dykes and training walls.

The wilderness that the settlers first encountered held dozens of animal species. The plant life was diverse and abundant, offering resources that have still not been fully explored, while the wealth of the marine environment was without equal. Many aspects of this wonderful natural world still exist, although they are often disturbed and fragmented. Some species have gone, unable to adapt or compete, and many others are at risk.

It is estimated that at least 70% of the original wetlands of the Fraser estuary were dyked and drained by 1900, and since then the rivers through the watershed have nearly all been tamed. Wetlands, wet prairies and wildflower meadows have disappeared, although a ribbon of new marshes has flourished outside the dykes. Many plants and animals not native to the Pacific Northwest accompanied the influx of newcomers from across the world, some of which have become abundant, usurping the habitat of earlier species.

In the path of the pioneers and settlers came exploiters, taking resources to the brink of existence with enthusiasm and boundless energy, and absolutely no apprehension for the future. It was the start of the modern world and resulted in a total metamorphosis of the landscape.

WET PRAIRIES AND MEADOWS

The low-lying land in the Fraser delta was originally a vast wetland, a mix of tidal and freshwater marshes and bogs. In winter it was awash with water, yet by midsummer a waving sea of grasses spread across the delta, mingled with sedges, rushes and colourful wildflowers. Along slough and river banks, on the old beach ridges, and scattered through the meadows were shrubs and small trees. Willows, crabapple, black hawthorn, red-osier dogwood, Nootka rose, hazel and purple-flowered hardhack grew profusely on the fertile delta soil.

These beautiful wet prairies have almost vanished. Remnant shadows are only found on the foreshore outside the dykes and where old fields have lain fallow for three or more years, flooding and ponding with winter rains. Dyking and drainage for farming, roads and urban development of the estuary, substantially dried out the native prairies. New grasses and crops arrived with each wave of settlers, on cattle hooves, cart wheels, trains and planes, spreading rapidly and displacing local plants (page 95).

In the 1800s, wetlands were seldom inherently valued and were often dismissed as wastelands or swamp. They were a force to be conquered and tamed, and while a source of food from waterfowl and fish, only a few people appreciated them for their wilderness beauty. People living on the delta waged a constant battle against the river and tides. Their houses were flooded, dykes washed away and sweet drinking water was scarce. Most of the year, everything was either covered with mud or dust.

As soon as the lowlands became available for pre-emption, European and American immigrants hastened to settle and develop them, despite the problems; flat land for building and farming has always been in very short supply on the mountainous west coast. Farms soon spread out wherever the wetlands were drained and the forests felled.

Beavers were abundant in the valleys in these early days, chewing on willow bark and demolishing trees to make their dams. John Work who travelled the Nicomekl and Salmon River valleys in 1824, found "the appearance of beaver being pretty numerous." The ponds and pools they created made excellent nesting areas for waterfowl, provided spaces for amphibians and fish, and acted as a natural flood reduction scheme by slowing the rate of water drainage. Fur trade hunting dramatically reduced the beaver population. As trees were felled, a call for dyking and flood prevention went out as several large floods subsequently occurred in the lowlands.

FRASER DELTA SURVEYS

In 1979, Margaret North, Mike Dunn and Jan Teversham used the journals of the Royal Engineers' survey of the Fraser delta to create a map of mid-nineteenth century vegetation and calculate the amount of wet prairie habitat lost to development (see page 85). The researchers found that in 1858 wet prairie covered 12,500 ha, denser shrub land accounted for about 2,500 ha and there were about 5,000 ha of other cattail and bulrush marshes, a total of 20,000 ha (1 ha = 2.4 ac). Bogs were not included in the calculation.

By 1978 these habitats had shrunk to less than 6,500 ha within the survey area. The wet prairie was reduced to less than 2,500 ha and the shrub landscape shrank to barely a tenth its original extent, at 250 ha. Dyking, draining and land development for agriculture and housing accounted for most of these irreversible losses.

A more recent study by Michael Church and Wendy Hales, that examined 75 years of air photographs, showed that marshes at the mouth of the Fraser outside the dykes expanded rapidly from 1894 to 1954, and increased more slowly from 1955 to present. On the active delta front they grew from 1,644 ha in the 1930s to 2,440 ha in 2004. However, where hard structures such as causeways prevent sedimentation, these gains are slowing or reversing. Delta marshes are important as the first buffer against storm surges and rising sea levels due to global warming, as well as being essential wildlife habitat.

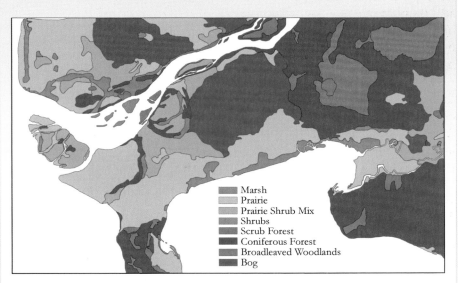

Historic delta habitats around Boundary Bay based on the map by North, Dunn and Teversham 1979.

Beach Grove houses buffeted in the storm of 4 February 2006. (Photo by Anne Murray.)

DYKING, DITCHING AND DRAINAGE

Early in the morning of 4 February, 2006, strong winds and exceptionally high tides unleashed a torrent of water on the west shore of Boundary Bay. Two hundred undyked homes in Beach Grove and the village of Boundary Bay were flooded. The sea overtopped sections of dyke in Boundary Bay Regional Park, swamped the beach and tossed huge logs around like matchsticks.

Flood has always been a real threat in the delta. The land is at or below sea level, and the sea level is slowly rising. The Fraser River that created the delta landscape is awesome in its might. It broke its banks disastrously in 1894 and again in 1948. Strong freshets (flow due to snow melt) occurred in 1972, 1999 and 2002. Other rivers in the Boundary Bay watershed have flooded repeatedly. The water table is very high in the delta and every winter many of the agricultural fields are inundated.

Early homesteaders tried to protect their properties by surrounding them with earth berms and ditches. They started with a series of drainage ditches emptying into Mud Bay in East Delta. This dried out the boggy land at the foot of the Panorama hills, and created fields for growing potatoes, hay and corn and rearing dairy cattle. The landowners still did not feel secure, and to guard against the risk of flooding from winter tides they cooperated in building a 10 km (5 mi) stretch of dyke along Boundary Bay. This 1892 dyke, built by the McLean brothers, reached from John Oliver's land at Mud Bay, west to 60th Street, Delta. Although Delta was relatively unscathed by the great Fraser flood of 1894, storm surges washing through Ladner in January 1895[1] prompted the newly-formed Delta Municipal Council to vote for complete dyking of the south bank of the Fraser through the delta, an enterprise begun in 1895 and completed by 1898. Much of this dyke was built by contractor Henry Benson, while the stretch around Brunswick Point was built by the Swenson family. The total length of Fraser estuary dykes measures 620 km (300 mi), creating the scenery visible today across the flood plain.

Ladner's Waterways

Prior to dyking, the mighty Fraser River flowed in many channels through the delta. The pattern of sediment deposition shifted regularly, and the Main (or South) Arm of the river moved too. At the time of the 1858 gold rush, the deep water channel swooped south to run along the bank past Ladner's Landing, a harbour for boats coming across the Strait of Georgia. An early resident, Leila May Kirkland, described the scene: "a rather fearsome world made up of a deep river, open prairie

with only one small house in sight; no marshes or sand bars could be seen and the river stretched in an unbroken expanse from the Delta side right across to Lulu Island in the distance."

Chilukthan Slough, Ladner, was then a wide creek flowing south from the Fraser and entering the sea at the Tsawwassen Reserve. A punt ferried settlers across it. After the 1894 flood, the Fraser River Main Arm changed course northward, silting up the old channel and creating the South Arm Marshes. Control structures built by the federal Public Works Department for the port at New Westminster, starting in 1866, contributed to this change in the river. Chilukthan Slough was choked off, and a flood box then reduced its flow so much that the southern stretch disappeared. The winding road of Arthur Drive, south of Ladner, follows the remnant waterway (page 192). Canoe Pass is also silting up and the Roberts Bank marshes are becoming increasingly saline as the Fraser flows northwest.

A break in the new dyke, Ladner 1899.
(BC Archives)

---

1. See photograph on page 78

Surrey's changing rivers

Flooding from the Serpentine and Nicomekl rivers in Surrey was an annual event during the second half of the nineteenth century, with dramatic washouts occurring in 1882, 1894 and again in 1895.

The Serpentine River was then tidal as far up as 176 Street and the Fraser Highway, north of Cloverdale. Approaching Mud Bay, its wandering channels merged with those of the Nicomekl River, which drained 180 sq.km (30 sq.mi) of watershed and was a major transportation route between the bay and the Fraser River. The two rivers emerged in the sea through a peat and salt marsh environment, the haunt of river otters, beavers, muskrats, short-tailed weasels and hundreds of thousands of waterfowl and shorebirds. Massive logs and tree stumps were buried deep in the peat and mud of the estuary, making land drainage and clearance a daunting task for new settlers. As late as the 1940s, Surrey farmer Allan McKinnon pulled out 12 m (40 ft) cedar and fir trunks with seaweed underneath them from land south of 64 Avenue.

After several disastrous attempts, during which everything was washed away, a sluice gate operated by water pressure (a sea dam) was finally constructed on the Serpentine River in the early 1900s. It prevented the salt water intruding into the boggy lowland and was followed by dyking and drainage of the floodplain. The estuary was considerably reduced in size by the dam and the sharp transition between salt and fresh water prevented incoming salmon adjusting to the river system. The Serpentine Canal was cut laboriously by hand through the bog and marsh of the floodplain; this was the work of Chinese labourers in the 1920s. It was originally 2 m (6.6 ft) wide, with cedar plank drains laid through the peat channels, part of a series of taming measures for the river, which now has dyked and reinforced banks throughout the agricultural land.

The Nicomekl was a beautiful river in the 1930s ...

All the early drainage schemes were privately financed by bonds purchased by landowners under the auspices of the Surrey Dyking District. Government money was not involved until much later. Colebrook Dyking district was formed in 1923 and that of Mud Bay followed in 1946, extending the dykes along the foreshore and onto the Nicomekl. Certain dykes in this area remain closed to the public.

The Nicomekl was a beautiful river in the 1930s, flushed by the tides, full of fish and oysters, and a favourite swimming location. For much of its course it is still a pleasant river, flowing through shady woodlands

Entire-leaved gumweed flowers in a remnant marsh at Elgin Heritage Park.

and grassy wet meadows, its banks lined with willows, salmonberries and Pacific ninebark. Sadly, it competes with increasing urban and agricultural development. The lower reaches are dyked on the northern bank and crossed by several highway bridges. A sea dam operated by tidal pressure prevents sea water from intruding far upstream.

For a brief time there was a canal across the large river bend near Elgin, as a short cut to log booming grounds in the estuary. It created Huckaway's Island, a boggy stretch of land full of cranberries and blackberries. The canal has since been filled in and the bog is now a golf course, although a remnant marsh exists at Elgin Park.

After the second World War, the B.C. government took responsibility for dykes along the Fraser River and around Boundary Bay, but not for the smaller rivers. This proved to be a problem for local communities. Heavy rains, gales and high tides in November 1951 tore a huge gap in the Serpentine dyke and flooded acres of farmland. More floods occurred in 1968, 1979 and 1982. The Nicomekl had major floods in 1966, 1967, 1972 (when a long section of dyke gave way) and in 1983. The government eventually relented and helped with the cost of repair and maintenance. With the pumps, dams, sluice gates and dykes in place, there was still the problem of getting salmon upriver. In 1990 the Erickson pump station

EXPLORING THE DYKES

The Boundary Bay dykes are a great place for walking, cycling, birding or horse-riding. A continuous 20 km (12.5 mi) trail extends from Mud Bay Park in the east to Beach Grove in the west, with further trails through Boundary Bay Regional Park. A dyke trail also leads from River Road, Ladner, around Brunswick Point at the mouth of the Fraser River. Richmond has many kilometres of dykes along Sturgeon Bank and the north bank of the Fraser River.

Dyke trails tend to be open and unshaded, so can be cold and windy in winter and hot in summer. The rewards are wide open views of sky, sea, marsh and fields and good wildlife watching, particularly in spring, fall and winter. Dunlin create smoke-like clouds in the sky as they wheel across the bay to escape peregrine falcons, and flocks of wintering snow geese and trumpeter swans descend on Brunswick Point.

The life of the river is fun to watch, with tugs and barges moving to and fro, the occasional sighting of harbour seal, beaver, or muskrat, and interesting communities along the banks - Canoe Pass float homes, beside the old Westham Island bridge, Ladner Harbour, and Finn Slough, Steveston.

Dyke access varies - consult *A Nature Guide to Boundary Bay* by Anne Murray and David Blevins for more detailed information.

was constructed using an Archimedes screw pump, designed to allow fish safe passage.

Soon upland forests were logged. Water then flowed rapidly down hillsides, causing new erosion and drainage problems. Ponds were dug to collect run-off water, but as development increased they proved ineffective. More recently, a large stormwater retention pond was constructed just east of 152nd Street, and its surroundings made into Surrey Lake Park.

Water quality is affected when shorelines are degraded

The Little Campbell River, Dakota Creek and California Creek were not dyked. Long stretches of them still remain as natural rivers, winding their way through gently rolling countryside. These streams are very sensitive to land use along their banks. Riparian vegetation, such as cottonwoods, alders and willows, are important wildlife habitat. Surrey pioneer Margaret Stewart, born in 1876, remembered seeing as many as fifty ruffed grouse perched in a cottonwood tree in the area now occupied by the town of Blaine. Ruffed grouse were also common throughout White Rock, Surrey and Point Roberts. They are seldom seen in the watershed today. The pastoral landscape on these rivers is now changing. Residential developments are more common, and wildlife habitat is disappearing, particularly on the Little Campbell River. Water licenses consume more than the flow of the river and water quality suffers as trees are cut, shorelines degraded and impervious ground surfaces increased.

Farmland ditches

Throughout the low-lying delta farmland, ditches run alongside field margins, draining away winter rains. These ditches need to be regularly dug and cleared, otherwise they rapidly silt up and become choked with vegetation. The ditches create wildlife habitat similar to that of natural sloughs; great blue herons, mallards and red-winged blackbirds are a common sight. Some animals are not welcomed by farmers, such as the muskrats and beavers that dig holes in the banks. Salmon use the ditches, creating arguments about dredging and destruction of habitat.

Dredge "Beaver No.2" in Richmond, ca.1908. (City of Richmond Archives)

LOST STREAMS

Unknown to their residents, some homes and yards around Boundary Bay are built over streams, ponds and even fresh water springs, now hidden in the landscape. These are the lost streams of the Fraser valley, dating back to the end of the Ice Age, when every bluff and bank drained water, eroding the glacial gravels and collecting in boggy pools at the foot of the uplands.

Many of these streams were filled or culverted when the land was developed, which greatly reduced the clouds of mosquitoes and other insects that hung over the delta in summer, much to peoples' relief. However, the lost streams had been vital fish nurseries, and fish, frogs, songbirds and waterfowl relied on the myriad of insects. As the drainage continued, many small runs of coho, chinook and chum salmon disappeared, together with birds like the purple martin, western bluebird and nighthawk, that feed on flying insects (see page 158).

At least three streams flowing into the Nicomekl and several into the Little Campbell have been lost. In Surrey, tributaries of the Serpentine disappear for long stretches through urban developments and many upland ponds have been drained. In White Rock, fresh water springs once emerged all along the foot of the hills where shops and houses now stand.

Streams poured down the bluffs into Semiahmoo Bay, flowing through deep ravines, thick with ferns and shade trees. A few survive, as they do in parts of North Delta, where streams down the northern escarpment flow towards the Fraser. Artesian springs also bring freshwater to the surface in Watershed Park (Map D3).

South of the border, California and Dakota Creeks, flowing into Drayton Harbor have silted up considerably since pioneer days partly due to the extension to Blaine Harbor. Both creeks are now little more than meandering streams. They drain the farmland, yet have been less affected by culverting or realignment than others in the watershed. Both have a full complement of feeder streams.

Freshwater is scarce in Point Roberts, which is supplied with water from Vancouver reservoirs. Streams and draws from the Tsawwassen uplands drain towards the ocean. The sea once lapped at the "dyke" of Boundary Bay road, and the area which is now regional park was foreshore marshes and tidal sloughs. Construction of the 12th Avenue dyke at Beach Grove effectively dried out this marshy area. Freshwater from a stream flowing through the neighbouring golf course is pumped into an intertidal lagoon on the north side of the dyke. The sandbars and channels of the lagoon constantly reshape under the influence of tides and storms, and slowly the area is drying out. A new spit is building further south along the beach, and creating a new lagoon behind it. This change in coastal morphology is a natural part of the ecosystem's development, involving cliff erosion and longshore drift. The resulting lagoons are used by many migrating and wintering birds especially at high water (Map C4).

As the ecological importance of streams is better understood and valued, some of the old culverts should be opened up and their banks restored, allowing fish to return. This is happening in other locations around the world.

Cain Creek empties from a culvert and flows across the mud flats near Blaine.

Wind throw and large downed trees were common in the old growth forests that surrounded the bay.

LIFE IN THE FOREST

Two hundred years ago the hills around Boundary Bay were covered with massive Douglas-fir, western redcedar and western hemlock, in a lush, oldgrowth, coniferous forest, typical of the Pacific Northwest. The forest floor was a dense tangle of fallen trees, shrubs and ferns. Streams fell in untamed torrents down the hillsides, drying to a trickle in late summer. Down on the flood plain, water-tolerant black cottonwoods, red alders and paper birches lined the rivers. Wildlife was abundant. Wolves, bears and cougars were all common and feared by the settlers.

According to one story, the trapper Sam Hall and his wife, who lived on Hall's Prairie for thirty years, were chased off their land by timber wolves. Wolves harassed the newly introduced herds of cattle brought in from Oregon by the Hudson's Bay Company to stock the Fort Langley prairie. They were still a problem for ranchers twenty years later, attracted by the easy pickings of domestic animals. Eventually they were shot out. The last wolf recorded in the estuary was at the end of the 1800s. Coyotes, in contrast, benefited from forest clearance and the lack of competition, proving to be both adaptable and persistent. Despite constant persecution and bounty hunting, they expanded their range across North America in the wake of the grey wolf, and are now common around Boundary Bay.

Oral histories of the Stò:lo and Semiahmoo recall the presence of grizzly bears historically in the region, but we never see them today. Black bears were common up until the early 1900s, although very few still live in the Boundary Bay watershed. Franklen Brunson described how "in the early days at Custer they were afraid to send children unattended to school on account of the numerous bears and cougars that infested the timber then." The last wild bear in Richmond was shot on No. 6 Road, Lulu Island, about 1948. The same year saw the last one in White Rock, although a black bear came down McNally Creek on the Surrey side of town one September evening in 1990. It stayed around for two weeks, fishing salmon out of the Little Campbell River and resting in the woodland of the Semiahmoo First Nation Reserve. Bears have also occasionally been seen in the lower watershed, especially when habitat upstream was destroyed.

Roosevelt elk were the largest ungulate in the Boundary Bay area. John Work noted them in his 1824 diary. "Immediately we put ashore Pierre Charles went to hunt and shortly returned having killed 3 elk and a deer....The great number of tracks seen by hunters indicated that elk are very numerous about this place." Further up the Nicomekl he noted "elk have been very numerous here some time ago but the hunters suppose that since this rainy season they have gone to high ground."

Elk were a regular sight in Burns Bog before 1900, where they were hunted by the Halkomelem-speaking Coast Salish. Game abounded in the forests of the Nooksack Valley and along California Creek. Their presence was recalled by George Kinley, a Lummi elder, in 1949: "My grandfather, Chanique,[2] I well remember; he was a great hunter of bear and elk, which he successfully pursued with bow and arrow. The low lands along California Creek were his favourite

hunting grounds for bear, as the salmon ran heavily up that stream to spawn and salmonberries were plentiful also...For elk he generally went to the uplands around Lake Terrell and what we now call Mount View, where *hias mowich* (big deer) were very numerous." The elk left this area around 1855, perhaps spooked by increased hunting pressure.

Columbian black-tailed deer were also very plentiful throughout the region and some remain. They were common in large acreages and woodlands in Tsawwassen until the 1960s and 70s. In South Surrey, residents remember them at Chantrell Creek and Crescent Beach, and recall seeing them lick salt from the rocks on Ocean Park beach. There are still deer living wild in Langley, Burns Bog, Point Roberts and other wooded areas of the Boundary Bay watershed.

LOGGING THE FORESTS

Oldgrowth forest in the Boundary Bay watershed was composed of trees of all ages, including some that were hundreds of years old, grown to an immense size. Douglas-fir regularly occur as forest giants yet all modern champion "big trees" would have been surpassed by one felled by William Shannon in 1881 near Hall's Prairie. The fallen trunk was 109.1 m (358 ft) long, the tallest conifer ever measured by a forester in

---

2. Pronounced Cha-hay-nack

British Columbia. The stump of the 1,100 year old tree was 3.5 m (11.5 ft) across! At first, pioneer settlers did not attempt to fell these giants, but cleared the forest in gradual stages, by removing shrubs and ring-barking the trees. The "ring-barking" method cut the bark away in a broad band so that the tree eventually died and came crashing to the ground. Another dangerous approach was to set fire to the roots and wait for the tree to fall. Joseph Figg, a Surrey pioneer, was probably not the only man to lose his life when a fired tree came crashing to the ground.

Logging at that time was all hand work. Trees were felled from spring boards, planks driven into notches on either side of the trunk, usually a metre or two from the ground. This avoided the stump's close-grained wood and provided space for the saw. It was slow, dangerous and difficult work, conducted under primitive conditions and in all types of weather, but there was no shortage of labour. Many settlers, both men and women, earned money for building cabins and getting established, at the logging camps, whether it was felling trees, hauling logs or cooking for the workers.

Old stumps, with their springboard notches still visible, are found in many upland parks. They are a poignant reminder of the beauty and wealth of the historic forest environment.

Two loggers are dwarfed by a giant stump at Point Roberts. (Point Roberts Historical Soc.)

Rivers were the major transportation arteries, so ditches and chutes were excavated to reach them more easily. Once logging began in earnest, teams of oxen and horses were brought in to skid the logs to the nearest river, where they were floated downstream to the booming grounds and then towed to the mills. "Booming grounds" are stretches of river, close to shore, where free-floating logs are stored in the water, confined by a chain of connected, floating logs, or booms. After 1887, steam donkey engines and railroads were incorporated into the operations.

Lowland forests and those on the banks of the rivers were the first to be removed, including one of the best

Elgin Douglas-fir today.

stands of Douglas-fir the loggers had ever seen, on the north-facing bank of the Nicomekl River, near Elgin. Tall second-growth conifers grow in this location today (Map E3). Across the Nicomekl valley, in Colebrook, the Burr family logged Panorama Ridge and the Mitchells worked Surrey Centre, both operators floating the logs along ditches at the foot of the ridge to the Serpentine River. The White Rock area was first logged by John Roper, using the Little Campbell River to store logs.

Once the towering trees had been felled, the stumps had to be hauled out using chains and horses or blown up with powder, caps and fuses. This highly dangerous work often had calamitous results. Harry Bose remembered a dreadful accident in the 1920s when Chow Wing, part of a Chinese labour crew clearing his land, was blown apart when he went to check on a reluctant detonator.

The stumps were often massive, far surpassing the stumps of modern trees. Even in 1947, when few giant trees were left, Allan McKinnon measured one oldgrowth stump in Surrey Centre at 3 m (9 ft) diameter. The holes they left were great craters, that had to be back-filled with soil scraped from the surrounding area before being plowed. Brush piles were burnt and the smoke of a thousand fires drifted over the Boundary Bay hills as the appetite for land and logs grew.

Herbert Gilley Logging Co. crew with steam donkey engine, Surrey, ca.1890 - 1910.
(Surrey Archives)

In the early 1900s, there was so much timber being cut that sawmills were built at Elgin, Tynehead, Cloverdale, Hazelmere, Sullivan, Blaine, Drayton Harbor and the mouth of the Little Campbell River. The Campbell River Lumber Company, which worked the woods throughout White Rock and Crescent Beach, was among the first to use a logging railroad for moving logs over to the Nicomekl estuary booming grounds. Railroads increased efficiency and sped up forest clearance, and were soon in general use for the industry.

Where the land was not developed, second-growth forests grew quickly in the moist coastal climate, softening the barren clear cuts. Even as second-growth, the conifers are magnificent trees, enormously tall by global standards. Mere youngsters, these trees lack the density of lichens, mosses and ferns of the oldgrowth trees. They will need to be allowed to grow for a good few hundred years to approach the maturity and richness of the forest they replaced. Sadly, thousands are still being cleared from Boundary Bay uplands today.

THE GREEN TIMBERS STORY

"How can a city not have a forest?" ~ Peter Maarsman, Green Timbers

By 1930 most of the forests had been logged and Surrey's premier industry came to a close. Green Timbers on King Creek, a tributary of the Serpentine River, was the last major area to be clear cut and it has a chequered history.

The busy Old Yale Road, connecting New Westminster to the Fraser Canyon, was built in 1874 through dense forest at the height of land in Surrey. The towering 60 m (200 ft) trees lining the road became a famous landmark. Their fame and citizen protests did not prevent them being logged through the 1920s.

In 1929, Premier Tolmie legislated the protection of Green Timbers forest, and creation of the first reforestation project in B.C. Plantings continued until the war and no further timber was cut until 1966, when two of the plantations were removed from the forest reserve and cleared. Many more trees were cut through the 1970s and 80s, with roads being widened and land being donated for other uses.

By the late 1980s, infuriated citizens insisted that the remaining forest be protected as Green Timbers Urban Forest Park, which is what it has now become. (Map E1- E2).

Further information
Green Timbers Heritage Society
www.greentimbers.ca

WHATEVER HAPPENED TO WILDFLOWERS?

Archibald Menzies, the botanist on George Vancouver's voyage of exploration, was delighted with the fragrant spring flowers blooming in Birch Bay. Early settlers recalled how every sunny bluff and prairie grassland was covered with chocolate lilies, blue larkspur, Hooker's onion, orange blossom and calypso orchids. Woodlands, dense with moss and ferns, were full of flowers: delicate white and pink fawn lilies, the showy blossoms of trillium, purple and white violets, the pink bells of bleeding heart and the flamboyant wild red currant.

Beautiful woodland plants can still be enjoyed in parks such as Hi-Knoll and Sunnyside Acres, yet many wildflowers once common around Boundary Bay are now scarce, and the clouds of lilies and larkspurs have become just a memory. Of the several hundred species of flowering plants growing wild in the Lower Fraser valley, a quarter of the native plants are now rare.

Wildflowers of the coastal grasslands are a hardy group, however, and many native species can still be enjoyed around the bay. The glorious yellows of gumweed and Canada goldenrod, and the purple of fireweed and Douglas' aster, brighten the late summer days. These plants bloom freely in disturbed ground

Entire-leaved gumweed blooms profusely on the Boundary Bay foreshore.

and share the foreshore with yarrow and Pacific silverweed. On the west side of Boundary Bay there is an unusual micro-habitat where relict beaches border the coast. No longer covered at normal high tides, these sand dunes are being stabilised by big-headed sedge and a crust of tiny lichens and mosses. In spring they are a mass of blue-eyed mary flowers, followed by lupines, a red carpet of sheep sorrel, white clumps of field chickweed and the pungent growth of lomatium. Lomatium has medicinal properties and is also known as Indian consumption plant.

Bogs have many beautiful flowering plants, and this habitat once covered much of low-lying Richmond and Delta. Today, bog plants can be admired in Burns Bog (Delta Nature Reserve) and Richmond Nature Park. Western bog laurel, bog cranberry, Labrador tea, cloudberry and marsh marigold are a few of the exquisite flowers to be found in these acidic landscapes. Also found in bogs are carnivorous sundews. These strange plants are able to obtain nutrients from insects that they enmesh in sticky traps.

The Coast Salish traditionally went into bogs to pick berries and herbs. Moss was collected for baby diapers and wound dressings, and reindeer lichen yielded a dye. These activities continued into recent times. Kim Baird, Chief of the Tsawwassen First

Pink fawn lily - a beautiful native wildflower.

Nation, picked medicinal teas in Burns Bog as a little girl, following in her great grandmother's footsteps. Many of the Coast Salish were happy to share their knowledge with the pioneers. Ted Wade, whose family settled near Boundary Bay in 1917, remembered walking with his mother over the railway trestle into Burns Bog to collect blueberries and herbs she had learned about from her Coast Salish neighbours.

Some native plants were actively harvested for thousands of years and suffered from lack of attention when traditional practices ended. Blue camas fields were carefully nurtured for their edible bulbs by Coast Salish women in fields near Semiahmoo Bay and out on the nearby Gulf Islands. The fields were regularly weeded, thinned, and the bulbs of poisonous plants, such as death camas, removed. At harvesting, blue camas bulbs were lifted and separated, and for each plant a bulb was replaced for sustaining the crop.

Without this attention, camas are often grazed by deer and rabbits, or choked by other plants. Similarly, berry grounds used to be cared for by selective harvesting and burning, which kept prairie areas open; this provided further habitat for other plants. Such interdependence of plant life and Aboriginal activities is a subject now beginning to receive academic attention.

Most wildflowers disappeared when they were dug up to make way for developments. Settlers were attracted to the same sunny meadows and forest clearings that flowering plants preferred, avoiding low, damp areas of bulrush, cattail and sedge. As a consequence flower meadows were the first to go. Subdivisions, towns and industry followed the pattern of settlement and as the small pioneer homesteads were demolished, the meadows and orchards surrounding them were bulldozed and cleared. Flowers such as the endangered Henderson's checkermallow, which grew at the water's edge, were lost when streams were culverted. Today, only the names of such places as Hall's Prairie, Kensington Prairie and Cloverdale hint at the flower fields that grew there.

Alien Plants

Many wildflowers were displaced by hardy invasives that travel with humans around the globe, including such common Eurasian plants as dandelion, hemp nettle, pineapple weed, common tansy, oxeye daisy, white clover, sheep sorrel and Himalayan blackberry (see page 176). Grasses and forage crops came with farmers, English ivy seems to have arrived with the railways, and Scotch broom was deliberately introduced to the Victoria area by a homesick settler. It has since spread through much of southern British Columbia and northern Washington.

Hardy invasives flourished because they evolved to take advantage of disturbed soil; their multiple reproductive strategies rapidly out-compete other species. Dandelions grow quickly in poor soil along road margins, have seeds that float easily on the wind and when cut down will grow up again from the root. Ivy and bittersweet twine over other plants, stealing their light and moisture while producing copious berries. Many wetlands are now clogged with purple loosestrife, a garden runaway that not only seeds profusely but also spreads by underground rhizomes.

At least 40% of the flowering plants are non-native

At least 40% of wild flowering plants in the lower Fraser Valley are non-native, introduced either intentionally or accidentally in the last 150 years. In addition, there are many introduced and naturalised tree species. The local landscape is irrevocably changed by the addition of all these newcomers.

Further information on local plants and habitats

Plants of Coastal British Columbia by Jim Pojar and Andy MacKinnon

A Nature Guide to Boundary Bay by Anne Murray and David Blevins

Georgia Basin Habitat Atlas: Boundary Bay www.georgiabasin.net

What a catch! Fraser River chinook salmon ca.1890. (Delta Museum & Archives)

HARVESTING RIVER & SEA

"It was said in those days you could walk across the Fraser River on the backs of the salmon." ~ William J. Dennison

The Fraser River is still one of the world's great salmon rivers and a source of provincial pride. The salmon runs astounded early explorers, cannery owners and fishermen, and still provoke amazement today; every British Columbian child is taught about the salmon life cycle and the glories of the sockeye runs. In the waters of Boundary Bay, fishing has been a way of life ever since someone first trod the sandy shores and pulled a salmon from the Point Roberts reef. This historic resource is threatened in many ways, and its future is strongly linked to clean, fresh water in the Fraser River and healthy wetland habitats in the estuary.

When the great glaciers of the last Ice Age finally retreated, fish recolonized the coastal rivers along with the inflowing sea. In geological terms, this was a recent event, occurring a mere 11,000 years ago. A relatively small number of species moved into the lower Fraser. Enormous white sturgeon, giant halibut and lingcod, massive schools of salmon, herring, surf smelt and eulachon occurred historically in great abundance, although often in fluctuating population cycles as part of the normal course of nature.

From the late 1800s, huge salmon runs encouraged the use of more efficient traps and boats, and there was a rapid rise in the number of canneries operating along the coast. Within a few decades, runs had collapsed, yet fishing fleets continued to grow as the catch shrank. Other fish species were targeted, often with unexpected ecological consequences and diminishing returns. This led to competition and controversy as different sectors battled for their share of fish. At the same time, the Fraser estuary was changing ecologically, as flow patterns altered and habitat was lost. Since the mid-1990s, local fishing has undergone a transformation. Many stocks are in peril, mixed stock commercial fisheries have declined, and the rules for Aboriginal fisheries have changed to reflect traditional rights. With the future of global fish populations at extreme risk, local conservation efforts have centred on fishing closures, improvement of water quality on fish-bearing rivers and habitat enhancement of spawning areas. Measures to mitigate climate change effects are also needed.

FRASER SALMON RUNS

Sockeye

Some sockeye populations, including the famous Adams River run, have a peak spawning year or "dominant run" every four years. Sockeye fry rear in fresh water but move quickly out to sea once they reach the estuary.

There are 4 seasonal spawning runs:
The "Early Stuart" (Stuart River)
Early summer
Summer
Late Fall (includes Adams River, Cultus Lake, Chilko runs)

Chinook

Also known as spring salmon, chinook spawn in the mid to upper reaches of the Fraser River and in the Lower Fraser watershed at Harrison River. Harrison fish are unusual in being white-fleshed, not pink, and fry rear in the estuary rather than lingering in freshwater before heading out to sea.

There are 3 main seasonal runs:
Early/Spring before July 15
Summer July 15 - Sept. 1
Fall (Harrison) Sept 1 onwards.

Pink

Pinks have a two year life cycle, and spend up to 18 months in the ocean. Odd year lines (dominant) and even year lines do not meet and are thus genetically distinct.

There are two Fraser runs in the fall of odd years, one to the Hope-Mission area and the other spawning above the Fraser Canyon.

SALMON FISHING: REEF NETS & CANNERIES

Everyone who has lived in British Columbia for any time at all knows about salmon. The Pacific salmon is king of fish, with its firm, succulent flesh and beautiful shimmering body. It presents an extraordinary natural spectacle as it leaps upriver, changing colour and shape, to die in an orgy of spawning, its carcass hauled from the river by grizzly, black bear and eagle. Salmon are anadromous: part of their life is spent in fresh water and part in the ocean. They normally return to their stream of origin to spawn, and some species have cyclical runs, which are poorly understood (see sidebars).

The First Fishers

The first signs of local fishing are fish bones found at the Glenrose Cannery archaeological site from over 8,000 years ago (page 10). Fish hooks and harpoons have been unearthed in abundance from Boundary Bay middens; perishable nets and lines are far less common.

Analysis of human skeletons from Crescent Beach shows that 3,000 years ago virtually all protein consumed came from the sea. The Coast Salish, whose ancestors relied on salmon for hundreds of generations, worked out methods of coping with its seasonal abundance. These included maintaining broad

Chum salmon.

family networks, participating in potlatches and other gift exchanges, and trading dried, smoked and fresh salmon. Oral history celebrates the salmon and sturgeon as founder people and many traditional stories relate to fishing and the sea. Cultural and language affiliations of the Coast Salish are intertwined with fishing methods (page 43).

One specialist method of fishing was the use of reef nets by Northern Straits Salish speakers, notably at Point Roberts (page 53). The reef net harvest there could be phenomenal. In August 1881, at least 10,000 salmon were taken in three Lummi reef nets within six hours. In 1889, sixteen reef nets were in operation, each catching up to 2,000 salmon a day, with probably the same number in use by Lummi elsewhere. Annual consumption was estimated at over 250 kg (550 lb) per person.

Halkomelem-speaking people, such as the Tsawwassen, Musqueam and Cowichan, set up summer camps to fish incoming salmon at the mouth of the Fraser and in the lower river.

FRASER SALMON RUNS

Coho
Coho spend much of their life in surface waters near the coast, both in the Strait of Georgia and off the west coast of Vancouver Island. Ideal spawning habitat is in low gradient streams within 100 km (62.5 mi) of the coast.

Coho return to rivers in late summer and fall, spawning between October and December.

Chum
Chum spend between 2 and 7 years at sea, travelling as far as the Gulf of Alaska during their oceanic lifetime. They return in two major runs to the Fraser River system, spawning in tributaries such as the Harrison, Chehalis and Chilliwack. Enhancement has been used as an important strategy to rebuild stocks.

Summer run: June to August, spawning September/October
Fall run: September to November, spawning October to January.

Steelhead & Coastal Cutthroat
There are two steelhead runs:
Winter run: November to May
Summer run: April to October
Both runs spawn Jan to May, and may spawn more than once.

Steelhead move rapidly offshore after reaching the ocean. In contrast, coastal cutthroat remain in the estuary, moving only a few kilometres out to sea. They can also spawn more than once.

Prior to 1873, all major Fraser sockeye runs seem to have had four year cycles with the same dominant year. This pattern changed after 1913 (page 125). Fishing methods included gillnets, traps and weirs. Weirs, such as those on the Nicomekl and Little Campbell rivers, were wooden fences that funnelled fish into holding pools. They were in use until the mid-1930s. The Stò:lo estimate that between four and twelve million salmon were caught from the Fraser River each year, based on a regional Aboriginal population of 20,000 to 60,000 people. The sustainability of such a catch is explained by the fact that the fishery was in-river and run specific, and sufficient salmon were left to go up river to spawn (known as "escapement"). Once the fish were landed, it was a challenge to process them before they spoiled - a huge investment of labour. Dried salmon kept the family going through winter and the surplus was traded, together with shellfish and eulachon oil. Salmon and sturgeon were traded to the Hudson's Bay Company at Fort Langley, and later to the "hungry people" who came in search of gold.

The Dominion Fishery Act took effect in British Columbia in 1877. It limited First Nations' access to fish by banning some traditional methods. Around this time, Native Americans were also excluded from the resource; ineligible for full citizenship, a requirement for commercial licenses, they were confined to fishing in reservation waters. The Lummi, who for years had caught salmon off Point Roberts, landed there in 1880 to find their village demolished, their reef nets destroyed and two squatters on their land. Later a large cannery opened at Lily Point.

The Alaska Packers Association (APA) cannery at Lily Point, Point Roberts, seen from the cliff top in 1899. (Delta Museum & Archives)

Fishing on the Fraser River, ca.1890-1900. (Delta Museum & Archives)

Fraser River Canneries

The Hudson's Bay Company exported salted salmon to Hawaii but it was not until the 1870s that salmon became really big business. The timing was linked with important technological advances, notably the introduction of safer canning techniques, invented a few decades earlier in Europe, and the arrival of fast boat and rail links across the world. Soon, thousands of factory workers in England were consuming canned salmon. The first cannery on the Fraser River was opened at Annieville in North Delta. Huge salmon runs in 1893 and 1895, together with the growth of markets, encouraged more people to get into the business. By the early 1900s, there were 33 canneries on the Fraser and at least nine on the Blaine-Semiahmoo shores. The land for Steveston on the river's Main Arm was purchased by William Herbert Steves in 1886, and the town became a centre for the fishing industry. The deep water channel had shifted here from the original port of Ladner's Landing on the south bank.

The canneries' mainstay were sockeye and pinks that sliced to the right size for cans. Coho, chum and halibut were also used, especially in "off" years, when sockeye runs were low. As the railroads opened up, halibut was sent to market using the new innovation of refrigerated carriages. Non-native cannery owners owned the boats, controlled work allocation and marketed the salmon.

The fishing industry brought multi-racial immigrants, from far flung countries such as Iceland, Finland, Norway, Greece, India, Japan and China, who worked alongside Coast Salish and other locals in boats and canneries. A tradition of multi-culturalism began, exemplified in later years by men such as Homer Stevens from Port Guichon, a leader of the United Fishermen and Allied Workers' Union who had Greek, First Nation and Croatian ancestry.

Several thousand Chinese men left wives and families on the coasts of Guangdong to follow the gold rush to British Columbia. They found work in the canneries and lived in camps or bunkhouses and later in "Chinatowns." Coast Salish women gutted and cleaned the fish, then packed the tins, while Chinese men were responsible for cutting the salmon and capping the tins. Huge profits were made by the canneries on the basis of their inexpensive labour. A prohibitive head tax, raised to five hundred dollars in 1903, discouraged further immigration from China for many years.

Steveston was also an attractive destination for immigrants from Wakayama Prefecture, Japan. Gihei Kuno was the first to arrive in 1887 and he was soon joined by other countrymen. Within a decade there were about 1,300 Japanese men and 15 women living in Steveston. In the off-season, Japanese men went to the southern Gulf Islands to make charcoal which was needed to fuel the canning process. Charcoal burning continued until 1908 when canneries switched to coal. Japanese boatbuilders developed the fishing fleet and Japanese brides joined their countrymen in Canada, taking jobs in the canneries and replacing the Coast Salish women, who moved north.

Women with babies worked the packing line

Women with babies strapped to their backs and toddlers at their skirts worked the packing line. In total, about 4,000 people moved from Wakayama to the Fraser River. In 1924 federal restrictions on Japanese fishing licenses for herring, salmon and cod forced families to take up farming. Yet another blow to the community came in 1942, when all their fishing boats were confiscated and families were sent away to internment camps for the duration of the Second World War. Despite the many economic losses and personal setbacks, Steveston maintained the heritage of its founders for many generations, and even today has a Japanese school and cultural centre.

Canadian canneries faced strong competition from the fish trap operations off Point Roberts, and many small, uneconomic canneries closed or merged in 1902. This was only a year after the biggest recorded sockeye run on the Fraser! In 1913 and 1914 railway construction caused massive rock slides that blocked the Fraser Canyon. Unable to get upriver to spawn, the dominant sockeye run was destroyed, and stocks remained low until the 1950s. Canneries continued to operate, switching to chum, pinks, coho and herring. Huge numbers of salmon were taken from the Fraser between 1950 and the 1970s, serving an insatiable demand for canned fish around the world.

Canneries had a high environmental impact, not just from the number of fish caught, but from wasteful practices. Thousands of fish that were too big for processing or exceeded the capacity of the cannery lines were dumped. These surplus fish, offal from butchering, and the bycatch of unwanted species, were thrown from wharves into the water. Even at that time this waste was recognized as unsustainable.

Fish Traps

In 1888 British Columbia imposed further restrictions on First Nations people, restricting them to a "food only" fishery and banning sales. The use of traps and weirs was forbidden even for personal use. In Washington State, tidewater traps were allowed up until 1934, provided they cut off no more than three quarters of a spawning stream. John Waller, a Point Roberts pioneer, set traps in

Coast Salish fishermen using Columbia River skiffs, Garry Point Cannery, Steveston, in 1891. (City of Vancouver Archives)

the late 1870s on reefs traditionally used by Coast Salish fishers. The location became the site of Wadham's cannery, later owned by Alaska Packers Association (APA).

By 1905, there were 47 fish traps lined up very closely around the Point Roberts' shore. Even as late as 1928, the APA owned 13 traps off Lily Point. These traps consisted of giant v-shaped nets, up to 1 km (0.63 mi) long, fastened to pilings on the ocean floor and narrowing at the far end, where they led to a spiller, a huge bag net that could be opened to remove the captured fish. Up to 35,000 fish could be caught at one time. The traps stretched across Boundary Bay to the east and south of Point Roberts, making an inpenetrable barrier for the incoming schools of salmon. In early summer a few traps caught chinook (kings) for markets from Vancouver to Seattle, then from the first of June to Labor Day, hundreds of thousands of sockeye were trapped and sold to local canneries. Dozens of other species, including now scarce lingcod, were caught as bycatch and discarded. Trap workers were armed with rifles to shoot seals, sea lions and eagles that pirated the fish.

For several decades, fish bound for the Fraser River were heavily intercepted, reducing the British Columbia catch and creating cross-border tensions.

American tidewater fish trap off Point Roberts ca.1890-1900. (Delta Museum & Archives)

FISHY HISTORIES

Life for the fishing families could be hard. Jan Hrutfiord wrote how her Icelandic forbears, Eythor and Hannes Westman, recalled fishing from Blaine in 1932, mostly at fish traps but also in late summer on seiners (a type of fishing boat): "It was a tough, seven day a week, 24 hour a day job, that earned $45 a month, from which we paid food and board."

"we're addicted to fishing..."

Gillnetting can be a dangerous and solitary pursuit for relatively little reward. Today, especially, there is the frustration of trying to survive on short openings and limited opportunities, yet the gillnetters still working the Fraser "are eternal optimists." Terry Slack, who grew up on the Fraser River, described the adrenaline rush of going after fish: "we're strange people, we're addicted to fishing, it's like a gold rush."

Georgia Strait fisheries have come and gone. Salmon, halibut, rockfish, herring, herring roe, sole, surf smelt, flounder, sablefish, Dungeness crab, opal squid, sea cucumber, oysters, sea urchins and shrimp have all made their way to the dinner table. There has even been a dive fishery for geoducks and spearfishing for cabazon.

Until recently, commercial salmon fleets operated a mixed stock fishery, using gillnets, trolling gear or purse seines in open water. This meant that all salmon were taken at the same rate, whether they were from an abundant stock, like the Adams River, or a very limited one, such as Cultus Lake sockeye. In time, small stocks disappeared and overall genetic diversity decreased. Coho were particularly affected by mixed stock fisheries. A sudden population decline saw catches go from 1.55 million in 1995 to almost none in 2001. Wild coho remain in very bad shape, and Fraser steelhead and sockeye are also in severe trouble. Today, closures of the fishery are enforced to prevent the incidental take of endangered stocks.

During openings, salmon are fished with gillnets in the lower Fraser River, and gillnets or trollers in the Strait of Georgia (see page 131). Salmon are not the only fish caught locally, and many other fish also have interesting histories.

Herring

A local Pacific herring fishery was started during the Second World War to feed the troops; it later became a reduction fishery, for conversion to livestock feed and for oil used in paints, cosmetics, fertilizers and lubricants. Huge quantities of herring were harvested, but it was unsustainable. Just four years after a record B.C. catch in 1963 of over 268,000 tonnes, the fishery came to an end. Subsequently herring have been fished for their roe, long a favourite of the Coast Salish, and prized as a delicacy for export to Japan, where it is an ingredient in sushi.

Herring populations have made a partial recovery, but many former spawning areas are no longer in use. Herring spawn erratically, choosing one area for years and then suddenly shifting elsewhere. This happened in Boundary Bay, which was a major spawning area between the 1950s and early 1980s. The urge to spawn is strong and when eelgrass habitat is lost, roe is deposited on pilings, boat bottoms, or other unsuitable places. The current decline in herring is not just due to changed habits but also to the loss of local stocks. Many marine bird and mammal populations have declined in consequence.

Eulachon (Oolichan)

These members of the smelt family are very important for First Nations, supplying both food and oil (grease). Eulachon are also called candlefish, since their dried, oily bodies can be used to provide a slow-burning light. The routes taken by First Nations trading them to the Interior were known as "grease trails." Another name for eulachon was "fathom fish" because they were sold strung on lengths of rope or sinew. (A fathom is 2 m (6 ft) long.)

Eulachon are anadromous, meaning they live most of their lives in the ocean, returning to freshwater to spawn. Entering the river in three distinct waves, the little fish move upstream on high water. They swim into the estuary by predictable routes, yet the spawning areas shift from year to year, with preferred locations being at the confluence of fresh and salt water. Schools spawn in April and May.

Eulachon are a vital part of the food chain with at least 26 other species highly dependent on them, including sturgeon, loons, sea birds, sea lions, and killer whales. Eulachon have undergone a huge decline in recent years. The last commercial gillnet fishery was in 1995 and there was no Aboriginal fishery in 2007. The cause of this decline is not understood; it may be the result of shrimp dragging off the West Coast, the loss of spawning habitats in the lower Fraser River, or a combination of effects.

Sturgeon

The extraordinary white sturgeon has survived many changes to its environment since its ancestors first swam the oceans 290 million years ago. This enormous fish was once found not only in the Fraser River, but also in Boundary Bay and the estuary, where green sturgeon also occur. Captain Vancouver named Sturgeon Banks, Richmond, after the fish he purchased there from

A 4.12 m, 411 kg (13.5 ft, 905 lb) Fraser River sturgeon caught in 1912 by Jim Burgess.
(Delta Museum & Archives)

the Musqueam. John Keast Lord described in 1866 how "sturgeon are continually leaping" during high tides. The Kwantlen and others later engaged in a trade of sturgeon to Fort Langley.

Sturgeon are long-lived fish and very slow growing, facts that the huge commercial fishery between 1885 and 1900 did not take into account. The population was rapidly depleted, almost beyond recovery levels. Terry Glavin documented the sad story in his book *A Ghost in the Water*. After fishing out the Columbia River sturgeon, entrepreneurs moved north and started a major commercial enterprise on the Fraser River. The catch quickly went from about 100,000 kg (220,000 lbs) in 1887, to over five times that amount in 1897. Only two years later, with the stock nearly exhausted, the catch fell dramatically.

As early as 1894, a Stò:lo preacher, Captain John of Soowhalie, protested the sturgeon slaughter. He recognized that it was unsustainable, and 158 Coast Salish people from along the river signed a petition against the commercial fishery. Fishing continued for some years, and photos from the early 1900s show men posing with giants over 4 m (13 ft) long (see previous page). More sturgeon losses were incurred when Sumas Lake was drained. This shallow lake in the lower Fraser Valley was a significant wetland and sturgeon habitat, but

it was converted to farmland in the 1920s. Sturgeon numbers in the Fraser River remained low for decades. In the 1990s, a mysterious die-off in the lower reaches saw huge carcasses washed up on river banks and the Roberts Bank causeway. The Stò:lo banned sturgeon fishing in 1994 and the Canadian government followed their lead, closing the fishery and classifying sturgeon as endangered in 2003.

The days of abundant giants had come to an end. It was only when fishing ended that local biologists first observed sturgeon actively spawning. Today, volunteers like the Fraser River Sturgeon Conservation Society are working to restore sturgeon populations on the river.

A quick-thinking detective filled the bath tub

Poaching can be a serious problem, and there is an incidental catch of sturgeon in gillnets. In 2002, police officers conducting a "grow op" raid, discovered a 1.5 m (5 ft) long, 38 kg (84 lb) white sturgeon in a house refrigerator. Surprisingly it was still alive. A quick-thinking detective filled the bath tub and saved the sturgeon by plunging it into cold water! Fisheries officers were able to return it safely to the river. The story gives credence to tales of fish found alive in muddy fields, days or weeks after floods.

The Fraser River fishing fleet at Steveston.

FISHING METHODS

A variety of commercial fishing methods are used on the northwest coast, including purse seining, gillnetting, trolling, long-lining and trawling. Recreational fishing also takes place.

Ocean-going seiners have a circular net, or purse, up to 400 m (1,320 ft) long, which they set around schooling fish; after a short wait, the lead line around the bottom of the net is drawn up under water, trapping the fish in the purse. The invention of hydraulic drives in the 1950s enormously increased the power of the haul-up drums allowing huge quantities of fish to be taken. Depth sounders and scanning sonars have further increased the efficiency of the fleet, even as the number of boats has declined. Purse seiners are used for catching salmon, sardines, tuna, and mackerel.

Gillnetting is an ancient method of salmon fishing, used originally by the Coast Salish, and by European commercial fishers since the 1890s. Small vessels haul a fence-like net, that is suspended in the water by floats, and weighted at the bottom by a lead line. The net lies about 500 m (600 yd) behind the boat, at a depth of about 10 m (33 ft). It is hauled in by a rotating drum. Gillnetters work in the river channels or close to shore in the estuary. Gillnets need to be invisible to salmon. In the old days, gillnetters used to often fish at night and their natural fibre nets were

THE ALBION BOX

A favourite location for old-time gillnetters, the Albion Box is an old main channel of the Fraser River.

The Fraser River estuary is a critical habitat for many species of fish. Salmon linger in the mix of tidal and fresh water at the mouth of the river allowing their bodies to adjust to salinity differences as they undergo their migrations. Chinook, coho and chum salmon use the estuary as a nursery and rely heavily on side channels, sloughs and marshes, especially where the water is shaded by over-hanging trees and other riparian vegetation.

Over the centuries the river mouth has changed enormously. Channels have filled and shifted, marshes have grown or been built over, sand banks have formed or dispersed. The waters around Westham Island and Woodward Island have some of the most productive fish habitats. This is the location of the Albion Box, used as a gillnetters' drift as salmon staged here before going into the river.

Eulachon and starry flounder enter the river through Canoe Pass, another one of the old channels of the Fraser, between Ladner and Westham Island. This is a good area to see birds like western grebe and great blue heron, that feed on these small fish.

Such productive areas are becoming rare along the river and the Albion Box is silting in so that fish can no longer access it.

dyed blue. Today's nets are made of "Alaska twist" - a multistrand mesh that is effective for fish yet can also be lethal to seabirds. One notable kill in Boundary Bay saw over 1,000 dead seabirds washed ashore, and in 2005, hundreds of common murres drowned when caught in gillnets off Point Roberts. Set nets and drift nets in the river can also be harmful to fish; if nets are not checked regularly spawning fish can wear themselves out avoiding them, or perish and wash out of the nets. Juvenile sturgeon are sometimes bycatch in these nets.

Trollers take salmon, halibut, tuna and other pelagic migratory fish with leaders and lures on long steel lines paid out from huge outrigger poles. The lines are supported by floats (pigs) and manoeuvered by hydraulic winches (gurdies). Fish are handled individually after capture and put on ice in daytime trollers; ocean-going boats have refrigerators aboard and stay offshore for weeks at a time.

Long-lining is used to catch halibut, employing a main horizontal line and multiple short lines, bearing hooks baited with octopus or herring. Trolling with a hand-held line from a small boat is called handlining and was especially popular in the 1920s and 1930s. For many years, rockfish were a favourite handlining target for sport fishers in the Strait of Georgia; these fish are now virtually extirpated from local waters.

Shrimp and sole are caught in trawls - cone-shaped nets towed behind the fishing vessel and drawn along the sea floor. Trawling is unselective, so creates a bycatch problem.

Local recreational fishing is catch and release only, and this year-round activity is focused on salmon, halibut and sturgeon. On the Fraser River, recreational fishing methods include bar fishing from gravel bars, casting from a dock, or driftfishing on a slow current.

CHANGING RULES

Present-day allocation of salmon is a huge challenge and management of the resource is the responsibility of the Pacific Salmon Commission, a joint US-Canadian agency. The Commission was created after nearly a century of disputes over salmon allocations, and followed from the Pacific Salmon Treaty of 1985.

Many Coast Salish people participate in the commercial fisheries but have had restricted access to catching and selling fish independently. By 1970, there was increased public interest in social justice and a desire to address some of the inequities of previous years. Aboriginal fisheries became the subject of legal challenges in both the US and Canada. The Lummi and other Native Americans gained the right to 50% of the salmon harvest in the 1974 Boldt Decision. Following this, the Lummi fishing fleet rapidly grew to be the largest in Puget Sound, with 600 gillnet boats and 40 seiners at its peak in the mid-1980s. When the salmon industry later crashed, the fleet was significantly reduced.

Fishing policy was slowly modified in Canada, and the First Nations were able to pursue their rights in the Supreme Court of Canada. In 1990 the Musqueam did just that, successfully challenging the Crown for "the Aboriginal right to fish for salmon for food, social and ceremonial purposes." The Court made a stipulation that conservation was the priority, but otherwise no restrictions on gear or openings were allowed. Fisheries and Oceans Canada subsequently launched the Aboriginal Fisheries Strategy to conform with this ruling. Access agreements were worked out with bands along the Fraser, with mixed success. The Aboriginal right to sell fish is still in dispute.

These fish are eaten and enjoyed by many

The most marketable salmon in North America is firm, oily, red-fleshed sockeye, in its ocean form. However, ocean catches are non-specific, increasing the risk of stock extirpation, and fishing the run offshore decreases potential for in-river catches. Salmon staging in the Fraser estuary before heading upriver are the target of commercial,

DECLINING SALMON

Since 1992, Fraser River salmon runs have suffered serious declines. The 2007 sockeye runs were the worst in 40 years. Salmon populations are only a tiny fraction of historic levels.

Possible causes for decline include: global climate changes resulting in ocean mortality, higher than normal river temperatures leading to instream mortality, unreported legal or illegal fishing in the river, counting errors and/or mismanagement. Estimates of run-size have frequently been much greater than the actual numbers, both for ocean returns and for spawning escapement.

Ocean conditions seem to be an acute problem: fish are having to travel further and are failing to find enough food. This results in smaller fish and poor survival rates. Timing is another factor. In some years, late summer sockeye arrived too early and then failed to stage in the estuary. Many were stricken with a kidney parasite, *Parvicapsula minibicornis*, and died in the river.

The health of the river and estuary are important factors. Good fish habitats are becoming more and more scarce along the lower Fraser. Cool, shady, gravel stretches are lost by bank reinforcement and riverside development. Urbanisation has created an overlit, overhot, polluted channel that the struggling fish must now face on migration, first as fry and then as spawning adults.

recreational and Aboriginal food fisheries. As they swim upstream, salmon lose texture and colour, becoming less popular for external markets, though these fish are eaten and enjoyed by many.

Expectantly waiting along the Fraser River banks are many First Nations people, including the Tsawwassen, Musqueam, Stò:lo and Cheam, and further upstream people such as the Nlaka'pamux in the Canyon, the Ts'ilhqot'in, the 17 nations of the Secwepemc in the Shuswap and the Stl'atl'imx of the Fraser-Lillooet valleys.[1] Each Nation is entitled to a share of this finite resource.

Allocation is not so difficult when runs are large, but a real challenge since the mid-1990s, when many stocks have diminished (see sidebar). Everyone hopes the fish will return and satisfy all the sectors, but the future looks far from rosy.

SALMON HATCHERIES

With many salmon stocks becoming endangered, enhancement programs were started about thirty years ago. In Canada, the Pacific Salmonid Enhancement Program (1977 to 2003) aimed to double the annual catch; later the goal became the restoration of salmon populations to historic levels, although experts could not agree on what these were.

1. Also known as the Thompson, Chilcotin, Shuswap and Lillooet respectively.

Little Campbell River at the fish hatchery.

Hatcheries were started for rearing juvenile salmon, in the hope that by decreasing fry mortality and putting fish into the system, stocks would regain their former abundance. It was soon realized that in order to raise fish, river habitat had to be improved.

Serpentine Enhancement Society volunteer, Lynn Price, described what was necessary to make the Tynehead hatchery successful: "we had to get people to think of the river again as a living viable habitat…once you clean up the water quality, you bring back all sorts of different species." Tynehead volunteers raised over 1.5 million coho, chinook and chum fry from 1987 to 1997. Other hatcheries operate on the Nicomekl and Little Campbell Rivers. To keep track, hatchery fish are marked with fin clips or nose tags.

The rearing program came under some criticism following a 1995 study of coho in the Columbia River. Salmon raised in hatcheries were genetically weaker and had lower survival rates than wild salmon; it was questioned whether hatcheries were actually benefitting fish stocks. On the other hand, thousands of volunteer hours have been spent restoring habitat, educating the public, involving school children and marking storm drains, all of which have certainly contributed greatly to fish and wildlife conservation. Fisheries conservation efforts are now increasingly focused on habitat enhancement, such as slowing water flow with strategically-placed logs, or planting vegetation to shade streams and encourage insect life that will feed the fish.

Further information on fish and fishing

The Jade Coast by Robert Butler

The Last Great Sea and *A Ghost in the Water* by Terry Glavin

Steveston Cannery Row by Mitsuo Yesaki, Harold Steves & Kathy Steves

Salmon without Rivers by James Lichatowich

Clam Gardens by Judith Williams

Fraser River Sturgeon Conservation Society www.frasersturgeon.com

Fish Stocks of the Pacific by Canada Dept. of Fisheries & Oceans www.dfo-mpo.gc.ca

Tynehead Hatchery www.tyneheadhatchery.ca

Nicomekl Enhancement Society www.nicomeklhatchery.com

Swaren Singh, Prahim Singh and Oudam Singh of the Crescent Oyster Company, sorting oysters, ca.1940-1965. (Delta Museum & Archives)

SHELL FISHERIES

Oysters, clams and crabs are some of the favourite seafood found in local waters. The sand flats of Boundary Bay are the best environment for oyster cultivation in British Columbia. From 1904 to 1961, they were the scene of successful oyster growing operations, centred around Mud Bay at the B.C. Packers cannery, the Crescent Oyster Company and the Olympic Oyster Company.

In the past, people had sometimes collected the small, native oyster from Blackie Spit. An attempt to culture Atlantic oysters in Mud Bay was made in 1903, but it was not until the introduction of the large Japanese Pacific oysters, from 1929 onwards, that the industry really got going. So successful was it, that Mud Bay oysters soon accounted for half the British Columbia product.

When the water warms in spring, oysters establish their gender, which can change from year to year. They spawn in July. Larvae are free-swimming, but while still very tiny, they settle on clean, hard adult shells (or in oyster culturers' language, they "spat on the cultch.") Oysters are harvested in winter when they are fat and tasty. In the early years, fresh oyster spat was imported annually to allow for a long growing season. A host of Asian shellfish accompanied the spat, including Manila clam, oyster drill (a predatory whelk) and mudflat snail, all of which established in Boundary Bay. The oyster fishery was successful, yet by the early 1960s, agricultural contamination, together with poor tidal flushing in summer, led to its closure. A similar fate overcame the Community Oyster Farm at Blaine when Drayton Harbor was closed to harvesting in 1999. To keep the farm going, watershed residents are cooperating to address pollution from the Little Campbell River, California and Dakota Creeks, and the sewage outfall at Blaine. Pacific oyster beds can still be seen at the mouth of the Serpentine River and their shells are found all around the bay.

The traditional Coast Salish method for digging clams and cockles was to watch for the show of water from the siphons and then thrust with a stick into the mud alongside the clam. If done accurately, the clam would bite and be drawn up to the surface. Clam digging was done by women, and the clams were steamed, dried or smoked for winter food and trading. Some coastal First Nations cultivated butter clams in "clam gardens," carefully made and tended areas of enhanced clam habitat. A retaining wall, many metres long, was made of stones rolled into place to create a flat terraced area. Judith Williams, one of the first researchers to describe the clam gardens, called them "living artifacts." The wide sand flats of Boundary Bay were already rich in growing areas, however, so

B.C. Packers oyster plant on Boundary Bay, ca.1940-1965. (Delta Museum & Archives)

clam bed enhancement was probably unnecessary, except perhaps for some careful tending. Butter clams and native littleneck clams were picked in early spring, although when commercial clam canning was introduced, it took place in winter. Introduced Manila clams became the most popular for canning, but poor water quality ended this lucrative fishery. Today clam beds are closed to harvesting in Boundary Bay.

The commonest crab in Boundary Bay, the Dungeness crab, grows to a large size and is very succulent; as a consequence, crab fishing has long been an important part of Boundary Bay life. The throb of the crab boats setting out at dawn was a familiar sound to several generations of Crescent Beach families. Crabbers still operate from Crescent Beach, and Blaine is the centre of a regional crab fishery. Only large male crabs should be taken, and there are regulations regarding crab pot construction. Minimum size limits are intended to allow male crabs to reproduce at least once before being harvested; female crabs seldom reach legal size.

Poaching is a problem, especially in September when crabs move into deeper water over the border, and boats follow them. So valuable is the harvest that drug dealers have been known to target crab fishers, trying to exchange drugs for shellfish.

Doug Sands, crabber, with fishing boat and crab traps, 1965. (Surrey Archives)

True Oliver with brant bagged at Beach Grove, ca.1910. (Delta Museum & Archives)

WATERFOWL HUNTING

Fish and shellfish are not the only harvest from the ocean: waterfowl, seabirds and marine mammals have all been taken. Coast Salish and early settlers alike relied on hunting to feed their families. Aboriginal duck hunters placed fine nets on high poles across the marshes to catch birds as they came into land on open water. Another approach was to light a little fire in the front of a canoe and, drifting with the current and obscured by smoke, creep up on sleeping ducks in the darkness.

Waterfowl hunting was a traditional and sustainable activity for thousands of years. Once rifles were used,

however, there was an immediate effect on bird populations. Tens of thousands of birds were shot during unregulated market hunting in the late 1800s. Mallard, green-winged teal, wigeon, sandhill cranes, brant, Canada geese, white-fronted geese and tundra swans were targeted. As a result some migratory swans, geese and cranes stopped landing at traditional staging areas.

Brant are a species of small sea goose, that nests in the Arctic tundra and winters from Boundary Bay south to Baja. They are only now recovering their numbers from this historic onslaught, while white-fronted geese, tundra swans and sandhill cranes are still scarce in the lower Fraser valley.

As millions of migratory birds were shot across North America, hunters became very concerned about the declining populations. In 1916 Canada and the USA cooperated on the Migratory Bird Act, legislation to regulate hunting seasons and bags. Organizations were set up to protect wetlands and other duck habitat.

Hunting shacks used to line parts of the beach, such as Beach Grove spit and Westham Island, but these were mostly gone by the late 1980s. Today waterfowl hunting is permitted only in limited seasons and in designated areas within Boundary Bay and the Fraser estuary (see also page 168).

STRAIT OF GEORGIA WILDLIFE IN PERIL

Changes to the marine environment are difficult for most of us land-dwellers to notice. It is only now becoming apparent that the undersea world, once seemingly limitless in its wealth, has seen profound changes during the last 120 years. Globally and locally, many fish species have been brought close to extinction either by wasteful overfishing or habitat destruction.

The Fraser River estuary and the Strait of Georgia have had their share of exploitation. For example, Pacific halibut and lingcod populations in the Fraser estuary are less than 3% of what they were in the 1890s, and modern fish are much smaller. Once

Above and right: humpback whales ...

abundant rockfish species, such as the yelloweye, are now rare in the Strait. Some rockfish species can live to be over 90 years old, so even if fully protected, recovery of populations will be painfully slow. Basking sharks were once common on the Pacific Northwest coast, including in the Strait of Georgia. They were extirpated from the 1950s onwards by the Department of Fisheries, because their habit of basking on the water surface interfered with fishing nets.

Marine mammal populations have also suffered. In the late 1800s, commercial whalers arrived, eager to capitalize on what appeared an unlimited resource. They hunted out the offshore Pacific right whales, then set their sights on chasing down the much faster humpback whales, once common in the Strait of Georgia. Commercial whaling operations, between 1866 and 1873 and particularly in 1907, soon led to the humpbacks' demise. Whales have a slow reproductive cycle and could not withstand the pace of the slaughter. During the whaling boom, 97 whales were harpooned in one season off the mouth of the Fraser River, the last humpback whales to be seen for many years in the Strait. A few recent sightings hold promise that this magnificent animal could make a comeback.

Grey whales, in contrast, are regularly seen in the Boundary Bay area, especially in late spring when they make their long migration north. Never hunted as assiduously as other whales, their populations are more stable. These slow-moving giants are at risk from boat collisions and dead ones are sometimes washed ashore. They may also beach themselves when they come into shallow water.

Killer whales and porpoises are a great attraction around our shores. The southern resident population of killer whales, or orcas, is now seriously imperilled. It is on both the Canadian and US endangered species lists. In the 1880s, people took a more utilitarian view and caught killer whales in a reduction "fishery." The whales continued to be harassed in salmon fishing areas up to about

1964, after which public interest in them increased. Sadly, this led to the capture of many for aquariums, and consequently, local pods of killer whales now lack sufficient individuals of breeding age. They are also badly affected by marine pollution.

Some species are more resilient than others and not all culls have led to extirpation or rarity. For over fifty years between 1914 and 1969 harbour seals were culled because they ate fish and interfered with nets. There was also a market for their hides. However, populations rebounded when the slaughter ended and numbers remain strong in the Fraser River and Boundary Bay.

Bald eagles are also survivors: they were the subject of a bounty in Alaska in the 1950s and 20,000 were shot. Some Fraser River fishers even shot local eagles and took the claws north to Juneau on their next fishing trip, to collect the reward! Pesticide poisoning of eagles has now declined, contributing to the eagles' revival (page 158).

.... a rare sight in the Strait of Georgia.

A potato field behind the Mud Bay dyke, with a South Surrey forest and Mount Baker in the background.

THE RICHES OF THE LAND

"Ours to Preserve ~ by Hand and Heart"
~ Delta municipality motto

The flat farmlands of the Fraser delta lie under wide skies just a short drive south of Vancouver. In winter the horizon beyond the brown rain-washed fields is lined by snow-capped mountains, with the majestic peak of Mount Baker rising prominently to the southeast. The fields dry out and become green as spring progresses, and the mountain view fades away into summer haze. This is rich arable land, some of the best in Canada. In just a few short months, the fields produce potatoes, corn, pumpkins, silage, blueberries, cranberries, strawberries, greens, barley, wheat and many other crops. Cattle, horses, llamas and other livestock are reared on lush pastures in watershed hills and valleys.

Farmland provides our food, but it also serves many other purposes. Everyone benefits from fresh air and wide open views. Low-lying delta fields are floodplains, absorbing heavy winter rains and redirecting run-off. Many of the wildlife in the Boundary Bay watershed would not survive without the presence of fields, ditches and hedgerows. Farmland feeds large numbers of wintering and migrating birds, including snow geese, trumpeter swans, dabbling ducks, shorebirds, hawks, owls, harriers, herons and flocks of finches, blackbirds and other songbirds.

Agriculture was a predominant way of life for a hundred and fifty years, yet during that time it has undergone a series of transformations. The future is likely to bring many more changes. Throughout the lower Fraser Valley, the Agricultural Land Reserve is under constant pressure from competing uses, including residential subdivisions, golf courses and playing fields, highway construction, port infrastructure and the need to protect areas as wildlife habitat. Only the last of these is even remotely compatible with farming.

Climate change and the need to reduce dependence on fossil fuels heighten the need for locally-grown food and better food security, while presenting new problems such as crop suitability and pest invasions. Modern, health-conscious consumers are pushing for organic products, more variety and consistently safe food sources. This all amounts to quite a challenge for local farmers, but facing change and challenge has always been part of farming's history in the area.

A Delta corn crop.

YESTERDAY'S FARMS

Different types of farming have been practised around Boundary Bay through the centuries. Traditional Aboriginal farming practices of the region included cultivation of camas bulbs and wapato, and burning to promote growth of blueberries, cranberries and other native fruit (see pages 45, 116). Around 1835, the Semiahmoo and Stò:lo people adopted potato growing from the Hudson's Bay Company at Fort Langley. Potatoes became a major crop all along the coast within just a few years; the surplus was sold to settlers and itinerant miners. Several varieties were grown, with brown, white, yellow or purple skins. The

Boundary Bay prairies, where local potato agriculture began, were noted for a smooth, white-fleshed, orange-skinned tuber, called "no eyes."

The first Hudson's Bay Company farm at Fort Langley comprised 12 ha (28 ac) around the old fort and half as much again on bracken-covered Langley Prairie. Pigs, cattle and oxen were raised and potatoes, wheat, peas, barley, corn and oats were cultivated, with mixed success. Poor growing conditions were one reason for the move to a new site in 1839. HBC Governor George Simpson, on a visit, described the new fort as "very well regulated" with "an excellent farm in the immediate neighbourhood, the produce of which with fish and venison maintains the establishment and assists in provisioning others on the coast."

As European pioneers arrived, they felled trees and ploughed the prairies as required under the homesteading agreement. These improvements for agriculture were a driving force in the newcomers' lives, as can be seen from their personal accounts. Yet they tended to preclude understanding of, or adaptation to, the landscape in which they had arrived. The new arrivals were from many different backgrounds, some seasoned farmers and others with little or no actual agricultural experience. They were also a mobile population, having already travelled across oceans or the continent, or come north from

California. Often they continued on to other areas in search of destiny or fortune. They came alone or in groups, young men seeking gold or families with women, children and cattle. Those with some experience of livestock rearing and homesteading were generally more successful and became leaders of the developing communities.

John Oliver rose to become Premier of British Columbia

John Oliver described himself as a "dirt farmer" having lived on farms in Derbyshire, England and Wellington County, Ontario, before pre-empting land in Surrey at the age of twenty. He later moved to Mud Bay and married Elizabeth Woodward, another English pioneer. Oliver became a prosperous farmer and rose to become Premier of British Columbia, embodying the rural conservatism of the delta and holding the strong belief that agricultural life was "the cornerstone of a healthy and prosperous society."

He was one of the main proponents of the Sumas Lake drainage scheme. This ambitious and costly project provided thousands of hectares of farmland, yet caused the loss of a vast wetland habitat in the lower Fraser Valley used annually by tens of thousands of migrating waterfowl.

John Oliver's farm in East Delta is now the site of a sports park.

The Booth sisters, who married the Ladner brothers of Delta, were daughters of William Booth who had been raising stock in California for some years before he immigrated to British Columbia. William and Thomas Ladner, originally from Cornwall, England, had tried their hand at mining and then run packhorse trains supplying the gold fields before pre-empting land on the delta shores of the Fraser River in 1868. The families began by raising livestock, especially oxen and horses for the transportation of goods, and when this proved successful went on to other forms of farming as well as entering the cannery business.

One of the first requirements of the pioneering farmers was hay and other fodder to keep their livestock through the winter. Native grasses were well liked for this purpose, but as time went on, introduced grasses such as oats and timothy were sown. The open lowlands, wet in winter but dry in summer, provided ideal growing conditions. Today a riot of intermingled native and non-native

Haying in Delta. (BC Archives)

grasses - delicate hair grasses, fescues, orchard grass and the silky plumes of brome - cover roadsides, fallow fields and slough margins, spreading out in profusion wherever they can set down roots. To make way for hay fields, shrubs were pulled out and drainage ditches dug.

Only horses could be turned loose to feed on the "saltings"

Land clearance was painfully slow. Huge tree stumps lay buried in the soil, with unexpected layers of peat and sand, and incursions of sour saline soil along the coast. Salt was often a problem close to Boundary Bay, as it still is now, and only horses could be turned loose to feed on the "saltings" or foreshore. The stumps broke plows and backs, and the sodden mud of winter turned rock-hard under the summer sun. Despite this, fields were gradually cleared, drained and plowed.

Dutch farmers played a key role in developing the dairy industry. They settled on the floodplain, especially in the Serpentine and Nicomekl valleys and East Delta, and raised black and white Holstein cows, a familiar breed from the Netherlands. By the 1890s, cooperative processing facilities, such as a creamery, were being set up. Farmers felt secure enough to invest in solid and elaborate houses, some of which still stand today.

Fresh water for livestock and human consumption was a problem on the delta coast. Animals usually drank from streams that ran through the floodplain, but in the 1890s water pollution from cannery waste became a big problem. Human drinking water was obtained from rainwater, and washing water from the river, until the artesian springs in East Delta were finally tapped and piped west in 1912. The springs can still be seen today in Watershed Park.

The heavily timbered watershed hills in Surrey were only slowly opened up for farming during this pioneer period. The McDougalls, Chantrells and Stewarts were among families homesteading near the Nicomekl River; the Stewart farm at Elgin still stands today (page 193). Most farms were in forest clearings, such as Langley, Hall's and Kensington Prairies and Cloverdale. These first farms raised beef and dairy cows, horses, hogs and bees. As the logging boom ended, farming became Surrey's premier industry, and remained so for many years.

A Delta cabbage crop.

Surrey landscape, ca.1940. (BC Archives)

The days were full of hard work. Pioneers picked fruit, grew garden vegetables, dug clams and took summer road-building work to augment their meagre finances. The cattle they raised were a constant preoccupation. Those left to graze in forest clearings would become mired in swamps, while others took off - at Birch Bay a wild herd roamed the beach until 1900.

Each individual effort to gain a living from the soil served to emphasize human frailty, yet collectively these efforts marked a huge change for the ecosystem. Dykes, drainage and forest clearance gradually allowed farm crops and pastures to cover the valley bottoms from Boundary Bay to the height of land at Langley, as well as many of the hillsides. Farm fields replaced wild prairies and marsh, ditches supplanted winding tidal sloughs, and roads and rail lines carved up the landscape.

While life was tough, with endless demands on resourcefulness and endurance, many pioneers enjoyed their experiences. George Triggs recalled with pleasure the living to be made from nature's bounty: salmon fishing on the Nicomekl, and selling crabs for a "fair day's pay." Dolly Wade wrote of once beautiful fields of chocolate lilies and lady slippers at Crescent Beach, and the "togetherness" that came from really needing one another.

Many farm buildings from this period have fallen into disrepair, but a number remain standing around Boundary Bay, especially in Delta. Ladner Trunk Road, 34B Street and Arthur Drive have some fine examples (see pages 189, 192).

POTATO CULTURE

In 1914, the first potato processor in Delta opened in an old fish cannery. Potatoes were shipped from Ladner to Victoria and Nanaimo, as well as being sold in the markets of New Westminster and Vancouver, the new city at the terminus of the Canadian Pacific Railway. The newly-constructed roads and railways were a great asset in getting crops to market. There were still very few people living near Boundary Bay at this time. The settler population of Delta was only about 2,500, which was mostly farming families. The rich soil produced vegetables in abundance including peas, sweetcorn, giant cabbages and prize-winning potatoes. Early crop potatoes and seed tubers became local specialities.

Plowing, 1901. (Delta Museum & Archives)

Farmers, Asahel Smith and Stewart Wade were declared "champion boosters to potato culture" when they won a trophy for their hundred varieties of potatoes at a New York agricultural exhibit. George London's farm on Westham Island was a big producer, especially in the 1940s, and the Harris family have farmed potatoes for five generations near Crescent Slough, becoming some of the first to make the switch to organic produce.

Chinese market gardeners also grew potatoes. Chung Mor Ping (Chung Chuck) who farmed in Ladner for over forty years was famous for his maverick personality. He sold his crop without going through the marketing board and had many disputes with the municipality over water. Sadly, his riverside farm was converted to residential developments following his death in 1986.

In the 1920s, the Tsawwassen - Point Roberts peninsula was mostly swampy forest, with three sawmills but little residential development. A few acreages were cleared by Italian families, including the Spetifores, Novellis and Amatos. The Spetifore family farm in Tsawwassen went on to become one of the largest potato producers in the Fraser valley. Mr. Amato sold his property to William and Zoe Dennison, who had seven children. William's son, Bill, recalled "our three enemies were drought, frost and the California market." The family grew early epicures and warbas, a heavy yield potato which is still popular with local gardeners. Their farm is recalled in the name "Dennison Park" near the high school in Tsawwassen.

A marsh known as "the Tule" had been dyked

On the US end of the peninsula, farming was difficult as freshwater was scarce. The best land was on the south coast, where a marsh known as "the Tule" had been dyked and drained; this was where the Point Roberts marina was built in 1976. The tight-knit Icelandic community did their best to survive, raising sheep and cows, and growing hay, vegetables and fruit.

A small town grew up on Gulf Road, Point Roberts, clustered near the cannery and docks. Grass pastures supported dairy and beef cattle until the 1960s; where the lagoon is now, there were once herds of Black Angus cattle.

TENTS TO TOWNHOUSES

The growing communities of the twentieth century provided both the workforce and markets for farm produce, yet there were new demands on the land base. City people looked for places to relax: by the early 1900s, Boundary Bay was becoming a popular summer holiday destination. To reach the west side of the bay, they came in steam paddle-wheelers to Ladner's Landing, and then by livery coach.

Once at the beach, families camped in canvas tents at Beach Grove, Boundary Bay village and Maple Beach in Washington. The water of the bay was clean, warm and shallow, ideal for summer parties for large families. Picnics were a regular event.

On the other side of the bay, at White Rock, cabins were built on cleared forest land on the south-facing bluffs. The railway brought weekend crowds and piped water was introduced in 1912. As well as a station, the town had a couple of mills, two stores, two hotels, some market stalls and about 300 houses. The barons of the Great Northern Railway wanted to deepen Semiahmoo Bay and build a port, a project which would have ended White Rock's future as a summer resort. This mega-project came to

Thomas Ladner and family camping at Boundary Bay, ca.1903.
(Delta Museum & Archives)

Camping Semiahmoo, 1950: the peaceful camping experience was not what it had been! (Surrey Archives)

nothing; however, the idea for a pier for coastal steamships became a reality. After years of discussion, it was completed in 1915. The pier and the large white glacial boulder on the beach became focal points for the town, although the original pier burned down in 1935.

The local population was booming and the price of land soared. Delta farmland reached $750/ha ($300/acre), then the most expensive in British Columbia, and then as now, other developments competed for land. In 1928 a resort opened at Beach Grove, with a golf course, tennis courts, ballpark and skating rink. There was a beachfront boardwalk, with a platform into the bay leading to an oceanwater swimming pool. Gen Matheson, like many who spent their childhood summers there, recalled how the bay had "always been a constant of good times and family belonging." Bonfires on the beach, clamming and crabbing, rollerskating at the rink and dancing to an orchestra at the hall were fond memories for that generation.

Small cabins lined the beaches and gradually more permanent homes were built. Subdivisions were made at Grauer Beach, and 300 new families moved into bungalows there. On the 4 July holiday weekend in 1929, 3,600 cars drove the gravel roads leading to Beach Grove and Point Roberts.

On Roberts Bank, a few houses were being built along Tsawwassen Beach, but the name Tsawwassen still only referred to the First Nations Reserve. Ladner was also still very much a fishing village, surrounded by farmland, until the 1950s. From 1913 onwards, it was reached by ferry from Woodward's Landing, Richmond, across the Fraser River to the mouth of Chilukthan Slough. In 1933, as the Ladner marshes silted up, the ferry was moved to the end of Ferry Road.

Years of Change

A huge technological shift occurred in the thirty years between 1890 and 1920. This was the era when electric lighting spread across the continent. Steam-powered engines, briefly used, were soon displaced by gasoline-powered machinery. Farmers made the switch from horses to tractors and operations became increasingly mechanised. Motor cars made their appearance and people began to demand more and better roads. The landscape south of the Fraser nonetheless remained predominantly rural until the late 1950s.

During the Second World War, the Atlantic blockade led to a national shortage of sugar and other commodities. Delta farmers promptly switched to the production of sugarbeet seeds, for shipment elsewhere in the country, and fibre flax was grown for making fishing nets. This was in the days before plastic twine had been invented. The war was also a time of intensive peat mining in Burns Bog (page 153).

After the Second World War, cars replaced every other means of transportation and road building became a priority. Large quantities of glacial gravel were excavated from the shores of Semiahmoo Bay and sold by Surrey to road builders and cement companies. Soon there was strong pressure from the car-driving public for Vancouver and the land south of the Fraser to be linked. "Of course Delta needs a bridge or a tunnel" proclaimed a 1946 writer in the local newspaper.

A Ladner machine operator, George Massey, spearheaded the tunnel lobby, persuading the authorities to adopt Dutch technology. Massey envisioned the tunnel connecting with a major highway that would link Vancouver to the US, cutting across Mud Bay. The lobby group was successful and in 1959 the Deas

Sand for Highway 99 was taken from Boundary Bay, 1960.
(Delta Museum & Archives)

Island Tunnel (renamed the George Massey tunnel in 1969) was opened under the South Arm of the Fraser River. It soon connected with the new Highway 499, (now Highway 99), which was built using thousands of tonnes of deep, soft Boundary Bay sand, dug from the beaches around the bay. Despite the original plan, however, the route looped clear of Mud Bay. Ferry service to Ladner, across the South Arm of the Fraser River, ended the day the tunnel opened.

Land south of the river developed quickly once the tunnel was built. New subdivisions reflected the car-centred lifestyle and the demand for family-sized houses in pleasant rural surroundings. The general mood was optimistic, with a focus on progress and growth. Subdivisions and shopping centres sprang up like mushrooms. Beach cottage communities merged into suburbs, supplanting potato farms, orchards and forests, and industries spread along the Fraser River.

Construction of Highway 99, looking east, in 1960. (Delta Museum & Archives)

Electricity was now readily available and lit up the night skies, powered machinery and transformed the work of farming and fishing. Around the world, there was a move to capital-intensive farming, with increased emphasis on petrochemical inputs in the form of fertilizers and pesticides, conversion to larger properties, and technological advances in farm equipment and processes. Increased yields from chemical fertilizers, together with greater efficiencies and rapid transportation reduced food prices and increased competition for local farmers.

> *There was no political will to conserve farmland at this time*

Farmland around Boundary Bay was expropriated for port use, highways, electricity transmission and other infrastructure, slicing and dicing the agricultural land base (see page 164 onwards). With 120,000 ha (300,000 ac) of arable land within the lower Fraser Valley (Metro Vancouver and the Fraser Valley Regional District), there was no political will to conserve farmland at this time, although the expropriations permanently affected the profitability of delta farming.

Many landowners chose to retire and sell their land, rather than struggle along under the increasingly complex conditions.

10 m (30 ft) high stockpiles of peat in Burns Bog, 1947. (BC Archives)

PEAT HARVESTING

Burns Bog is bordered by fields of cranberries and blueberries, but the bog was once renowned as a major source of sphagnum peat moss. Harvesting began in the 1930s but intensified as part of the war effort in 1941. The peat was dug by hand and sent to the Basic Magnesium plant in Nevada, where it was used in refining metallic magnesium for tracer flares and incendiary bombs. At the height of the operation 1,600 people were employed, but towards the end of the war demand faded. The works were then sold to Western Peat Co. Ltd. who continued the peat harvest for horticultural uses and poultry litter. The peat was hand dug in winter, dried through the summer, then stockpiled, milled, baled and distributed all over the continent. After the mid-1960s, hydraulic methods were used to mine the peat 24 hours a day, seven days a week. Nearly three million bales were dug. In the final years of the harvest, discs and harrows were utilised for digging and the peat slurry was pumped through pipes using a vacuum system. Peat mining ended in 1984.

The old workings remain as deep pools, growing over with sphagnum moss, and the scarred soil is being reclaimed by bog flowers, mosses and lichens. Relics of old conveyor belts and small gauge locomotives used for hauling peat lie rusting in the bog.

BEE FARMING (Apiculture)

Eurasian honey bees supplied honey and pollinated crops in the Old World for thousands of years. In North America, the pollinator role was taken either by solitary bees, such as the orchard mason bee, which can pollinate over 2,000 flowers a day, or by native bumble bees, that use the frequency of their buzz to release flower pollen.

In the late 1800s, tame honey bees migrated from the States onto newly-cleared land in Langley, attracted by fruit trees and flowering plants, such as the foxgloves that Alexander Annand planted around his Campbell Valley farm. In 1913, the Forslund brothers made a profit by bee-tracking, looking for feral bee hives in hollow tree trunks.

Today, honey bees are essential for berry crop pollination and producing honey, and greenhouse operators use bumble bees to pollinate tomatoes.

Visit apiaries at Westham Island Apiary, Delta; Honeybee Centre in Serpentine Valley,Surrey; and Murray Creek Farm, Langley. The old Annand/Rowlatt farm is in Campbell Valley Park.

THE AGRICULTURAL LAND RESERVE (ALR)

A mere 5% of mountainous British Columbia is farmland and only 1.1% is prime farmland. "Prime" refers to land classes one to three, under the seven class, Canada Land Inventory classification system. Most prime farmland, including the Fraser delta, is in the south of the province. The delta has fertile, alluvial soils that are among the very best in Canada, a mild climate, compared with similar latitudes across the continent, and a long growing season. The frost-free period, from April 15 to Oct 21, is the longest in Canada. However, such mild, lowland areas are also in high demand for settlement.

Until the 1970s, prime agricultural land in the province was being lost to highways, residential and industrial developments at a rate of 4,000 - 6,000 ha/yr (9,900 - 14,400 ac/yr). Speculation raised the cost of land and created uncertainty about the future. The provincial government responded by passing the B.C. Land Commission Act in 1973, creating the Agricultural Land Reserve (ALR), a land use zone dedicated to agriculture. Provincially, 4.76 million ha (11.8 million ac) are in the ALR, about 5% of the province, and in Metro Vancouver there are 61,000 ha (150,500 ac). Within the Boundary Bay watershed, in Delta, Surrey and Langley, there are about 43,000 ha (106,200 ac) of ALR land.

The ALR was key to saving much valuable farmland from being lost to urban sprawl. For example, less than 5% of Delta's agricultural land was excluded from the ALR between 1974 and 1991, although some non-farm uses were allowed. However, the ALR is under continual pressure from land speculation in the Boundary Bay area.

MODERN FARMING

While the creation of the Agricultural Land Reserve cooled the rapid land speculation of the 1970s, it did not solve all the problems of farming viability. Crop processors gradually withdrew from the delta, citing rising costs and lack of access, and this caused many farmers to stop growing beans, peas, and strawberries. Some people switched to old staples like potatoes, corn, and pumpkins, but at 9 cents a pound, pumpkins did not pay the bills. Today the farming community is experimenting with a wider variety of products.

Vineyards and hop fields have been planted on sunny slopes in the rolling hill country east of the bay and fruit wineries have opened. Langley produces much of British Columbia's mushroom crop, and cranberries from the lower Fraser Valley are the province's biggest berry crop. Blueberries are also profitable and many Boundary Bay farmers have rushed to plant them. Whatcom County has the largest raspberry crop in the USA. Other crops grown in the watershed include tayberries, red currants and hazelnuts. Turf farms occupy land around Boundary Bay alongside enormous greenhouses, that were rapidly built to supply export tomatoes, peppers and cucumbers (see page 160).

Livestock diversification has also occurred: deer, goats, emus and sheep are being raised alongside the more conventional cattle, poultry and horses. Sheep are often kept with a companion llama in the field, as protection from coyotes.

Caine llama farm in the Little Campbell River valley.

Pressure on the land base continues to rise, with increasing demand for residential, industrial, institutional and transportational uses. Of all the communities around the bay, Surrey has seen the greatest changes in land use in the last two decades. The once attractive pastoral landscape has almost completely disappeared under tarmac and concrete, with a resulting huge loss in wildlife habitat. The remaining hillside farms, and the ALR land on the Nicomekl and Serpentine floodplains, are hemmed in on all sides by developments. Throughout the watershed, families that try and stay in agriculture must wrestle with the problems created by urban proximity. These include road access, vandalism, nuisance complaints and high land costs.

At one time, nearly everyone in the lowlands around Boundary Bay was connected in some way with agriculture or fisheries. That is no longer the case. Only about 3,000 people are employed on farms in Delta and Surrey, and there are about 1,230 farms in Langley. Agriculture is an important source of employment for the Punjabi community. Punjabis first arrived from India in 1906 and 1907, at which time most men worked clearing land or milling lumber. John Berry hired about twenty men to clear his Langley farm of timber at the rate of 50 cents a day. The Punjabis prided themselves on hard work and persistence: "we got established here with muscles and money." Soon many families were running prosperous businesses and farms, especially in the Surrey area. Later immigrants adopted the same attitude, accepting unpopular physical work as a means to improved circumstances, and as a consequence, they were often a predominant ethnic group working in the fields.

High quality, locally grown food benefits all of society

While no longer the primary local industry, agriculture is crucial to the wider community. The advantages that attracted the early farmers still hold true. Agriculture can be a sustainable industry that benefits all of society by ensuring a safe supply of healthy, high quality, locally grown food. It provides employment both directly and in added value services, and it promotes national security and environmental protection.

Intensive agriculture helps keep the the industry competitive in global markets, while the profit from open-field crops is often relatively small. Open-field farming, however, provides communities with less economically tangible assets such as the presence of wildlife, the views and quality of life experienced by open spaces, trees and hedgerows, and carbon dioxide absorption due to vegetation growth.

Boundary Bay farmland is among the very best in Canada yet is under constant threat from expansion of the cities around it.

For all these reasons, many citizens hold passionate views about the use of farmland and what constitutes valid activities for the region. The Agricultural Land Reserve has not prevented the use of farmland for other purposes. Only a massive public outcry in the late 1980s and early 1990s prevented a proliferation of golf courses and subdivisions on farmland around Boundary Bay (see pages 172-173). The issue became such a hot one around the province that it led to sweeping political changes, both locally and provincially. Despite this, valuable soil is still converted to non-soil based agricultural uses, residential developments, and highways. The agricultural landbase is squandered, with little recognition of its limited and essential nature.

Green choices

Organic farming is steadily taking hold, driven by consumers' demand for healthy food, with authentic quality and better taste. Certified organic farms do not use synthetic inputs, such as chemical fertilizers, pesticides, hormones or antibiotics and the land must be chemical-free for three years to permit certification. According to Fraser Valley Organic Producers Association, management practices on organic farms "restore, maintain and enhance ecological harmony" and include many positive activities such as crop rotation, green manure crops, composting and other soil improvements. Plants and animals grow stronger and healthier on good soil and pasture, and benefit from greater space. Many local farms are adopting "organic" practices, even if they do not go through the full certification process. It is sometimes forgotten that buying local produce, whether organic or not, can be a "green" choice - the food has not had to travel far, so carbon emissions are low.

FARMING & WILDLIFE

Ecological conflicts between modern farming methods and wildlife are particularly acute around Boundary Bay because of the international significance of the bird populations and the importance of the fisheries resource. Waterfowl and shorebirds, migrating on the Pacific Flyway or wintering in the area, use farm fields for resting and feeding. Songbirds and birds of prey use grasslands and hedgerows. Salmon and other fish are found in farm ditches. The Agricultural Land Reserve therefore provides much essential habitat for a diversity of wildlife.

Since farming began in the lower Fraser Valley, landscape changes have affected bird populations in a variety of ways. First there were the effects of dyking and draining of marshes, as discussed on pages 102 to 108. The resulting paddocks and hay fields were a reasonable substitute habitat for some species, but as dairy farms declined and were replaced by arable crops, grassland habitats became scarce. By 1981, burrowing owls and horned larks were extirpated as local breeding species.

Agricultural pesticide use increased in the 1960s and 70s, and combined with habitat loss had a cumulative impact on songbirds and other insect-eaters. Yellow-billed cuckoos were extirpated in 1950, and western bluebirds by 1971. Nighthawks used to nest even in downtown Vancouver, yet are now scarce visitors, and swallows and swifts are declining. Since pesticide spraying has been reduced, some species may recover. Purple martins, that had gone from the region, now nest again at Blackie Spit. Bald eagles, at the top of the food chain, were particularly affected by pesticides such as DDT, and their numbers plummeted in the 1950s and 60s. However, since these chemicals were banned, both breeding and wintering eagle numbers have rebounded.

Homeowners use ten times as many chemicals as farmers

Many local farmers have switched to integrated pest management with biological controls, although spraying against weeds continues for some crops. Today, blueberry and cranberry producers are among the heaviest agricultural pesticide users,

Bald eagles are again a common sight.

acreage has decreased, particularly in Richmond. Snow geese flocks there have been forced to feed on school playing fields! While ducks and geese are the chief culprits for crop damage, muskrats, beaver, deer, gulls and songbirds may also cause problems. Berry farms are targeted by flocks of birds, such as starlings and blackbirds, necessitating the use of propane cannons to scare them off. Since this creates a high noise level for neighbouring communities, some growers use fluttering kites in the shape of hawks instead.

Another factor in modern farming is the need to maximize field size. Hedges prevent the manoeuverability of farm machinery so are often removed to clear ditches or make fields larger. Yet hedgerows and the long, rough grass in field margins are excellent habitat. At least 45 bird species use hedges for breeding, and over 70 species rest there during migration. Many of these birds are pest predators, as are some native insects living in field margins. These species contribute to natural control of infestation outbreaks. Hedges also provide livestock shelter, optimise local microclimates and prevent soil erosion.

yet suburban homeowners use, on average, ten times as many chemicals per acre on their lawns and gardens as are used on farmland. Herbicides and pesticides lower the water quality of streams and rivers running into the bay, and can be extremely dangerous for human and wildlife health. It makes sense for home gardeners to avoid the use of such substances for cosmetic purposes, especially since pulling weeds is such good exercise!

Wildlife populations and agricultural operations may conflict in other ways. One night of intense feeding by a flock of geese or ducks such as wigeon, can reduce a field of newly-emerged winter wheat to a sea of mud and stalks. Snow geese have increased in recent years while delta farmland

Hedgerow planting is one of the programs undertaken by the Delta Farmland and Wildlife Trust, a non-profit organization formed in 1993 to address farming and ecological issues. The Trust plants winter cover crops and establishes rotational set-asides. Set-asides are grass fields that lie fallow, without being cut, for two to five years. Once a traditional part of the agricultural cycle for restoring soil nutrients, such fields are also important habitat for voles: small mammals that are prey for hawks, owls and herons. Trust programs reach 80% of farms in Delta, as well as educating the public about farming's role in society.

GROWING UNDER GLASS

Much of the land adjacent to the bay in the 1990s was optioned by developers, who speculated that the floodplain would inevitably undergo a transition to urban development. When golf courses were outlawed as an outright use of farmland, the same land saw an influx of large greenhouses, a permitted agricultural use but one with cumulative impacts on the local environment. There is now more land under glass in Delta than any other municipality in British Columbia: about 2% of the areas's farmland is covered.

Industrial greenhouses are large glass structures designed to grow plants such as tomatoes, cucumbers and peppers, in hydroponic conditions, that is, without soil. Nutrients are supplied by fluid-bearing pipes. The ability to control temperature and other conditions allows for uniform crops and a long growing season, providing lucrative exports. Land

around Boundary Bay is flat, the area has climatic advantages such as good light conditions, and farmland has relatively low tax assessments. All are excellent conditions for greenhouse farming. Local greenhouses were a massive investment, with building costs of up to $2.5 million/ha ($1m/ac). Over 90 ha (222 ac) were erected in Delta between 1996 and 2006, with some covering up to 23 ha (57 ac). This far surpassed the investment or revenue from other agriculture in the region, and is so large that a return to soil-based, open-air farming on these sites is most unlikely. Conservation concerns relate to the importance and sensitivity of the Boundary Bay ecosystem. Large site coverage obscures natural soil and reduces the amount of available habitat for waterfowl and other birds. Birds are displaced to neighbouring arable farms, increasing crowding and the risk of disease, reducing feeding opportunities, and aggravating crop predation. Air pollution and night lighting by some operations are other environmental impacts (pages 179-181).

Part of a massive greenhouse on the bay.

A plume of water irrigates a farm field.

FARMING'S FUTURE

A seminal study of agriculture in the Boundary Bay area was completed in 1992. It recommended better soil management, longer leases on Crown land, a provincial right-to-farm act and the establishment of a wildlife-agriculture committee. These goals were achieved yet local farmers continue to struggle competitively in an increasingly tight global market. While growing conditions, access to markets, interest in value-added products and consumer demand are all positive, only a few processors remain in the area because crop volumes are not large enough.

According to the B.C. Ministry of Agriculture, current farm problems include a lack of seasonal field workers, insufficient research, and increased international competition, especially from the southern states and Mexico where growing seasons are longer, the weather more settled, and wages and other costs considerably lower. More food is also being imported from China,

where wages are even lower and environmental regulations less strict. The future will likely see more local demand for produce, especially niche items, but also increasing concerns about food safety and quality. This was illustrated in spring 2004 when an outbreak of avian flu in the Fraser Valley caused enormous problems for the poultry industry. Health issues will likely increase interest in organic methods, yet rising populations will also push the drive towards closed systems, genetically modified organisms and highly intensive methods. New challenges could arise as a consequence of climate change, including emergent diseases, alteration in rainfall patterns and season length, and rising sea level.

For many years, the farmers' lands around Boundary Bay provided food for the whole community, sustained migratory and wintering birds, and served as a buffer between urban centres and the wetlands. With the changing face of agriculture in the region, it is essential that some way is found to permanently protect soils, agricultural values, and farmland biodiversity and sustainability.

Further information on local farming

Delta Farmland & Wildlife Trust
www.deltafarmland.org
BC Ministry of Agriculture
www. agf.gov.bc.ca
Agricultural Museums - see page 185

Roberts Bank - an ecologically-critical area, at risk from structures such as the port, ferry terminal and causeways, and the associated habitat deterioration.

CONSERVATION OF BOUNDARY BAY

"A chance to see wildlife under natural conditions is one of the important rights of man. It is a real enrichment of living."
~ *Roderick Haig-Brown*

Boundary Bay is a very special place, for thousands of years a source of abundant food and shelter for people and wildlife. What will the future bring? This final chapter examines the rapid social changes that have occurred in the last fifty years, their impact on the landscape, and the efforts that have been made to mitigate the resulting environmental damage. It also looks at current and future threats facing the Boundary Bay ecosystem, in the wider context of accelerated habitat loss, biodiversity declines and global climate change.

Wildlife declines and extinctions are usually precipitated by either habitat loss, pollution, over-hunting or alien species introduction, or a combination of these causes. Effective conservation must therefore address all these factors, as well as even greater challenges. The complex effects of the changing climate will include altered plant communities, shifts in bird and animal populations, declines in fish survival and invasion by new pests, viruses and diseases among the many unforeseen events. The Pandora's box of possibilities opened by the ability to modify genomes and construct new organisms also awaits us in the coming decades. Industrialised agricultural monopolies, greater efficiencies and higher profits, are driving the increasing predominance of domestic and genetically modified stocks over wild and natural species. If current trends continue, genetically modified organisms and patented life will become the norm, even if it means the loss of individual rights and the privatisation of our common natural heritage. We may soon see a vastly changed landscape.

These possibilities are deeply troubling for all those who love nature or delight in our priceless heritage of wildlife and landscape. Humans need nature to survive - it is not a luxury but a necessity. It is well known that ecosystems provide many services essential for human health and increasing numbers of studies report on nature's immense curative, cognitive and psychological benefits. Nature conservation is not just about wildlife, but about our own future on this planet, and the actions we take locally have a global impact.

There have been many ideas for reclaiming all or part of Boundary Bay. Pilings mark the site of an old cannery.

RESHAPING THE BAY

Grandiose ideas for "improvements" have been a constant theme around Boundary Bay, including reclaiming the entire bay for an industrial site, and constructing a deep sea port in Burns Bog! Development projects that came to completion often had unintended ecological side effects. A brief catalogue of projects around the bay illustrates their variety.

After the Second World War, the shallow nature of Boundary Bay encouraged the idea of reclaiming the intertidal areas, and in 1958 a giant industrial site was proposed within the bay, with oil refineries, steel mills, petrochemical plants and

a deep sea harbour. George Massey, of tunnel fame, was a sponsor of this proposed "Southport" which would have destroyed 4,800 ha (12,000 ac) of fish and wildlife habitat. The huge project was vociferously opposed by local citizens and came to nothing. A few years later, back came more plans for marinas, canals, industrial uses, and residential developments. Between 1958 and 1971, at least six mega-development proposals were advanced by promoters. However, in each case, communities united in strong opposition and defeated them. Advocates for the bay included the Save Our Beaches Association, who not only wanted to protect the wildlife of the area, but also proposed creating a recreational resort along the lines of Mission Beach, San Diego, with artificial sand spurs and islands in the bay. Financial support for that was not forthcoming, however, though the group successfully rallied the community to clean beaches of accumulated trash (see page 170).

Some projects went ahead. A make-work project to ease the 1930s Depression, changed the Drayton Harbor ecosystem. The shore at Blaine had been lined with mills and canneries, but when logging came to a halt, disused piers, buildings and waste lumber lay in the sand. A breakwater constructed out of this jumble of waste reduced water flow into the harbour and the enclosed bay almost silted up. Sedimentation near Peace Arch Park, on Semiahmoo Bay, was also increased. In the forties, Surrey used some of this beach sediment for road fill. Sand from along the north side of Boundary Bay was later used for building Highway 99 (see page 151-152).

Roberts Bank

Two railway companies envisioned port terminals on Roberts Bank in the early 1900s, to serve the needs of the growing population. Premier W.A.C. Bennett also promoted a superport there, so it is not surprising that the edge of the bank was chosen for the 1960s construction of both a B.C. Ferry terminal and the first Roberts Bank port. Two 4 km long (2.5 mi) causeways were cut across

the ecologically rich intertidal flats, linking the deepwater berths to the mainland and permanently altering the mixing of river and ocean water on Roberts Bank. Currents generated by the barriers scour the banks, creating large eroded gulleys. Sediment deposition from the Fraser River is obstructed, reducing the intertidal banks and eroding beaches along Point Roberts' western shore.

The port expanded from its original 20 ha (48 ac) to five times that size by 2007, resulting in the loss of intertidal sand flats, essential habitat for shorebirds migrating up and down the Pacific Flyway. Massive deep sea ships now pass through waters used by endangered killer whales, porpoises and other marine mammals, all sensitive to disturbance. Marine transportation is a source of water and air pollution, and can introduce alien species; several such aliens have arrived, despite the 1994 instigation of ballast water controls. Causeway powerlines serving the port are lethal for migrating birds, and increased night lighting is likely adding to the disruption of critical fish and wildlife habitat (see page 181).

As part of the industrialisation of this area of Delta, 1,700 ha (4,000 ac) of agricultural floodplain was expropriated by the B.C. government in 1968. Known as the "back-up lands," they were leased back to the original farmers, under short term leases that gave no incentive for capital investments in either soil, machinery or maintenance. Some of this land was sold back to farmers in 2003 and the rest was retained for treaty settlements, container depots and railyards. A whole new port terminal is proposed even though current impacts remain unresolved. Habitat compensation in this area is difficult: an attempted eelgrass planting project adjacent to the ferry terminal turned into a salt marsh, which, while popular with Caspian terns and other birds, was not the anticipated outcome.

Western sandpipers dramatically declined

Channelisation of the Fraser River made changes on both Roberts and Sturgeon Banks. The latter area had also been proposed through the years for land reclamation, port structures and industrialisation, although none built. The river's freshwater plume is being pushed north, increasing salinity to the south and expanding eelgrass beds at the expense of unvegetated mud. Sedge and cattail marshes are spreading on both banks. These changes are good for some birds, like waterfowl, but detrimental to migrant flocks of shorebirds that rely on open sand. Numbers of western sandpipers recorded on the delta have dramatically declined since 1994, perhaps in response to reduced habitat or greater disturbance.

Airfields

Boundary Bay airport, located on farmland just north of the bay, first opened during the Second World War as a Royal Canadian Air Force flight training base. After the war, the land was used for non-aviation activities until the airfield was reactivated in 1983. It is now a busy small plane airport. Reactivation was criticized by Environment Canada because of the high risk of bird strikes. The airfield lies between Boundary Bay and Burns Bog, an area heavily used by birds, and right on the dawn and dusk flight paths of roosting gulls. Despite this, the airport has plans to enlarge.

Further east along the bay, the Delta airpark was set up by the Embree family in the 1960s as an uncontrolled aerodrome for private owners of small planes. It was purchased as parkland in 1995 by Metro Vancouver, which maintains a license agreement with the operating organization. Part of the land is farmed; the remainder is airpark and dyke trail car parking. The farmhouse dates to 1918.

Construction of Vancouver's first passenger airport began in 1930 on Sea Island, Richmond, at the mouth of the Fraser. The airport steadily grew, and by the late 1960s nearly the entire island was federal airport reserve. When yet another runway was built in 1991, important riverside wildlife habitat was lost. Federal

Peat excavation created ponds in Burns Bog.

safety regulations also now require the control of bird flocks.

All these airport developments have meant a loss to the quality of our environment. Habitat is destroyed, planes and helicopters disturb bird flocks, and noise and air pollution increase. Unfortunately, this seems to be a price many people are willing to pay, even as the beautiful natural world declines.

Burns Bog

Burns Bog has been the subject of all sorts of development ideas since peat harvesting ended in 1984 (page 153). In 1988, community outcry, led by ardent bog supporters, squashed a $10.5 billion proposal for a city of 125,000 people built around a port complex. Proposals for a racetrack, fairground and sewage plant were also defeated. However, nothing could stop the construction of the Alex Fraser bridge and Highway 91, which split the bog from the North Delta uplands. Private landfills were a precursor to industrialisation of the riverside, and in 1964, the Vancouver City landfill opened on about 260 ha (624 ac) in the southwest corner of Burns Bog. It now takes in about 475,000 tonnes of waste a year. In response to environmental concerns, recent improvements to the landfill have included collecting recyclables and compost materials, trapping and using methane gas, and restorative landscaping. The footprint has also been reduced by piling the waste higher.

HUNTERS AND CONSERVATIONISTS

Boundary Bay has a history of use by food and sport hunters. In historic times, even many naturalists hunted; in fact, bird identification had to be confirmed by obtaining a specimen.

The frontier was an unregulated place in the 1800s, with fish and game free for the taking. When wildlife disappeared, it was tempting to blame neighbours. According to one account, the Little Campbell River was once full of trout but "the people of Blaine came across and cleaned most of them out." The Canadians blamed the Americans for taking too many sockeye salmon at Point Roberts and the Americans blamed their northern neighbours for intercepting them in the Strait. Everyone blamed everyone else for the loss of the trumpeter swans.

The impact of unregulated hunting affected bird populations along the Pacific Flyway, the migration route followed by millions of waterfowl. As more people settled in the west, game was shot for commercial sale and recreation on a huge scale (see page 139). Stocks plummeted, causing first concern and then the call for restraint. "Conservation" became the word of the day after it was promoted by President Theodore Roosevelt in the early 1900s. He established standards of scientific management for game animals, and set up institutions and government wildlife departments to administer the new approach. Scientific knowledge grew with the collection of zoological data. In British Columbia, the Provincial Museum was established in 1886, with ornithologist John Fannin as its first curator. Numerous aspects of the province's wildlife were researched and published.

A view prevailed for many years, of game animals as "good" and all predators as "bad." Destruction of predators was thought to increase game, with even some naturalists believing that animals such as the cougar should be "hunted down and destroyed, regardless of cost." Bounties were placed on coyotes, cougars and wolves until the mid 1950s. Opinion swung slowly in favour of predators, especially among the increasingly urban public. Sadly, by that time wolves and cougars were long gone from the Boundary Bay watershed, although coyotes have held their own in suburbs and cities. Even so, there are still those who think they should be shot.

The Christmas Bird Count, an annual event in both Canada and the USA, had its beginning in 1900 as a gentle alternative to the widespread Christmas hunt in which teams competed for the "biggest pile of feathered and furred quarry." Today, the Ladner Count is often the top one in Canada, averaging about 140 different bird species found by teams of observers. The early years of the century also saw the formation of local naturalist clubs, such as the Vancouver Natural History Society (now Nature Vancouver), ushering in a long tradition of volunteer involvement in nature observation and record keeping.

A more ecological perspective gradually took over, under the guidance of renowned naturalists like Ian McTaggart-Cowan, who in 1941 became Canada's first teacher of a university course on wildlife management, and Vernon "Bert" Brink, who worked throughout his long life to foster the appreciation and protection of nature.

Hunters on Boundary Bay.

When waterfowl stocks declined, rod and gun clubs rallied to protect them. Wood ducks in the estuary marshes were restocked in the 1960s, aided by the provision of nest boxes. The Reifel Bird Sanctuary was set up by a group of conservationists in the 1970s, followed by Serpentine Fen Wildlife Area, adjacent to Mud Bay. Both sites were at first used to breed exotic waterfowl, in retrospect a very mistaken idea. A non-migratory subspecies of Canada goose raised there multiplied rapidly and is now far more abundant than the migrant Canada geese that stop only briefly in the delta. Eventually, interest in introducing these non-native wildlife waned, replaced by the concept of natural habitat protection.

Hunting still takes place around the bay, but over the last 30 years it has declined as other outdoor leisure activities developed. Today wildlife viewing and photography are greatly enjoyed by many people, and birdwatching is considered to be among the most popular recreational pastimes in North America.

Further information on hunting and conservation

Waterfowl on a Pacific Estuary by Barry Leach

Our Wildlife Heritage by the Centennial Wildlife Society

Local nature clubs: www.bcnature.ca

B.C. Waterfowl Society:
www.reifelbirdsanctuary.com/bcws2.html

Wrecked vehicle bodies were used to shore up the Boundary Bay dyke in 1965.
(Surrey Archives)

CHANGING VALUES

As higher education became more common, salaries rose and leisure time expanded. Interest grew in travel, viewing natural landscapes and watching wildlife, rather than shooting it. Concerns about pollution and the biosphere's health led to the birth of environmentalism and greater support for conservation.

The advent of television and nature shows fostered this interest while also creating certain expectations as to how nature should look. The secretive world of mud flats or the untidy expanse of marshes, lacking large, charismatic game animals, did not match up to these expectations. It took many years for people to appreciate the beauty and value of natural wetland areas. By the 1970s, however, recognition was growing of the Fraser estuary's ecological importance. Studies on waterfowl habitat, fish populations and geology emphasized the area's unique attributes. Public attitudes to the environment began to change, and this led to local actions. People rallied around to clean up the old cars and trash that had been dumped on the Boundary Bay foreshore, and to remove the debris of the forest industry - driftwood logs that were choking the life out of marshes. Cars were excluded from the beach at Blackie Spit, and parks were opened at Deas Island, Centennial Beach and

Ladner Harbour, the latter adjacent to an old sewage lagoon that was replaced by new regional facilities. Government biologists shifted focus, expanding their interest to non-game animals, biodiversity and endangered species. Concern about pesticides led to control of their use, so that bald eagles and peregrine falcons were once more a common sight.

"Greenbelt" lands were purchased on the shore of Boundary Bay

Political will also changed, if only temporarily. For example, in 1972 the B.C. Green Belt Protection Fund Act was passed. "Greenbelt" lands were subsequently purchased on the shore of Boundary Bay with the intent of protecting them in perpetuity for their agricultural and conservation value. Similar concern was shown for the environment in 1979, when recommendations of an Environmental Assessment Panel reviewing Roberts Bank port led the federal Minister of Environment to conclude that "full expansion of the port would present an unacceptable threat to the ecosystem." Sadly, these progressive statements did not influence the actions of later governments. Greenbelt lands were sold off in 1999 and the port on Roberts Bank expanded well beyond the original 1977 plan (page 165).

IN DEFENCE OF HABITAT

Despite increased public knowledge and concern for the environment, habitat loss continues to accelerate. The urge to develop our wonderful coastal landscape is just too great. All too often, political commitment to conservation fades in the face of economic interests, even though an insatiable demand for growth is unsustainable in our finite world. Other community goals, such as quality of life, clean air and clean water, become superceded. Nature, with no voice at the bargaining table, does not get a chance.

Living creatures need habitat with food, water and shelter to survive. Boundary Bay, in the heart of the Fraser River estuary, is remarkable in having many different habitats. Land and water merge here in a dynamic and life-sustaining way. In 1987, ornithologists Rob Butler and Wayne Campbell produced the definitive report, *The Birds of the Fraser River Delta*. They found that migratory and wintering birds using the delta travel

A male hooded merganser.

between three continents and 20 countries, and that "no comparable sites exist along the Pacific coast between California and Alaska." The delta exceeded the criteria for designation as a "Ramsar" site or Wetland of International Importance by more than 30 times for waterfowl and 60 times for shorebirds. Butler and Campbell stated that habitat loss was becoming critical, and that land should immediately be set aside to conserve internationally important bird populations. Yet, within a year of this report, farmland adjacent to Boundary Bay was proposed for many controversial developments.

A 1988 provincial Order-in-Council allowed golf courses as an outright use of farmland, and a string of golf courses, condominiums and club houses were planned for the shoreline. Virtually all the land just north of Boundary Bay was optioned by developers. If every proposal had come to fruition there would now be 44 golf courses around the bay! Community conservationists rallied once more to make politicians aware of the bay's ecological importance, and even went to court to prevent farmland being rezoned. Simultaneously, another housing and golf course proposal for the Spetifore Farm (Southlands) roused the South Delta community to action. The potato and dairy farm had been removed from the Agricultural Land Reserve some years previously, and housing developments had been proposed in 1971 and 1983. Rezoning was required for the Tsawwassen Development Ltd. (TDL) proposal to go forward, and this time the public hearing was the longest in Canadian history, involving hundreds of people and a hundred hours of testimony over twenty-five evenings. When it was eventually terminated, citizens arranged a public plebiscite to convince Delta Council to turn down the rezoning bylaws and say "No to TDL," as the bumper stickers proclaimed. A high turnout on voting day was 98% against the proposal. The bylaws were defeated, the community having chosen overwhelmingly to maintain open farmland rather than build more subdivisions.

This strong public outcry against the continual loss of farmland spilled over to other areas of the province, provoking a debate that carried into the next municipal and provincial elections, at both of which politicians espousing "greener" viewpoints were elected. The first action of the new provincial government was to repeal the offending Order-in-Council.

Public participation in community planning continued to be strong in the early 1990s. Rezoning bylaws were challenged, and the Greater Vancouver Regional Green Zone was approved, confining development to core areas and protecting the ALR. Richmond, Delta and Surrey municipalities were all taken to court over controversial developments.

The old Spetifore farmland was the subject of the longest public hearing in Canadian history.

Conservationists successfully urged governments to conduct a series of studies of Boundary Bay and these confirmed the area's importance to wildlife, as well as analysing rural and farming issues. In 1992, a volunteer environmental group, the Boundary Bay Conservation Committee, advanced the idea of a local biosphere reserve. Although supported by conservationists and three municipalities, this idea was not upheld by senior governments. In time, however, many of the group's goals were achieved: Wildlife Management Areas were designated at Boundary Bay and the South Arm Marshes, a farmland trust was created (page 159), and about a third of the Spetifore property was purchased for Boundary Bay Regional Park through the provincial Nature Legacy program. With thousands of people calling for protection of Burns Bog, Delta taxpayers voted strongly in favour of its purchase as an Ecological Conservancy Area. In Whatcom County, land planning considered ecological and heritage values when waterfront recreational facilities were expanded. As Surrey rapidly urbanized, groups scrambled to protect remaining forests. Parks had previously been set aside at Tynehead, Green Timbers and Sunnyside Acres; now a new park was designated at Mud Bay. Much rural land was converted to urban use, however, with resulting loss of habitat.

BIODIVERSITY

The satyr comma butterfly.

Biodiversity is the complete variety of living organisms found on earth.

The incredible richness of form, colour, texture, sounds, movement, reproduction, feeding habits, growth patterns and a myriad of other life variations is what makes our planet such as special place.

This great variety means that every habitat can be used to advantage by some especially adapted species. Where conditions such as vegetation and climate combine to provide a rich habitat, wildlife will proliferate. Conversely, a degraded habitat will be marked by a paucity of species and a preponderance of opportunistic, adaptable ones, such as rats, crows, rock pigeons, and starlings.

Biodiversity in the Boundary Bay watershed is high. Among vertebrates alone, there are over 60 species of mammals, reptiles and amphibians, 330 recorded species of birds, of which 250 or so occur annually, and over 90 species of freshwater and marine fish. In addition there are thousands of invertebrates and smaller species, and hundreds of native plants. Biodiversity fluctuates as some species become locally extirpated while others move in.

Although biodiversity is high, many species are declining in B.C. and Washington. For example, B.C. breeding bird surveys show there are fewer songbirds than two decades ago. Numbers of relatively common species such as Bewick's wren, barn swallow, golden-crowned kinglet, rufous hummingbird and yellow warbler are down. Belted kingfishers are seen less often and wintering seabird numbers have also declined. These trends are part of a global problem; in the US, for example, the National Audubon Society has estimated that one in four birds is at risk.

Birds are not the only ones in trouble. Examples of federally listed "Species at Risk" found in the Boundary Bay watershed include killer whale, mountain beaver, Pacific water shrew, western toad, white sturgeon and Henderson's checkermallow.

Further information on Boundary Bay's biodiversity

A Nature Guide to Boundary Bay by Anne Murray and David Blevins www.natureguidesbc.com

The Georgia Basin Habitat Atlas: Boundary Bay www.georgiabasin.net

PARKS AND WILDLIFE

Some people might think enough has been done for wildlife, with several new parks now set aside. However, parks are insufficient to protect our wildlife populations. We need a much more holistic approach to habitat protection, right across the landscape.

One problem with parks is that recreational activities often conflict with wildlife use. People enjoy taking their dogs walking in the park. On weekends and holidays, large numbers of people picnic, hike, jog, mountain bike, kayak, boat, kite-board or windsurf in parks around the bay. Some of these activities can cause serious disturbance problems for certain species of birds and animals. Shorebird flocks, which stop for just a few days on migration, have to feed almost continuously at low tide to fuel-up for the next stage of the journey. Being constantly interrupted and put to flight by, for example, a loose running dog, can cause shorebirds to starve and die at sea. Kite-boarders like windy winter days and high tides; these are times when large flocks of waterfowl are also present along the bay shoreline. It is easy to notice the big, bold birds, like the eagles, and not realize that other less confident species are being scared away. Sadly, there is often nowhere else for them to go, as parks are frequently just islands of habitat in a sea of development.

Public preferences for parks differ. Some people want very manicured areas, with no natural "untidy" vegetation of the kind required by wildlife. Some parks are even entirely given over to sports, with buildings, carparks, artificial fields, and bright lights.

Another issue is genetic diversity in parks. When wildlife, large or small, are sequestered in protected areas their chances of breeding successfully are diminished. This may be less of a problem for most birds, with the great mobility of flight, but for other species it can be a major drawback. How long will bears continue to live in Burns Bog, with their movement corridors through the Surrey forests increasingly shut off? Where will long-toed salamanders and western toads replenish their populations if wetlands are drained, culverted or polluted? We have already seen the loss of western spotted skunks and red foxes from most of the region; reducing habitat to a few protected areas leads to declining biodiversity (see page 174).

An American bittern hiding in a marsh.

ALIEN INVADERS

The introduction of non-native plants, invertebrates, and even larger animals to the Boundary Bay area has resulted in significant changes to the local ecosystem. In most cases, these alien species arrived accidentally in association with human activities, such as agriculture, shellfish farms or ports. In some cases, animals and plants were introduced deliberately. Several have been mentioned earlier in this book (see pages 95 and 117).

American bullfrogs grow to the size of a dinner plate

Of the larger introduced animals, American bullfrogs have been among the most destructive. They escaped and proliferated after a failed attempt at farming them for frog's legs, and now occupy many freshwater wetlands around the bay. They can grow to the size of a dinner plate and will even eat ducklings!

Black rats and Norwegian rats were also unwelcome additions to the local fauna. Like many other alien species, they arrived by ship. Marine invertebrates travel in ballast water or on the hulls of deep sea vessels; some came with oyster spat from Japan. Dozens of species now live in Boundary Bay. Purple varnish clams and invasive English cordgrass (*Spartina*) are two recent marine introductions.

Most Boundary Bay birds are native species but common starlings, rock pigeons and house sparrows were introduced. European starlings were originally brought to New York City but spread across the continent in just a few years, aggressively displacing other hole-nesting species in disturbed habitats. Homing rock pigeons were used for racing and sending messages, yet have now adapted to city streets, as have confident English house sparrows, a species unrelated to native North American sparrows.

Not all aliens cause major problems; ring-necked pheasants, originally from India via the countryside of England, were regularly reared and released by rod and gun clubs; some still survive. California quail brought to the delta in the 1890s have since died out. Eastern cottontails, that seem to have come with early settlers, now provide prey for raptors.

The ecological impacts of the hundreds of alien animal and plant species now in the landscape is not well understood. Their interactions within the local food web have hardly been studied, and it would be impossible to find local control habitat unaffected by foreign species. Restoration of habitats therefore presents a number of problems. Enhancement and stewardship of habitats require detailed knowledge of the behaviour and needs of local wildlife. Merely pulling out invasive

Rock pigeon - an introduced species.

plants is not enough, and could even cause harm. Many introduced plants provide critical food or shelter for species that have adapted to these new habitats. Berry-bearing alien shrubs, such as Himalayan blackberry, holly and English ivy, are used extensively by wildlife because so many native shrubs, such as cranberry, blueberry, Pacific crabapple and salmonberry, have been destroyed. Blackberry thickets can provide essential shelter for birds and small mammals in winter and during the breeding season. The untidy bramble tangles replicate shrubby areas found on the delta prior to dyking and draining.

If alien vegetation is removed, quick replacement of adequate food and shelter is vital, otherwise animals will disperse or die. Only with local ecological knowledge and adequate planning is it possible to successfully restore habitats and enhance chosen species' populations.

Climate-induced migrants

New species for the bay may arrive in response to the changing climate. Animals and plants disperse because of factors such as food availability, rainfall and temperature cycles, and territorial needs. Others may move into new areas as part of a general range expansion; for example, cattle egrets and Caspian terns are two southern bird species now regularly occuring in the Boundary Bay area.

Climate change is already known to be altering bird distributions and disrupting migration patterns elsewhere on the continent. Nature Canada reports that many birds are laying eggs days, or even weeks earlier in spring, compared with a few decades ago. Many migratory species are arriving earlier or leaving later. In some places birds are completely failing to migrate. Annual observation records kept by birders and naturalists are going to be of increasing importance in studying the long-term impacts of climate change.

Birds are very visible indicators of ecosystem disruption. Microscopic bacteria and viruses will be among more subtle invaders; West Nile virus is only one of many southern pathogens we may play host to in the near future. Other indicators include increased "red tide" occurrences and the spread of unusual fungi, such as the disease-causing *Cryptococcus*.

KEEPING IT CLEAN

For 10,000 years, the Boundary Bay watershed enjoyed the cleanest water and air that nature could provide, with glacier-fed rivers and fresh ocean breezes. Pollution problems began more than a 100 years ago: in 1892, dairy farmers, William Ladner, Henry Benson and John Kirkland, spoke up about cannery waste contaminating their cattle's water supply and were successful in getting a clean-up of the sloughs. Family connections have never meant unanimous view points: William's brother, Thomas, spoke for the canneries, claiming that fish offal never hurt anyone!

Drinking water in the western delta was originally derived from aquifers beneath the East Delta escarpment. Delta, Richmond, Surrey and Point Roberts residents are now supplied with water piped from reservoirs on Vancouver's North Shore, while White Rock, Blaine and parts of Langley access groundwater from

Fresh artesian water flows from an aquifer below the North Delta hills.

aquifers. Waste water and sewage on the Canadian side of the border is mostly handled by Metro Vancouver, at secondary treatment plants in Northwest Langley, Annacis Island, and Lulu Island. Blaine has its own system and rural properties upstream in Langley and Whatcom County are on septic systems. Poorly operating septic systems are a major source of aquifer pollution in rural districts, while significant problems in urban districts include storm drain abuse, urban run-off, garbage dumping and lack of streamside trees in protective riparian buffers.

Urban run-off is a mix of oil, garden chemicals, soil, air-borne particles, animal waste and rainwater that flows into streams, ditches and storm drains. Described as "non point source pollution" it is now more chronic and harder to control than direct industrial pollution. It is worst where the ground is covered by asphalt, concrete or other impervious materials, a situation common to the modern landscape of vast parking lots, acres of highways and driveways, extensive residential areas and industrial complexes.

The critical point for marked degradation in water quality comes with an impervious covering of 7 to 10% of watershed land. In 1997, the Serpentine River watershed reached 12.8% coverage, and since then urban development has escalated dramatically throughout Surrey.

It is amazing what gets into rivers. Besides the regular toxic mix of oil, grease and antifreeze, the Serpentine River has had soap spills, wine spills, garbage dumps, salt and PCBs. Fish were killed in a chlorophenol spill in a tributary of Hyland Creek, and landfill leachate contributed to fungal growths in the Nicomekl.

These problems were described in a study that looked at how well Boundary Bay and its tributaries adhered to provincial water quality standards. Unfortunately, the report found that none of the rivers were trouble-free. Most suffered from poor bacteriological quality, high temperatures and high fecal coliform counts, particularly in the lower reaches and during the summer. Agricultural sources raised fecal coliform, ammonia and phosphorus levels, making the bay unsafe for shellfish harvests although not bad enough to prevent swimming.

Drayton Harbor also suffered poor water quality because of agricultural, urban and sewage contamination. Recent shared watershed efforts have led to considerable improvements. A similar cooperative project should be initiated for all of Boundary Bay's watershed. It would soon lead to improved water quality for the benefit of both people and wildlife. Just imagine, pristine, healthy streams of sparkling water flowing through our neighbourhoods and into the bay!

Good air quality is essential to human health.

THE BREATH OF LIFE

An indisputable necessity for life, air is a mixture of gases, including nitrogen, water vapour, carbon dioxide, oxygen and methane. The flow of air across Boundary Bay is generally from the southwest. Rain clouds are carried across sunny Tsawwassen, Blaine and White Rock and deposit precipitation on the North Shore mountains. Air quality is generally quite good around the bay, compared with denser urban areas or the Fraser Valley, yet smog continues to be a problem and a cause of respiratory diseases.

Smog results from the presence of air contaminants such as ground level ozone and particulate matter. Thanks to the regional "Air Care" program, smog-causing emissions from light trucks and cars have declined over the last two decades. However, smog is predicted to once again be on the rise, with marine sources becoming the largest contributor, followed by emissions from agriculture.

The farming culprits are livestock ammonia emissions, especially from poultry. Poor air quality from smog is a different problem from that of "greenhouse gas" emissions, although many of the sources are the same. Greenhouse gases, such as carbon dioxide and methane, contribute significantly to climate change. An overall increase in greenhouse gas emissions is forecast, unless people become much more proactive in prevention. Space heaters (buildings), cars and light trucks were the top sources of greenhouse gas emissions in Metro Vancouver from 1985 to 2000, with the percentage from heavy goods vehicles predicted to rise.

This is bad news. Local consequences of escalating climate change are already being observed. Sea level is slowly rising as glaciers melt: it is now 4 cm (1.57 in) higher than a hundred years ago, and the rate is increasing. A storm surge could deeply flood the Fraser delta lowlands. Sea surface temperatures in the Strait of Georgia are rising at three times the global average, and very high Fraser River water temperatures have led to salmon mortalities. The growing season is longer, spring comes earlier and the first frost is later. Winters are less cold, yet extreme storms may be more frequent, with intense winds and heavy precipitation. Rapid climate change causes disruption in marine and terrestrial food chains, with resulting complex effects on wildlife populations and ecosystems (page 177). The possibility of more fires would be particularly dangerous in Burns Bog (page 97).

Some of the simplest solutions to cutting emissions, such as leaving the car at home, are not always the easiest, given the infrequency of public transport to much of the bay area. Combining fitness goals with a greener lifestyle is appropriate for more and more people. Walking and cycling promote health; constant car travel is linked to high body-mass indices, a measure of obesity.

Leaf blowers are smoggy polluters

Lawn and garden equipment, such as power mowers and leaf blowers, can be a significant source of smog-forming pollutants in the airshed. Improved technology can help energy conservation while saving money. Turning down the heat a notch, using double-glazing, energy-smart appliances, and a washing line, even just for some of the time, will cut fuel bills and help reduce emissions.

Further information on climate change

An Inconvenient Truth directed by Davis Guggenheim

The Weather Makers by Tim Flannery

Metro Vancouver www.gvrd.bc.ca

Night falls over Boundary Bay, the shores lit up by the surrounding city lights.

GUARDING THE DARK

Excessive night lighting emanating from settled areas has increased dramatically in the last few decades and has become an issue in the Fraser River estuary.

Much night lighting is proven to be inefficient, unnecessary and expensive; in effect, it is a form of pollution. Typically across North America, one third of all artificial lighting shines upwards or sideways, wasting $1 billion a year in energy costs and causing environmental and sociological effects. Sky glow is at its worst during periods of rain and fog, conditions that prevail markedly in the Fraser delta during winter.

Artificial light disrupts the natural cycle of daylight and darkness that activates the hormonal regulation of many human and wildlife biological functions. Darkness stimulates the production of melatonin, a key factor in circadian rhythms like sleep cycles and body temperature, blood and urine chemistry, immunity to disease and seasonal behaviour patterns. True darkness is needed for a range of biological activities, including deer giving birth and owls hunting successfully. Locally, night lighting of greenhouses has been observed to change the way some falcons hunt.

Bright lights at night disorient flying birds and moths, and are particularly dangerous for the hundreds of bird species that migrate at night. They are drawn to the light, becoming confused and blinded, and collide with structures or fall to the ground exhausted. Across North America, tens of thousands of song birds die every year, crashing into floodlit smokestacks, transmission towers or lighted buildings. The death toll from night lighting is calculated to be over 100 million birds a year. A reduction in some moth populations has also been linked to excessive light kills.

Salmon migration has been observed to peak during the darker nights of the monthly lunar cycle, so these fish could be particularly vulnerable to artificially bright nights. It is difficult these days for fish at the mouth of the Fraser River to find an area uncontaminated by bright lights.

Light pollution also destroys our ancestral right to gaze at the beauty of stars and planets in the night sky. Furthermore, studies are finding links between artificial daylight and hormone-related health problems such as breast cancer, pineal gland disfunction and depression. For all these reasons, thought is now being given to restoring the darkness in some jurisdictions. Abbotsford has created a Dark Sky Preserve in the Fraser Valley, and Delta is considering a bylaw to prevent light pollution. Energy conservation suggests that turning off unnecessary night lighting makes sense from all sorts of perspectives.

THE NEED FOR NATURE

People are part of the web of life although we often act as if we were independent. We truly need nature to survive. Healthy soil and vegetation provide us with food, fresh air, water and other vital services. Natural biodiversity stores genetic opportunities and survival for the future. Yet it goes even further than that.

The intuitive sense that the sights and sounds of nature are essential for health is now being proven. A childhood spent playing in nature strengthens physical coordination and concentration, while nurturing imagination and creativity. Children with attention deficit disorders improve cognitive ability and powers of observation and lower depression levels, when they play regularly in natural surroundings.

Not only children benefit. Seniors in care homes have better mental states, and people of all ages heal better, when they are able to see plants and animals. Outdoor exercise is healthy and stress-relieving for everyone, especially when taken in a natural setting. We are part of nature, and it enriches our lives.

We have every reason on earth to conserve the wonderful natural area that is Boundary Bay. We must now just take the necessary steps to achieve that goal.

Relaxing in natural surroundings and watching wildlife are among life's greatest joys. This playful river otter family was found off Point Roberts.

Britannia Heritage Shipyard, Steveston - one of many heritage destinations in the area.

HERITAGE DESTINATIONS

This part of the book presents ideas for destinations where the long history of Boundary Bay can be appreciated and enjoyed. Although the region has many heritage sites, they are not collectively well known. They come under different jurisdictions and are managed under a variety of programs. Nonetheless, the area possesses several British Columbia and Washington National Historic Sites, as well as exceptionally fine museums. I have included some destinations which are a bit further afield, as they are relevant to the topics in this book.

Many archaeological sites and middens have been covered over by development or are on private land. For most of the remaining ones, security and cultural concerns require that their exact location is not revealed (see page 40). Artifacts from sites are scattered in museums in several municipalities; a cultural centre to describe and celebrate the local Coast Salish history would be a wonderful addition to the region. Totem poles have been erected at various locations, and although house posts and mortuary poles, not totem poles, are the traditional Coast Salish art form, the poles are of interest as an example of the carvers' skill and the stories they reference. European and Asian exploration, pioneer history, and the story of farming and fishing are recorded in many museums and old photograph collections. Some historic buildings have been protected and restored, although not always in their original setting. Note that some museums are only open in summer.

Archaeology and First Nations - museums, pages 188-191, totem poles, pages 191-192, Xa:ytem Longhouse, page 187. See also pages 51-59.

European heritage - Fort Langley page 186, 194, museums, pages 188-191, heritage buildings, page 192-196, exploration and the sea, page 190.

Fishing history - Britannia Heritage shipyard and Gulf of Georgia cannery, page 186, Plover ferry, page 187, Semiahmoo Spit, page 195, Steveston Museum, page 188.

Agricultural history - Delta Museum & Archives, page 188, B.C. Farm Machinery & Agricultural Museum, page 189, Lynden Pioneer Museum page 191, Stewart Farmhouse, page 193, London Heritage Farm page 195.

For information on public transport around the Boundary Bay area see Translink www.translink.bc.ca or Whatcom Transit Authority www.ridewta.com. Site numbers and map grid references refer to map on pages vi-vii.

NATIONAL HISTORIC SITES, B.C.

Several important sites related to the history and culture of the region are found just a short distance from Boundary Bay.

1. Britannia Heritage Shipyard

The site has ten buildings associated with fishing and the river, dating to 1885, located on 3.2 ha (8 ac) along the bank of the Fraser River, South Arm, at Steveston.

A waterfront cycling and footpath trail links the site with Steveston to the west and **London Farm** and **Woodward's Landing** to the east, with cross trails along Shell Road through Lulu Island, Richmond.

5180 Westwater Drive, Richmond, B.C.
(at the foot of Railway Ave)
Telephone: 604 718 8050
Website: www.britannia-hss.ca
Hours: summer: Tues. to Sun, closed Mon.
Winter: Sat. & Sun only. Admission free.
Bus numbers 402, C93, 490, 492
Map grid A3

Britannia Heritage Shipyard, Steveston.

2. Fort Langley

This reconstruction of the Hudson's Bay Company's trading fort is located near the confluence of the Salmon and Fraser Rivers. The fort includes historic buildings, artifacts and art work. Traditional skills, such as barrel-making and blacksmithing, are featured here. The **Fort to Fort Trail** runs along the Fraser River. It links Derby Reach Regional Park and the site of the original Fort Langley, marked by a commemorative cairn, with the National Historic Site.

23433 Mavis Avenue, Fort Langley, B.C.
Telephone: 604 513 4777
Website: www.pc.gc.ca
Hours: March to October.
Admission charges apply.
Bus number C62 from Langley Centre.
Map grid H2

3. Gulf of Georgia Cannery

Built in 1894, this is the only historic cannery left standing in the Fraser River delta. It features information on fish and fishing methods, a herring reduction plant and canning line exhibit. Site tours and films are available in both official languages. **Kuno Japanese Gardens** and the **Fishermen's Memorial** in Garry Park lie to the west of Steveston Village.

12138 4 Avenue, Richmond, B.C.
Telephone: 604 664 9009
Website: www.gulfofgeorgiacannery.com
Hours: summer only: April - Oct.
Admission charges apply.
Bus numbers 401,402, 407, 410, C93 etc.
Map grid A3

4. Xa:ytem Longhouse Interpretive Centre

Somewhat north of the Boundary Bay area, *Xa:ytem* (pronounced Hay-tum) is the location of Hatzic Rock Transformer Stone, a place of great spiritual importance to the Stò:lo. Artifacts dating back 9,000 years were discovered here. A replica of a pithouse and artifacts are on view at the museum. The centre is both a National Historic Site and a B.C. Heritage Site.

35087 Lougheed Highway, Mission, B.C.
Telephone:. 604 820 9725
Websites: www.xaytem.museum.bc.ca and
www.xaytem.ca
Hours: generally Mon.- Fri. afternoons in
winter, daily in July and August.
Tour bookings year round.

NATIONAL HISTORIC SITES, WASHINGTON

5. Boundary Marker Number 1.

Overlooking the Strait of Georgia, on west-facing bluffs in Point Roberts, this obelisk marks the beginning of the world's longest international border. The border was inaccurately measured in 1858, so the marker, the first on the 49th Parallel, is positioned too far north (see page 83).

West end of Roosevelt Road, Point
Roberts. Note: the Border crossing is at
56 Street/Tyee Drive, and not on 48th St.
Delta.
Nearest bus numbers 601, C84
Map grid B5

Plover ferry, Blaine.

6. Historic Plover Ferry

This small wooden boat is the oldest foot passenger ferry in Washington. It plies the narrow marine entrance to Drayton Harbor, between Blaine and Semiahmoo Spit. Built in 1944, it was used to ferry workers to the APA cannery on the Spit. It was restored in 1996, and now caters to foot passengers and cyclists on weekends from Memorial Day to Labor Day.

Visitor's Dock, Blaine Marina,
215 Marine Drive, Blaine, WA
Telephone: 360 332 4544
Hours: summer, Friday through Sunday
Map grid F5

7. Hovander Homestead Park

A short distance south of Boundary Bay, this 1901 farmstead is located at Ferndale. The park has hiking and cycling trails, fishing, a fragrance garden and picnic areas. Nearby Tennant Lake is a natural wetland.

5299 Nielson Road, Ferndale, WA
16 km (10mi) south of Blaine, exit 262.
Website: www.co.whatcom.wa.us/parks/
hovander/hovander.jsp
Park hours: 8 am to dusk; house open to
visitors in summer.

8. Peace Arch State Park - Peace Arch Provincial Park

This is an historic 40 acre (16 ha) park on the International Border. The 20.4 m (67 ft) **Peace Arch** stands amid flower gardens and spans the border. It was built in 1920 to commemorate the centennial of the Treaty of Ghent between the United States and Great Britain, that led to the subsequent peaceful coexistence of Americans and Canadians. No passport is needed to be in the park, which is neutral territory.

The park is at the Douglas border crossing, just north of Blaine (I-5 junction 276) and on Highway 99, South Surrey. Website: www.peacearchpark.org Nearest bus numbers for 8 Avenue at Stayte Rd, White Rock, C51, 354 Map grid F5

B.C. MUSEUMS

9. Delta Museum & Archives

Housed in the old **Delta Municipal Hall**, the museum collection has artifacts from the Whalen Farm archaeological site, as well as pioneer farm implements, cannery items, household memorabilia, and old photographs.

4858 Delta Street, Delta, (Ladner Village) Telephone: 604 946 9322 Website: www.corp.delta.bc.ca Hours: closed Monday. Admission by donation. Bus numbers 601, C86. Map grid B3

10. Richmond Museum

Located in the Richmond Cultural Centre, this museum holds special exhibitions on a range of heritage and environmental topics.

7700 Minoru Gate, Richmond, V6Y 1R9 Telephone: 604-247-8300 Website: www.richmond.ca Hours: vary, open daily. Admission free. Bus numbers: all Richmond Centre services. Map grid B2.

11. Steveston Museum

A small collection of late 19th and early 20th century artifacts from this interesting fishing community.

3811 Moncton St. Richmond, B.C. Tel. 604 272 6868 Steveston Village Community website: www.steveston.bc.ca. Bus numbers 401,402, 407, 410, C93 etc. Map grid A3

12. Surrey Museum & Archives

This museum recently relocated to a new building close to the old village of Cloverdale. It features historical Coast Salish artifacts and pioneer memorabilia from around Surrey, as well as special exhibitions. The Archives, situated in the renovated 1912 **Surrey Municipal Hall**, include many photographs relating to logging and farming activities

17710 - 56A Avenue, Surrey Telephone: 604 592 6956 Web: www.cloverdale.bc.ca/museum.htm Hours: vary; check website or call ahead. Admission charges apply. Bus numbers 320,341, C70. Map grid F3

The 1873 Eric & Sarah Anderson cabin, at the Surrey Museum & Archives, is the only one of this era in Surrey.

13. Langley Centennial Museum and National Exhibition Centre

The collection includes memorabilia, archival photographs, and works of art, presenting a window on the past life of the Langley community.

9135 King Street, Fort Langley
Telephone: 604 888 3922
Website: www.langleymuseum.org
Hours: daily, opening times vary.
Admission by donation.
Bus number C62 from Langley Centre.
Map grid H2

14. British Farm Machinery & Agricultural Museum

This is a museum for the farm enthusiast! It is located in the old village of Fort Langley on the Fraser River, adjacent to the Centennial Museum.

9131 King St. Fort Langley
Telephone: 604 888 2273
Website: www.bcfma.com
Hours: open daily in summer
Bus number C62 from Langley.
Map grid H2

15. White Rock Museum & Archives

Situated in the old railway station on White Rock waterfront, adjacent to the pier, this small museum has regular exhibits on topics of local interest, as well as a full collection of archival material from the White Rock area.

14970 Marine Drive, White Rock, B.C.
Telephone: 604 541 2222
Website: www.whiterock.museum.bc.ca
Hours: Museum daily 10 am - 5pm
Archives 10 am - 4.30 pm, Mon.-Thurs.
Admission by donation.
Bus numbers C51, C52.
Map grid E4

16. Museum of Anthropology, University of British Columbia

The museum has a major collection of artifacts and artwork, including local Coast Salish work. The totem pole gallery is not to be missed.

6393 NW Marine Drive, Vancouver
Telephone: 604 822 3825
Website: www.moa.ubc.ca
Hours: open daily in summer; closed Mondays in winter. Admission charges apply; entry by donation, Tues. 5 - 9 pm
Bus number C20 from UBC Loop.

17. Simon Fraser University Museum of Archaeology

The museum contains artifacts from around B.C.

8888 University Drive, Burnaby, B.C. (NE corner of the Academic Quadrangle, Concourse Level.).
Telephone: 604 291 3325
Web: www2.sfu.ca/archaeology/museum/
Hours: 10 am - 4 pm, Monday to Friday. Call ahead as times may vary.
Bus numbers 135,143,144,145, N35

18. Vancouver Museum

Permanent and travelling exhibits on the natural world, historic and cultural issues, are provided in this hundred year-old museum, located near the Burrard Street bridge.

1100 Chestnut St., Vancouver, B.C.
Telephone: 604 736 4431
Website: www.vanmuseum.bc.ca
Hours: closed Mondays in winter. Admission charge.
Nearest bus numbers 2,22, 44, N22, 32, 258 westbound.

19. Vancouver Maritime Museum

Located on the Kitsilano waterfront, this museum has displays on boats, explorers and maritime instruments. Visit the harbour in summer to see a fleet of heritage vessels.

1905 Ogden Avenue,Vancouver
Telephone: 604 257 8300
www.vancouvermaritimemuseum.com
Hours vary seasonally.
Admission charges apply.
Bus numbers: see Vancouver Museum.

20. The Maritime Museum of British Columbia, Victoria

The museum displays many aspects of B.C.'s maritime history including British and Spanish exploration.

28 Bastion Street, Victoria
Telephone: 250 385 4222
Website: http://mmbc.bc.ca/
Hours vary seasonally. Admission charges apply. Reached from Tsawwassen by B.C. Ferry to Swartz Bay and express bus (pay and board on ferry) www.bcferries.com

21. Royal B.C. Museum & Archives

The provincial museum showcases B.C.'s human and natural history, and exhibits from other countries and cultures.

675 Belleville Street, Victoria, BC
Telephone: 250-356-7226
Website: www.royalbcmuseum.bc.ca
Hours daily except Christmas. Admission charges apply. Reached from Tsawwassen by B.C. Ferry to Swartz Bay and express bus (pay and board on ferry).

WASHINGTON MUSEUMS

22. Burke Museum of Natural History and Culture

An important collection of Northwest Coast heritage items and ecological displays are held here.

University of Washington campus, 17 Ave. NE and NE 45 Street, Seattle.
Telephone: 206 543 5590
Web: www.washington.edu/burkemuseum
Hours: 10 am - 5pm daily (not stat. hols.)
www.greyhound.com; www.amtrak.com.

23. Lynden Pioneer Museum

The small farming town of Lynden, WA, celebrates its Dutch ancestry and pioneer roots at this memorabilia-packed museum. Worth a visit for the old carriages, farm implements and reconstructed street scene.

217 Front Street, Lynden, WA
Telephone: 360 354 3675
Web: www.lyndenpioneermuseum.com
Hours: Mon. to Sat. and Sun. afternoon.
Admission charge; free on Sundays

24. Whatcom Museum of History and Art

The ornate 1892 Town Hall museum has permanent collections of clocks, woodworking tools and toys, as well as special exhibits and art displays.

121 Prospect Street, Bellingham, WA.
Telephone: 360 676 6981
Website: www.whatcommuseum.org
Hours: Tues. to Sun. afternoons.
Admission free.
Bus to Bellingham www.greyhound.com.

POLES & PETROGLYPHS

Coast Salish tradition includes carved house posts and mortuary poles; however, a number of totem poles have been carved on commission or to commemorate anniversaries. Placing of totem poles can be politically and culturally sensitive to Coast Salish people.

As well as local poles, there are many fine ones in Vancouver, B.C. (e.g. Stanley Park) and Bellingham, WA.

Totem Pole at Delta Museum, Ladner.

25. Delta Museum poles

The taller of the two poles standing in front of the museum was carved in 1932 by Nanaimo carver, Wilkes James for his Tsawwassen wife. It references *Tsaatzen*, "First Man of Tsawwassen," his helpers black bear, eagle and beaver. The smaller pole was carved by Marvin Joe in 1974. Both poles are western redcedar.

4858 Delta Street, Ladner.
Bus number 601, C86.
Map grid B3

26. Pole, 12 Avenue

This colourful pole carved by Simon Charlie from Cowichan First Nation, was commissioned by Century Group in 1966, and stands in front of their Tsawwassen office.

5499, 12 Avenue, Tsawwassen. Delta Bus numbers 601, C84. Map grid B5

27. Blaine Marine Park Totem Pole

This pole was designed and carved by Phill Claymore and donated to the park by Phill and Virginia Claymore.

Blaine Marine Park, Blaine, Washington. Map grid F5

28. Totem Plaza, White Rock

Two totem poles carved by Haida master carver Robert Davidson, from thousand year old western redcedar trees, stand on Marine Drive in White Rock. The poles were commissioned as a commemoration of the 125th anniversary of the Royal Canadian Mounted Police. One was designed by Davidson and the other by Susan Point, a Coast Salish artist.

Marine Drive at Cypress Street, White Rock. Bus numbers C51, C52, 354. Map grid E4

29. Heron Park Petroglyph

This carved stone was moved from its original site and placed in a small park at Crescent Beach.

Location: Beecher Street and Gordon Avenue, Crescent Beach. Bus number 351   Map grid D4

HERITAGE BUILDINGS

The whole Boundary Bay area is scattered with houses, farms and other buildings dating from the late 1800s and early 1900s. Each municipality has information on the most historic ones. Here is a sample of what to see.

30. Arthur Drive, Ladner Village

Arthur Drive (Slough Road) in Delta is a picturesque route south from Ladner Village alongside Chilukthan Slough. It has period homes and mature trees, including copper beech and bigleaf maples. Starting at **All Saints Church**, just north of **McKee House** in Ladner Village, the road winds south for about 3 km (1.5 mi). **Hawthorne Grove** (Kirkland House), built 1911 for William Kirkland, lies in 1.6 ha (4 ac) of garden. The old Jubilee farm and McNeely house, built 1893, are now the **Augustinian**

The Heron Park petroglyph.

Monastery. Arthur Drive has a cycle lane and a return loop to Ladner can be made through quiet, rural roads.

Arthur Drive, Ladner, Delta.
Bus number 601. Map grid B3.

31. Deas Island

Deas Island Regional Park, a nature park by the Fraser River in Delta, has several pioneer buildings that were relocated here. They include **Burrvilla**, home of the Burrs, built in 1905, the 1899 **Old Agricultural Hall**, and 1909 **Inverholme Schoolhouse**.

River Road, Delta. Bus number 640.
Map grid C3.

32. Boundary Bay

Two Boundary Bay heritage houses have recently been restored. The 1884 **Alexander-Gunn farmhouse** was the home of four generations of Gunns, who had purchased it in 1897 from Robert and Margaret Alexander. It is now the Earthwise Garden and Farm. Nearby, the **Cammidge House** was restored and moved to Boundary Bay Regional Park by the local Lions Club.

Boundary Bay Road, Delta. Bus number 601, C89 (summer only) Map grid C4

Further information: Delta

Gwen Szychter's books describe the history of Ladner and Tsawwassen - see page 220.

Delta's Rural Heritage by Donald Luxton has illustrations of all the heritage buildings.

Burrvilla, built 1905, now in Deas Island Regional Park, Delta.

33. Old McLellan Road

Christ Church, founded 1882, is one of the oldest buildings in Surrey and the city's first church. Situated on a hillside above the Serpentine River valley this Carpenter Gothic style church has a peaceful graveyard with mature trees. One of these is the **Royal Oak**, planted to commemorate the coronation of King George VI and Queen Elizabeth in 1937.

16603 Old McLellan Road, Surrey
Bus numbers 341, 320. Map grid F3

34. Elgin

Elgin Heritage Park beside the Nicomekl River, is an attractive destination. The **Stewart Farmhouse**, barn and bunkhouse, date to 1894. Other historic buildings clustered in the Crescent Road area include the 1878 **Elgin Community Hall** and the 1921 **Elgin Schoolhouse.** The trailhead for the **Semiahmoo Heritage Trail** is at 144 St. and Crescent Rd. A trail also runs beside the Nicomekl River.

13723 Crescent Road, South Surrey.
Bus number 352 (rush hr). Map grid E3.

Further information: Surrey

Surrey heritage information available from Surrey Parks, Recreation & Culture: *www.city.surrey.bc.ca* 604 502 6456.

35. Green Timbers Urban Forest

This historic forest was saved from complete annihilation by the efforts of generations of conservationists (see page 114). A peaceful place for a walk in the woods.

96 Ave at 140 Street, Surrey. BC Bus number 502. Map grid E1, E2

36. Langley: portage to the Fraser

The **Nicomekl River** is located just south of Langley's shopping area and its tributaries wind among the surrounding suburbs and fields. The **Michaud historic house,** owned in 1888 by Joseph and Georgiana Michaud, is now used by the Langley Arts Council. It sits beside the Nicomekl in **Portage Park,** where a plaque commemorates James McMillan's 1824 journey to establish the Hudson's Bay Company fort at Langley.

204 Street, Langley, BC Bus numbers C63, C64 Map grid G3

37. Old Yale Road

In the 1870s, the **Old Yale Road** led from Brownsville, on the Fraser River across from New Westminster, to Fort Yale, the start of the Cariboo Wagon Road up to the gold fields.

At Murray's Corner in Langley, the **Traveller's Hotel** was built in 1887 and has welcomed travellers ever since. One notorious customer was Billy Miner, who stayed here the night before he stole over $8,000 in gold from the Canadian Pacific Railway, in Canada's first great train robbery!

Old Yale Road is mostly replaced by the Fraser Hwy 1A, though some old sections remain. Murray's Corner: 48 Ave & 216 St., Langley. Buses C60, C61. Map grid G3

38. Wark-Dumas House

This house was built in 1890 and restored by the Langley Heritage Society in 1987; it is now part of Kwantlen College Langley Campus.

20901 Langley Bypass, Langley, BC Bus numbers C64, C62, 502. Map grid G3

39. Campbell Valley Park

Campbell Valley Regional Park is a beautiful nature destination and the site of two heritage buildings. The **Annand/Rowlatt Farmstead** dates from 1898. The **Lochiel Schoolhouse** was built in 1924 and subsequently moved to the park.

Between 8 - 16 Ave, 200 - 208 St, Langley. Nearest bus number (Fernridge) C63. Map grid G4

Further Information: Langley

Township of Langley *www.tol.bc.ca*

Roads and other Place Names in Langley, B.C. by Maureen Pepin

Semiahmoo Spit cannery, Drayton Harbor.

40. London Heritage Farm, Steveston

Just upstream of Shady Island, London Farm is an 1890s pioneer homestead situated in a pleasant flower garden.

6511 Dyke Rd, near south end of Gilbert Road , Richmond. No direct bus service - use Steveston routes. Map grid B3

41. Blaine, Washington

Blaine was first a gold rush town, then a bustling, noisy place of mills and canneries, the hub of a fishery that reached as far as Bristol Bay, Alaska. Interpretive signs in the **Marine Park** show where the **E-Street wharf** stood in 1888, the site of Jasper Lindsay's sawmill, the Cain Brothers shingle mill, and the H.L. Jenkins Lumber Company. Only pilings in the mud remain and the noisiest inhabitants are the ducks, shorebirds and gulls.

West off Highway 5, just south of international border. Map grid F4

42. Semiahmoo Spit & Cannery

Across the water from Blaine, Semiahmoo Spit has an ancient history, dating back thousands of years (see page 54). The first cannery was built in 1891. Bunkhouses from this cannery were moved to **Semiahmoo Spit Park**, where they were restored and now feature in a small interpretive centre. Other cannery buildings were incorporated into the Inn at Semiahmoo.

Southwest side of Drayton Harbor, WA. Map grid F5

Gulf Road, Point Roberts, looking west ca.1910. (Point Roberts Historical Society)

43. Drayton Harbor watershed

Rural communities in the Drayton Harbor watershed feature old-style wooden barns and modest homes. This is farming country and there are fruit stands in season. **Custer,** one of the first homesteading areas, has several period buildings including the **G.** and **F. Brunson houses** and the 1904 **United Methodist Church**.

Whatcom County, WA.
Map grid F5, F6, G5, G6, H5,H6

44. Point Roberts - Lily Point

This little US peninsula has a quiet rural atmosphere. The coastline is excellent for wildlife viewing and **Lily Point** has an ancient history (see pages 52, 53, 74). A few pilings on the shore are all that remain of the giant Alaska Packers Association (APA) cannery. The peaceful **Pioneer Cemetery** hints at the Point's multicultural past.

East end of APA road, Point Roberts, WA
Map grid C5

45. Point Roberts - Gulf Road

Gulf Road was the centre of the old Point Roberts town, and the location of the George and Barker cannery. Only a few older buildings are still standing. They include **Brewsters Restaurant**, ca.1920, and the 1930 **Community Hall**, which was once the village school.

West side of Tyee Road, Point Roberts, WA
Map grid B5

TRACES OF THE PAST

In this brief guide we have travelled from the far distant past of the Pleistocene Ice Ages to the challenges facing us in the modern day, winding our way through a varied and ever-changing landscape. Every history is just one interpretation of events; in covering such a long era, this guide has only traced a brief personal perception of the past. It is hoped that this book has both stimulated an interest in the Boundary Bay area and encouraged readers to look with fresh eyes on the the surrounding landscape and the outstanding natural assets found here.

I hope too that the book will foster both respect and celebration for Boundary Bay's past: the long history of the Coast Salish, the Indigenous people of the land, and the interlocking thread of stories made by settlers from many other countries around the world, who came in search of something new, and different, and wonderful. By knowing about our past, we can learn from mistakes, set priorities, and find a path towards a sustainable and healthy future.

People work hardest to protect what they know and value. Nature and people coexisted in this rich environment for thousands of years; with care and consideration they could continue to do so long into the future. The challenge is to conserve the intricate web of life, in all its diversity, by keeping places for wildlife in a landscape that has been almost entirely appropriated and transformed to meet human demands.

If you would like to know more about Boundary Bay's story, visit the museums and heritage sites listed on pages 185 to 196 and consult the many written sources listed in the bibliography. Further information on the wildlife and nature of Boundary Bay can also be found in our companion book, *A Nature Guide to Boundary Bay*, available from book stores, bird shops and on line at:

www.natureguidesbc.com.

This website also has checklists of wildlife species and other information on the bay, a list of retailers for the books, and links to organizations and clubs interested in nature and local history.

To see more of David Blevins' photographs or to order prints visit:

www.blevinsphoto.com

NOTES & SOURCES

Direct quotes from external sources are given in *italics*. See page 222 for photo credits.

The Dawn of Time

Page 3. "*We have been here since time immemorial.*" Chief Kim Baird in Tsawwassen First Nation 2004. "2.6 million years ago" Dr. John Clague, pers. comm. Jan. 2008. "mammoth lived in the lower Fraser Valley...about 25,000 years ago." There were also bears on the coast 14,000 years ago and mastodon south of the Juan de Fuca Strait 13,000 years ago. Information from Roy Carlson, pers. comm. Jan 2007. Mammoths survived up to about 4,000 years ago on Wrangel Island, Russia, the nesting area of snow geese that winter in the Fraser estuary. "The ice extended as far south as Olympia" Clague and Turner 2003. "at least 9,000 years ago" Matson 1996.

Page 4. "was about 2 km thick" Blunt et al. 1987; Clague et al. 1998. "During the last Ice Age, sea levels fell to 125 m" Dr. J. Clague pers. comm. Jan. 2008. "The climate changed about 16,000 years ago" Clague et al. 1998. "ice free by 13,500 years ago" Ibid; Dr. J. Clague pers. comm. Jan 2008. "encroached all the way to Pitt Lake" Clague et al. 1998.

Page 5. "by 12,000 years ago the ice had finally gone." Arcas 1991. "Sea level fell rapidly reaching its present level shortly after 11,000 years ago..." Dr June Ryder, pers. comm. Jan. 2008.

Page 6. "Redwood Park" Dr. J. Clague pers. comm. Jan. 2008. "a growing series of sandy spits" June Ryder in Arcas 1991. "lodgepole pines" Mathewes & Heusser 1981. "red alder" Hebda 1977. "Herds of caribou may have attracted hunters" Roy Carlson personal communication, Jan. 2007. "Pollen grain studies" Mathewes & Heusser 1981. "Earthquakes" Mathewes & Clague 1994.

Page 7. "The weather became warmer and drier around 10,000 years ago" Mathewes & Heusser 1981; Pielou 1991. "Deer, elk, bear and wolves" Matson 1996. "Spikemoss, bracken and grasses" Mathewes & Heusser 1981. "The same distinct style of technology" Carlson & Dalla Bona 1996; Simon Fraser University Archaeology Museum. Halkomelem name courtesy of Jill Campbell and Musqueam Language Committee - this version is anglicized due to font restrictions. The u represents the semi-vowel known as schwa, generally pronounced like the u in but.

Page 8 to 10. Prehistory. Many sources were consulted including Suttles 1990; Arcas 1991; Wright 1995; Matson & Coupland 1995; Carlson & Dalla Bona 1996; Univ.of Tennessee 2005.

Page 8. "Stone age tools" info from Carlson & Dalla Bona 1996; SFU Archaeological Museum. "Genetic studies suggest" Crawford 1998; Carlson & Dalla Bona 1996. "Dyukhtai people hunted woolly rhinoceros" Wright 1995; Ames and Maschner 1999. (Spellings of Dyukhtai vary, e.g. Dyuktai, Diuktai, etc.)

Page 9. "Bering Strait land bridge" see Manley, 2002 for a geospacial animation of sea level changes in the Bering Strait area (Beringia). "Bison, mammoths and musk ox roamed" Yukon Beringia Interpretive Centre website www.beringia.com has details of wildlife found on the land bridge. "sites found in the Americas" Carlson & Dalla Bona 1996. "Clovis hunters" Flannery 2001. "One theory" Fladmark 1986; Crawford 1998; Smithsonian Encyclopedia online 2007.

Page 10. "The lowest layers are 9,500 years old" (8570±90 BP): R.G. Matson in Carlson & Dalla Bona 1996. Lower layers were originally radiocarbon dated at 8150±250 BP (Matson 1976). "R.G. Matson" Matson 1976. "on the Puget Sound coast" Stilson 2003.

Page 13. "*the place where the first man of this race was created*" Chief Harry Joe, quoted in Arcas 1991. "*stl'elup*" orthography, Jill Campbell, pers.comm."a number of older people from the 1930s to 1960s" Jenness 1955; Suttles 1987; Arcas 1991; Bierwert 1999. "Suttles conducted detailed studies" Suttles 1951; Suttles 1990. "Arcas Consulting Archaeologists" Arcas 1991. "The Stò:lo people published a history and atlas" Carlson 2001. "In the local traditions" see Jenness 1955; Arcas 1991; Carlson 2001. "*I want from now and everlasting*" Chief Harry Joe quoted in Arcas 1991. "Another First Ancestor" Jenness 1955. "The First Ancestor age" - clarification of the three stages of Coast Salish history, thanks to Don Welsh, archaeologist, Semiahmoo First Nation, pers. comm. 2008. "Transformers" see Jenness 1955; Suttles 1990; Carlson 2001; and First Nations' websites.

Page 14. "*Tsau'wuch*" on map for Birch Bay: based on Boundary Bay Survey map 1858, courtesy Don Welsh. "In the version of the story told by Old Pierre" and following quotation "*three brothers accompanied by 12 servants....*" from Old Pierre, Katzie, in Jenness 1955.

Page 15. "*you shall dwell among the clam beds for ever*" Ibid. "*centre from which the underground passages radiate*" Jenness 1955; Arcas 1991. "*power over all the underground channels that lead from Chelhtenem....*" Whatcom County 1992. "*people drowned at distant places...*" Ibid.; see also Jenness 1955 and Appleby 1961 for legends about the underground passages. "Ancestral name used as a place name: *Smakw'ets*" Arcas 1991. "*P'qals... a legendary site*" Don Welsh, pers. comm. *P'qals* is the preferred Northern Straits Salish transcription.

Page 16. "According to Old Pierre, it was *Smekwats*' son who was responsible" story from Jenness 1955. "*her tears turned into raindrops*" Old Pierre quoted in Jenness 1955. "*His vitality went into the deep water off Point Roberts....*" Ibid. "the popularised one ....of the Cowichan 'sea god' " see for example Hastings 1981, in conversation with Mrs Bernard Charles, May 1970. "*Sqwemay: yes* or "Kwomais"" Carlson 2001. *Sqwemay:yes* is Halkomelem transcription, *Kwomais* preferred by Semiahmoo, Don Welsh pers. comm. "*Sqwema:yes* is 'dog-face' " Simon Fraser University, Department of Anthropology: www. sfu.ca/halk-ethnobiology. "The climate 9,000 years ago was warm and dry" Pielou 1991; Carlson 2001. "The presence of bay mussels" Matson 1996.

Page 17. "wolverine at Beach Grove" Thom 1997. "Californian condor" Lord 1866. "reputed to carry off whales in its talons" Webb 1988.

Page 18. "A cataclysmic geological event occurred 7,700 years ago" Clague and Luternauer 1982. "By 4,000 years ago, these trees had achieved a position of dominance in the landscape" Carlson & Dalla Bona 1996. "the versatility of the cedar was fully recognised" Stewart 1984; Pojar and MacKinnon 1994; Mathewes 1991 and others.

Page 19. "The Tsawwassen, for example, cook the first salmon" Tsawwassen First Nation 2004. "The Lummi, traditionally address the salmon" Yates 1992.

Page 20. "For 4,000 years after the glaciers retreated" Luternauer et al. 1998. "Around 6,000 years ago" Thomson 1981; Luternauer et al. 1998. "The origin of the bog dates back to 6,000 years ago" Dr J. Clague pers.comm. Jan.2008.

Page 21. "about 3,500 years ago" Hebda 1991. "According to botanist Richard Hebda" Ibid. "as high as 5 m above the surrounding delta" Hebda 1977.

The Hidden Story of Middens

Page 23. "*In the neighbourhood of Boundary Bay....*" ~ Charles Hill-Tout, quoted in Maud 1978. "intertwined layers make interpretation difficult" Stein 1992. "ancient artifacts found around Boundary Bay"a selection of artifacts from the collection at Simon Fraser University Museum.

Page 24. Artifact Survival - much information from Stein 1992. Calibration of radiocarbon dates in this book are based on Prof. Richard G. Fairbanks' calibration curves from the University of Columbia www.radiocarbon.Ideo.columbia.edu

Page 25 - 27. Archaeology of Ancient Times: information on these pages taken widely from Matson & Coupland 1995; Carlson & Dalla Bona 1996; Thom 1997; Stein 1992 and other sources in bibliography, or listed by page as follows.

Page 25. "Fishing nets from this period have been found at Beach Grove" reported in Delta Optimist newspaper, June 25 1989, Delta, BC.

Page 26. "a traditional style that spanned millennia" According to Kathryn Bernick in Matson et al. 2003, chapter 10, Coast Salish baskets were originally woven not coiled, a tradition that spanned three millennia. Baskets found at the Beach Grove site resemble this style. "A rare find of skunk cabbage" Wright 1995. "It is assumed...that long, post and beam...." Stein 1992.

Page 27. "by 2,400 years ago" R.G. Matson, pers. comm. "Totem poles are not part of the Central Coast Salish tradition" Baird 1999. "Women were responsible for spinning wool" Suttles 1987. "The Coast Salish have serious concerns" Mitchell 1996, McLay et al. 2004. "The earliest graves..." R.G. Matson pers. comm. "A Tsawwassen adult and teenager.." Arcas 1991. "The health of the people…" Carlson and Dalla Bona 1996. Sources for Archaeological Sequence from Arcas 1991; Matson & Coupland 1995; R.G. Matson personal communication.

Page 28. "Professor Roy Carlson" Carlson and Dalla Bona 1996.

Page 29. "the shores of the estuary and of Puget Sound…" Charles Hill-Tout, quoted in Maud 1978. "Crescent Beach" Ham 1982; Matson & Coupland 1995; R.G. Matson pers. comm. Spelling of Halkomelem names for this and other sites mentioned in the book courtesy of Jill Campbell and the Musqueam Language Committee. Capitalization is not used. Due to the problems of font reproduction, spelling has been anglicized rather than written in the full Musqueam orthography. For information see http://fnlg.arts.ubc.ca and list on www.natureguidesbc.com/

Page 31. "Beach Grove midden" information on excavations, house depressions, from Matson et al. 1998; Matson & Coupland 1995; "Archaeologist Julie Stein" Stein 2000. Information on bird remains found at Beach Grove from Matson et al 1980. "*ttunuxun*" Jill Campbell pers.comm. "A stone sculpture of a human figure" Don Welsh pers. comm.

Page 32. *Stl'elup* information from Arcas 1991. Orthography: Jill Campbell, pers. comm. 2008

Page 33. "*c'a:yum*" orthography Jill Campbell, pers.comm. "Whalen Farm, Maple Beach" most information from Thom 1997; Delta Museum. Coast Salish dogs: "kept on islands" Lord 1866. "woolly gene was recessive" Crockford 2005.

Page 34. "Over 400 artifacts were collected … by Carl Borden in 1949 and 1950" Thom 1997. "Further salvage excavations in 1972" Ibid. "Archaeologist Dimity Hammon excavated in 1985" Ibid. and Delta Museum exhibit. "Semiahmoo Spit, *nuwnuwuluch* and *tsi'lich (s7luch, s'eeluch)*" Wayne Suttles' interview with Julius Charles and Lucy Celestine, Semiahmoo, reported in an unpublished document Semiahmoo Place Names, provided by Suttles to Don Welsh and quoted in Welsh 2004. "*s7luch*" Musqueam orthography, J. Campbell pers. comm. "A burial ground... subject of lawsuits" Henderson 2000. "The midden on Semiahmoo Spit is 4,200 years old." Welsh 2004. "Lummi policy" Lummi Nation website accessed 2003.

Page 35. *Chelhtenem* - Lummi spelling in Whatcom Co. 1992, *sc'ultunum* - Musqueam orthography (Jill Campbell pers.comm), *tselhtenem*: Arcas 1991. Island Halkomelem is similar (Brian Thom pers.comm.) "archaeological remains" Whatcom County 1992. Birch Bay middens: "European settlers were puzzled" Jeffcott 1949. "*Tsau'wuch*" a Northern Straits Salish name for Birch Bay. "*Strav-a-wa*, the place of clams" Chief Martin, Lummi, in Jeffcott 1949.

Page 36. "St Mungo Cannery" Calvert 1970, Boehm (neé Calvert) 1973. Names see note 29.

Page 37. "Red ochre." e.g. Beach Grove, Matson et al. 1980, Crescent Beach, Don Welsh pers. comm.; Glenrose and St. Mungo sites, Wright 1995. "Upriver Halkomelem story tells" Boas 2002. "Australian Aboriginal tradition" Finlay 2002. "Five people buried at Glenrose Cannery" Wright

1995. "Ochre was used" Wright 1995. "in the writings of Spanish and English explorers" see for example: Galiano 1792 in Espinosa y Tello 1930; George Vancouver quoted in Lamb 1984; Scouler quoted by Wayne Suttles in MacLachlan (Ed.) 1998. "One of the most renowned dyes" - information on dyes from Turner 2001; Kuhnlein & Turner 1991; Pojar and MacKinnon 1994.

Page 38. "One of the oldest carved objects …" Ames and Maschner 1999.

Page 39. "The Spanish explorers in the Strait of Georgia recorded…" Galiano and Valdes expedition, Galiano 1792 and descriptions in the anonymous journal Relacion del Viaje see Espinosa y Tello 1930. "trait also noted by the Russians at Sitka in 1860…" Ames and Maschner 1999. "Anthropologists listening to older members …" Suttles 1955; Suttles 1987.

Page 40. "At least 39 archaeological sites " Parsons 1981.

Page 41. "Human burial sites" McLay et al. 2004 for the Hul'qumi'num Treaty group perspective. "This can lead to misunderstandings" Henderson 2000 and other contemporary media reports.

Hallowed Ground

Page 43. "*The smoke from their morning fires...*" Old Pierre, Katzie Elder quoted in Jenness 1955. "Coast Salish " - some of the 54 Coast Salish Nations in the Pacific Northwest. "Halkomelem" - this is the anglicized version of a general word for dialects of this language, see page 56. Suttles also described the Halkomelem-speaking "Snokomish" who ostensibly died out in the smallpox epidemics (Suttles 1987). According to Don Welsh, archaeologist for the Semiahmoo First Nation, there is little other evidence for this band in the Boundary Bay area. The word may relate to people or a group living near Fort Langley. "based on published literature and conversations" - see following sources and bibliography.

Page 44. "The tradition of the Coast Salish is to live in family groups" Suttles 1987; Baird 1999; Carlson 2001. "obsidian from Oregon…. nephrite from Sumas Mountain" Carlson and Dalla Bona 1996. "Copper has been found" Ames and Maschner 1999. "*on the north side there is some flat country.....*" Pantoja's 1791 account quoted in Wagner 1933.

Page 45. "*another band of migrant Indians....*" Joseph Whidbey quoted in Naish 1996. "clearing clam beds" Williams 2006.

Page 46. "Each month was associated with specific activities…" information drawn from many spoken and written sources, e.g. Suttles 1955; Suttles 1987; Yates 1992; Boxberger 2000; Carlson 2001; Turner 2001. Cow parsnip (Indian celery) is another common edible plant, but it can easily be confused with similar poisonous species - White Plume Woman, Yvette John, Stò:lo, pers. comm.

Page 47. "Ethnographer Wayne Suttles interviewed elders" Suttles 1987.

Page 48. "Smallpox" - historical information from this section comes from Harris 1994; Harris 1998; Boyd 1999. "A form of inoculation...called variation" www.immunisation.org.uk/article.hp?id=347. "Edward Jenner's 1798 vaccine…" for more on the story of variation and vaccination see "*Smallpox ~ A great and terrible scourge*" on National Library of Medicine and National Institutes of Health website at http://www.nlm.hih.gov/exhibition/smallpox/sp_variolation.html. "90% of the Indigenous population died" Boyd 1999. "microscopically tiny virus" information on smallpox virus from Dr. C. Murray personal comm. 2007. "As historian Robert Boyd describes" Boyd 1999.

Page 49. "two thirds of the Stò:lo people died" Carlson 2001. "A story circulated … Several Lummi villages …" from Straits Salish informants, gathered by Suttles 1987. "*very fatal among them....*" George Vancouver quoted in Lamb 1984. "*the smallpox most have had...*" Peter Puget from Lamb 1984. "*the site of a very large village....*" Archibald Menzies in Newcombe 1923. "Joe Splockton described the pestilence as *stalacom*" - Don Welsh explained the meaning of this word and shed light on the quotation from Splockton, given in Appleby 1961; Boyd 1999; viz."*Pestilence was a giant called Stalacom.*" "*The people became frightened and began to shoot*" Joe Splockton, Tsawwassen, 1950s, quoted in Appleby 1961. "*and Tsawwassen, where many people had lived....*" Ibid. "Oral history and the archaeological record" Arcas 1991, Vol.1.

Page 50. "Botanist Dr John Scouler, noted in his August 1825 journal" Boyd 1999. "The 1836 -37 epidemic passed by the Lower Fraser" Ibid. "*When smallpox came to Birch Bay......one family who used the wood of a little tree called pcinelp for firewood. That killed the germs.*" Julius Charles Semiahmoo, born 1865, speaking to Wayne Suttles: Boyd 1999. "'*pcinelp*' or *put'thune7ilhp* is the strong-smelling Rocky Mountain juniper." Turner and Bell 1971. Saanich orthography from www.cas.unt.edu/~montler/Saanich/wordlist/index.htm "used by First Nations" Turner 2001.

Page 51. Information on family networks courtesy Don Welsh, pers. comm. Also E.C. Fitzhugh 1857 report in Jeffcott 1949. "Chanique" - his family story is in Jeffcott 1949. Pronounciation explained by Sharon Kinley, Lummi, Chanique's great granddaughter.

Page 52. "Lily Point" also Cannery Point, or as *Chelhtenem, sc'ultunum, tsehltenem, tsalhten, etc.*

Page 53. "Reef net fishery" Stewart 1977. "to hang salmon for drying" WA State Advisory Council 1992. "*incredible quantity of rich salmon ...*" Juan Pantoja quoted in Wagner 1933. "*on the White Bluff*" Peter Puget from Lamb 1984.

Page 54. "*a number of Indians in groups..*" Sir George Simpson, Chief Factor for the Hudson's Bay Company, in *Overland Journey around the World 1841-42* quoted in Gibbon 1951. "on August 20 1829, at least 200 canoes" Work 1945. "The Semiahmoo" Suttles 1951; Don Welsh, pers. comm. "it appears on the earliest European map of the region" Hayes 1999.

Page 55. "[The Semiahmoo] would travel as far as Waldron Island" Glavin 2000. "*Lekwiltok* (also known as the *Yukulta*)" Wayne Suttles in MacLachlan (Ed.) 1998. "the Semiahmoo had two forts" Thrift 1929. "The Lummi traditional territory" - information from Suttles 1987; Boxberger 2000; and Lummi websites: www.goia.wa.gov/tribalinfo/lummi.html and www.lummi-nsn. org. "Mount Baker, called *Kul-shan*" Jeffcott 1949. "seasonal activities" Boxberger 2000. "Recent generations of Lummi" Halliday and Chehak 2000.

Page 56. Salishan Languages: To hear spoken examples of the languages visit the Simon Fraser University website: http://www2.sfu.ca/halk-ethnobiology/. See also Timothy Montler's work on the Saanich dialect at www.cas.unt.edu/~montler/Saanich and the UBC Language Program http://fnlg.arts.ubc.ca/. "Chilliwack area" Smith 1950. Saanich information from Fort Langley Journals, MacLachlan 1998; Tsawout website www.tsawout.ca; Boxberger 2000.

Page 57. "A vibrant presence in the lower Fraser watershed, the Stò:lo" Carlson 2001; Stò:lo website www.stolonation.bc.ca; and individual band sources. Re-structuring of groups occurred in July 2004. "about 5,000 people" Stò:lo Statement of Intent, BC Treaty process, www.bctreaty. net/soi-2/soistolo.html.

Page 58. "On July 25 2007, the Tsawwassen..." for more information see Tsawwassen First Nation: www.tsawwassenfirstnation.com "*Kikayt*" and "*Skaiametl*"Appleby 1961; Arcas 1991; Baird 1999; Carlson 2001. "Land facing the sea" Tsawwassen First Nation 2004. "Katzie" www. katzie.ca "About 900 Musqueam registered" www.bctreaty.net/soi-2/soimusqueam.html

Page 59. "*Musqueam means the place of the muthkwuy*" Rose Charlie, Musqueam, quoted in http://public.sd38.bc.ca.8004/~timecapsule/Div6/Rose.htm and Simon Fraser University website: http://www2.sfu.ca/halk-ethnobiology/. "the Musqueam have a special affinity with the wetland grass" Musqueam website: www.musqueam.bc.ca. "Cowichan and Hulquminum Treaty Group" www.hulquminum.bc.ca. The Hulquminum Treaty Group comprises the Cowichan, Chemainus, Penelakut, Lyackson, Halalt and Lake Cowichan bands. "The Nooksack" Nooksack websites: www.goia.wa.gov/tribalinfo/nooksack and www.nooksack-tribe.org. "described as *Pekows* or *Quck-sman-ik*, the White Mountain" depending on sources: *Pekows*, literally "white top" is from an a 1966 interview with J.W. Kellerher quoted in Oliver N. Wells Oral Histories on Centre for Pacific Northwest Studies website: http://www.acadweb.wwu.edu/cpnws/wells/wellsdescription.htm; *Quck -sman-ik* is from Jeffcott 1949.

Sails in the Mist

Page 61. "*Two places in that direction...*" Peter Puget in Lamb 1984; "The Chinese...had a magnificent fleet of ships" Levanthes 1994.

Page 62. "Local writer, Samuel Bawlf" Bawlf 2003. "reached only 42 deg.north" Goetzmann and Williams 1992; Thrower 1984; Kelsey 1998; Hayes 2001; Williams 2002; etc. "*vile, thicke and stincking fogges*" and "*extreme and nipping cold*" Drake 1628. "*bad weather again obliged us to keep the sea*" James Burney's journal, 1778, quoted in Hayes 1999. "*exceeding tempestuous weather*" James Cook 1778 in Hayes 1999. "*the north wind in winter is extremely strong ....*" Mozino 1970. "*thick fogs and continuous rains*" Ibid.

Page 63. "*the dreary wilderness of the north west coast*" James Douglas quoted in Hayes 1999. "accurate timekeepers were invented by John Harrison" Sobel 1996. "Inaccuracies of one or two degrees were quite regular" Williams 2002.

Page 64. "Galiano, in 1792, had to correct..." Espinosa y Tella 1930.

Page 65. Fighting Scurvy: Naish 1992.

Page 66. "On a summer evening in June 1791" Pantoja's report in Wagner 1933. "*El Gran Canal de Nuestra Senora del Rosario la Marinera*" Ibid. The name given by the Spanish lingers on as Rosario Strait, between Bellingham Bay and the San Juan Islands. *El Gran Canal* was a year later named the Gulf of Georgia by George Vancouver and subsequently became the Strait of Georgia. "*no less than 30 vessels on the NW coast*" Menzies to Sir Joseph Banks, Jan 1793 in Naish 1996. "The Lummi told stories" E.C.Fitzhugh, Special Indian Agent, Bellingham District, Report of 18 June 1857, in Jeffcott 1949. "Jose Maria Narvaez Gervete" Kendrick undated; Hayes 1999.

Page 67. "Narvaez drew the position" Narvaez map in Wagner 1933; Hayes 1999.

Page 68. "*At the Isla de Zepeda....*" Narvaez' report, "*Description of the Estrecho San Juan and the Great Canal*" in Wagner 1933. "*which the foreign vessels have brought*" Ibid. "*An Indian boy obtained....*" Wagner 1933. "*Nothing will be acquired from the Indians...*" Conde de Revillagigedo, Viceroy of New Spain (Mexico), *Instruccion que el Exmo. Senor Verey dio a los commandantes de los Buques de exploraciones in California*, Historia y Viajes, tomo 1 MS 575 Museo Naval, quoted in Cutter 1991.

Page 69. "*the schooner being anchored two miles out they collected and drank sweet water*" Francisco Eliza's account in Wagner 1933. "*line of white water more sweet than salt*" from Narvaez' Description of the Estrecho San Juan and the Great Canal, in Wagner 1933. "*gulls, tunny fish and immense whales*" Ibid. "The Galiano and Valdes expedition": biographical data on Galiano, Valdez and crew members in Vaughn 1977; Cutter 1991; Kendrick 1999. For pictures see www.mmbc.bc.ca/source/schoolnet.exploration. "Jose Cardero" Vaughn 1977.

Page 70. "the account of the voyage" Espinosa y Tella 1930. "Sailing up the Juan de Fuca Strait" details of their voyage - Ibid; Kendrick 1997. "*would be full of interest*" Espinosa y Tella 1930. "*closed bay with trees all around it*" Galiano's journal in Wagner 1933. "*low land, marshy and full of trees*" Espinosa y Tella 1930.

Page 71. "*astonished at the difference in face ....*" Galiano's journal in Wagner 1933.

Page 72. "George Vancouver" - for more about his voyages see Lamb 1984; Hayes 1999; Hayes 2001.

Page 73. "Menzies introduced Europeans…" Justice 2000.

Page 74. *"A habitation of near four hundred people, but was now in perfect ruins and overrun with nettles and some bushes"* Peter Puget in Lamb 1984.

Page 75. *"in full bloom diffusing its sweetness that beautiful shrub the Philadelphus"* Archibald Menzies' journal quoted in Justice 2000. *"I departed at five o'clock on Tuesday morning.... we had passed around."* George Vancouver, in Lamb 1984. *"two places in that direction...."* Peter Puget in Lamb 1984. *"very low land, apparently a swampy flat "* Ibid.

Page 76. *"spoke a little English"* Ibid. *"this gentleman spoke English"* Peter Puget in Lamb 1984.

Page 77. *"by far the most pleasing place..."* Thomas Heddington, artist, quoted in David 1992. "The names of Tsawwassen and Semiahmoo..." for more about place names see Akrigg 1986.

Pioneers in a New Land

Page 79. *"The pioneer spirit lived strong in those days ...."* ~ Rebecca E. Jeffcott, pioneer settler, in Jeffcott 1949.

Page 80. "James McMillan and his multicultural team" Work 1824; Merk 1968; MacLachlan (Ed.) 1998. *"made a portage across Langley Prairie"* Sir George Simpson's journal in Merk 1968. Portage Park at 204 St. in Langley is the site of the portage stop on the Nicomekl. The expedition then went on foot to the Salmon River that flows into the Fraser near the present Fort Langley. A plaque at Portage Park commemorates the event. *"the Indians are very numerous....."* Chief Trader James McMillan, Hudson Bay Company, report in Merk 1968. *"Cowitchens"* - Halkomelem-speaking people lived on both the lower Fraser River and parts of Vancouver Island, where there were winter villages. There are various spellings of this old collective name for these members of the Coast Salish. Today the name refers to the Cowichan Tribes, based at Duncan (Hul'qumi'num Treaty Group).

Page 81. "The captain was too timid…" Information in this paragraph from MacLachlan (Ed.) 1998. "On Friday July 13 they anchored in the bay…" George Barnston's journal, in MacLachlan (Ed.) 1998. Barnston was a naturalist but says little about the wildlife in this account. "bringing in salmon for salting" - for stories of the fort see Work 1945; Merk 1968; MacLachlan 1998; Carlson 2001. "the Nooksack people trading furs" Jeffcott 1949; MacLachlan 1998. Fur trading virtually ceased by 1845 because the suppliers could obtain a better price elsewhere and were deterred by the raiding parties of northern tribes, see Work 1945. *"the sea yields an abundant supply of fishes of the most delicious kinds... every rivulet teems with myriads of salmon."* Scouler 1905.

Page 82. *"the woods on both sides of the river are all on fire ... met great numbers of Indians.."* Work 1945. "In 1818, the United States and Britain had agreed to define…" information in this paragraph and next from Hayes 1999; Hayes 2005.

Page 83. "The border line through the San Juan and Gulf Islands" - the dispute began as the so-called Pig War of 1859, in which the only casualty was a pig. International tensions escalated to a military stand-off, followed by occupation of San Juan Island by US and English camps. The dispute was settled in favour of the US in 1872. "Derek Hayes tells the story...." Hayes 1999; Hayes 2001; Hayes 2005. "A map of Semiahmoo Bay, Washington Territory, printed in 1858" Ibid. "A transborder shooting" Stewart 1959.

Page 84. "Zero Avenue" MacDonald 2004. "Captain James Prevost" - he was in HMS *Satellite* and Captain George Richards was in HMS *Plumper*. "Their maps show an intricate mass of depth records from Sand Heads to Derby Reach" Map 268, p 159, Hayes 1999.

Page 85. "Margaret North" North, Dunn and Teversham 1979. Information on Royal Engineers Camp, Jack Brown pers.comm. September 2007. "John Keast Lord" all quotes in box from Lord 1866.

Page 86. "Negotiations for this pivotal treaty" Hayes 1999. "Chinook Jargon"- this was a trade language, a mixture of English, French, American and Chinook dialect from the lower Columbia River. Coast Salish languages are difficult for foreigners to pronounce and speak, so this simple one was used instead, but it lacked the ability to communicate any nuances or complexities. "those living on the American side" Brown website. "The Stò:lo Atlas records..." Carlson 2001.

Page 87. "James Douglas...responded assertively" Ladner 1972. "This was the route taken by the Ladner brothers" Philips 2003. "Miners landed at Semiahmoo Spit" Jeffcott 1949. "Attempts were made to push through an overland route" Ibid.

Page 88. "Roadworks carried out along White Rock's beach road" Don Munro, personal communication 2003. "the checkerboard appearance of the landscape" - information in this section from Jordan 1982 and Taylor 1958. Other types of survey method in use at the time include "metes and bounds," a traditional British method, and "longlot" survey, used by French Canadians.

Page 90. "only about 300 immigrants.." Carlson 2001. "James and Caroline Kennedy" for more about Delta and Surrey's 'first family' see the website www.kencom.ca/kennedytimeline.htm. "The first men to get their name on the registry" - information on pioneers in following sections from Thrift 1931; Jeffcott 1949; Whiteside 1974; Ladner 1979; Hastings 1981; Philips 2003, etc.

Page 91. "1874 voters' list for the Elgin area had only fourteen men's name" McKinnon 1996. "The Ladner brothers" Ladner 1972. "Terrence Philips' book" Philips 2003.

Page 92. "Innes brothers came from Ontario" www.city.langley.bc.ca/History.shtml. "George Boothroyd pre-empted" Hastings 1981; Surrey Centennial Museum handout 1973 (origin of place names in Surrey); Akrigg 1986. "Dinsmores, Collishaws and Loneys" Whiteside 1974. "John Harris, his wife and four children" Jeffcott 1949. "the Harris family" Clark 1980.

Page 93. "a community of Icelanders" Thor 2002; Point Roberts Historical Society. "Only two of the squatters, Kate Waller and Horace Brewster" Clark 1980. Homesteads were confirmed in 1892. "Sarah Olson" Pt Roberts Historical Society

Page 94. "James Kennedy cleared a trail in 1861" - More information on local roads in Pepin 1998; Philips 2003. "In 1865, the Collins Overland Telegraph Line" - for an interesting true story about the telegraph line, read *The Woman who Walked to Russia* by Cassandra Pybus, Thomas Allen publishers, Toronto, 2002. "a branch line out to Ladner" - in 1906 the Victoria Terminal Railway and Ferry Company, a subsidiary of the Great Northern Railroad (and later the Vancouver, Victoria and Eastern Railway and Navigation Co.) went from Blaine to Colebrook, along the east shore of Boundary Bay, and out to Port Guichon in Ladner. To the east, a line linked with Cloverdale. See Philips 2003, map on page 61.

Page 95. "black slugs" Ladner 1979, Don Munro, personal comm. 2003. "Yellow-bellied marmots" Vancouver Port Authority pers. comm. "cats kill hundreds of millions of birds" Bird Studies Canada www.bsc-eoc.org, Audubon Society www.audubon.org, FLAP www.flap.org. "rabbits and turtles" e.g. European domestic rabbits and red-eared slider turtles bought as pets and then released when owners tired of them.

Page 96. "John Work" Work 1945. "burnt timber" and "burnt ground" John Fannin 1873, map in Hayes 2005, page 36-37. "Logging camp fire...in 1912" Hastings 1981.

Page 97. "Waters Pavilion" - the story of the fire from Don Meikle's letter to the All Point Bulletin, November 2003 issue.

Metamorphosis of Landscape

Page 99. "*It was nothing to see forty or fifty grouse...*" ~ Margaret Stewart, Surrey pioneer, in Hastings 1981. "at least 70% of the wetlands" Ward 1980.

Page 100. "The low lying land in the Fraser delta" see descriptions in Ladner 1972, Ladner 1979, Hutcherson 1982. *"the appearance of beaver being pretty numerous"* Work 1824.

Page 101. The Royal Engineers' Survey information from North, Dunn and Teversham 1979. "A recent study by Michael Church and Wendy Hales" Church and Hales 2007.

Page 102. The Fraser River "broke its banks" Ladner 1979.

Page 103. "they cooperated in building" Philips 2003. "total distance of 620 km (300 miles)" Bocking 1997. *"a rather fearsome world* ...." Leila May Kirkland in Hutcherson 2002. "a punt was used to ferry settlers" archival photo in Delta Museum & Archives. "Main Arm dramatically changed course" Philips 2003.

Page 104. "dramatic washouts occurring in 1882...." Thrift 1931. "through a peat and salt marsh environment" North, Dunn and Teversham 1979. "Alan McKinnon pulled out 12 m (40 ft) cedar" McKinnon 1996. "After several disastrous first attempts"- for the full story see Thrift 1931. "was the work of Chinese labourers in the 1920s" Ibid. "were privately financed by bonds" Ibid.; McKinnon 1996. *"The Nicomekl was a beautiful river in the 1930s...."* as remembered by K. Hardy in Hastings 1981.

Page 115. "there was a canal" Hastings 1981; Munro 2003. "Heavy rains, gales and high tides...." McKinnon 1996. "In 1990, the Erickson pump station was constructed" Emery 1997.

Page 116. "Margaret Stewart, born in 1876" Stewart 1959.

Page 107. "birds like the purple martin, western bluebird, nighthawk" see note page 158.

Page 108. "Streams poured down the bluffs" unpublished 1990 map by Ted Wade, made available by Don Munro 2003. "The sandbars and channels of the lagoon" - lagoon study, Page 1999.

Page 109. "Wolves, bears and cougars" Jeffcott 1949. "According to one story…". Thrift 1931, Hastings 1981. "The last wolf recorded in the Fraser estuary" Butler and Campbell 1987. "Oral history of the Stò:lo and Semiahmoo" Carlson 2001; White Rock Museum & Archives 2003 Semiahmoo museum exhibit. *"in the early days at Custer...."* Franklen Brunson quoted in The Weekly Blade January 13 1904, New Whatcom WA www.rootsweb.com/nwawhatco/newspapers/blade1904.htm accessed Jan. 2008. "The last wild bear in Richmond" Kidd 1973. "A black bear came down McNally Creek" Don Munro personal comm 2004.

Page 110. *"Immediately we put ashore..."* Work 1824. *"Elk have been very numerous..."* Ibid. "they were hunted by the Halkomelem-speaking Salish." Carlson 2001. *"My grandfather, Chanique..."* George Kinley quoted in Jeffcott 1949. *"The elk left this area around 1855"* Jeffcott 1949. "In South Surrey, residents remember them." Don Munro, personal comm. 2003 and others.

Page 111. "big trees" - for US State Champion Tree lists visit www.championtrees.org. For article on B.C. and WA State big trees see www.arthurleej.com/a-peninsulatrees.html. "One felled by William Shannon in 1881" - this Douglas-fir was felled on Hall's Prairie Road. A count of its rings gave its age as 1,100 years old. The trunk was measured after felling so a couple of metres could be added to its length for the height of the stump. Details from the Know B.C. website: www.knowbc.com/iebc/book/T/trees accessed November 2002. "Joseph Figg, a Surrey pioneer" Thrift 1931; Stewart 1959; Hastings 1981. Henry Thrift had a similar, but non-fatal, experience in March 1885 when a tree fell across his house, smashing everything, after taking three weeks to burn. "Many settlers" Jeffcott 1949. "Once logging began t" - information on early logging in the watershed from Hastings 1981; Surrey Centennial Museum.

Page 112. "The Burr family logged Panorama" - logging information, Surrey Centennial Museum. "Harry Bose remembered" Frank McKinnon in SHS 1995. "Allan McKinnon measured" Ibid.

Page 114. "The Green Timbers Story" Green Timbers Heritage Society. "Archibald Menzies, the botanist" Justice 2000. "calypso orchids"- also called fairyslippers or ladyslippers. "a quarter of the native plants are now rare." Gray & Tuominem (Eds.) 1998.

Page 115. "The Coast Salish people traditionally went into bogs" Tsawwassen First Nation 2004. "Moss was collected" Turner 2001. "Chief Kim Baird...." Tsawwassen First Nation 2004.

Page 116. "Ted Wade, whose family settled near Boundary Bay" Don Munro, personal comm. 2003. "Camas fields were carefully nurtured" Brown website http://members.shaw.ca/j.a.brown. "plant life and Aboriginal activities " - see for example Sewid-Smith 1999. Dr Daisy Sewid-Smith, a Kwakwaka'wakw elder explained in this talk how plants were traditionally tended by the Coast Salish people.

Page 117. "Scotch broom…" Pojar and MacKinnon 1994. "At least 40% of the flowering plants..." Gray & Tuominem (Eds.) 1998a.

Harvesting River & Sea

Page 119. "*It was said in those days you could walk across the Fraser River on the backs of the salmon.*" Dennison undated. "is still one of the world's great salmon rivers" according to Bocking 1997, the Fraser was the world's greatest salmon river, but the drastic decline in many salmon stocks now questions whether that statement still holds true in 2007. "11,000 years ago" Department of Fisheries and Oceans Canada www.dfo-mpo.gc.ca; Don McPhail pers.comm. Jan.08. "moved into the lower Fraser" - the Interior Fraser, was part of the Columbia River system at this time, since the Puget Sound ice lobe was slower to melt than the Okanagan lobe: Ibid.

Page 120. "Fraser salmon runs" - various sources including Fisheries and Oceans Canada 2001. Sockeye "dominant run" - 8 out of 20 Fraser River sockeye populations currently have dominant four year runs. "fish bones found...Glenrose Cannery" Matson 1996. "virtually all protein consumed" Carlson & Dalla Bona 1996.

Page 121. "Oral history" Carlson 2001. "At least 10,000 salmon were taken" - all figures in this paragraph from Whatcom County 1992; Boxberger 2000. "Halkomelem-speaking people" Baird 1999; Carlson 2001; Tsawwassen First Nation 2004.

Page 122. "Prior to 1873, all major Fraser River sockeye runs" Fisheries and Oceans Canada 2001. "The Stò:lo estimate" Glavin 2000; Carlson 2001. "hungry people" Stò:lo name for newcomers at that time - Carlson 2001. "The Dominion Fishery Act ...1877" Knudsen 2000. First Nations were denied the vote in BC in 1872, until 1960. "two squatters" - these were John Waller and Joseph Goodfellow: Ladner 1979; Clark 1980; Lummi 2004. See photo of Goodfellow's Camp page 52.

Page 123. "Fraser River canneries" - information from Gulf of Georgia Cannery Museum exhibits; Steveston Museum section on virtual museum website www.virtualmuseum.ca; Terry Slack pers. comm. August 2007, Jan. 2008; Gold Seal website www.goldseal.ca; and other sources as noted. "William Herbert Steves" for more information on Steveston see Community Memories on Virtual Museum of Canada: Steveston website www.virtualmuseum.ca.

Page 124. "Homer Stevens" BC Packers website www.intheirwords.ca/english/people_labour. html accessed Oct. 2007. "Several thousand Chinese" Yee 1988. "Gihei Kuno was the first Japanese" Ibid. "1,300 Japanese men and 15 women" Ibid. "Japanese men went to the southern Gulf Islands" Yesaki 2001. "Japanese boatbuilders" e.g. Kyuzo Kawamura (Phoenix Cannery) and Tsunematsu Atagi (Atagi Boatworks): Community Memories on Virtual Museum of Canada, Steveston website www.virtualmuseum.ca accessed June 2007; Yesaki et al. 2005.

Page 125. "biggest recorded run" - 1901: Terry Slack, pers. comm. Jan 2008. "In 1888, British Columbia imposed more restrictions" Following the restrictions of the Dominion Fisheries Act of 1868, which took effect in B.C. in 1877, the 1888 Fisheries Act confined First Nations to a food-only fishery. Knudsen 2000; Lichatowich 2001. "John Waller" see ref. page 122.

Page 126. "By 1905, there were 47 fish traps" Clark 1990. Traps information: Hrutfiord 2002.

Page 127. "Jan Hrutfiord" Ibid. "We're strange people..." Terry Slack in conversation, August 2007. "A sudden population decline" Fisheries and Oceans Canada 2001.

Page 128. "Pacific herring fishery...catch of over 268,000 tonnes" Gulf of Georgia Cannery Museum, Parks Canada. Herring spawn, see DFO website http://www.pac.dfo-mpo.gc.ca/sci/ herring/herspawn/GIF/0293.gif. "Eulachon" much information from Terry Slack, pers. comm. Aug. 2007.

Page 130. *"sturgeon are continually leaping"* John Keast Lord's journal, Lord 1866. "Terry Glavin documented"- information in this paragraph from Glavin 1994. "a Stò:lo preacher, Captain John of Soowhalie" Ibid. "classifying sturgeon as endangered" Committee on Status of Endangered Wildlife in Canada, 2003. "Fraser River Sturgeon Conservation Society www.frasersturgeon. com. "In 2002, police officers.." Tanner 2002.

Page 131. "Seiners" - info from Pacific Coast Salmon Fisheries website: http://collections.ic.gc. ca/pacificfisheries/ and Gulf of Georgia Cannery Museum, Parks Canada. "Gillnetting is an ancient method" info from Pacific Coast Salmon Fisheries website: http://collections.ic.gc.ca/ pacificfisheries/ and Gulf of Georgia Cannery, Parks Canada.

Page 132. "Albion Box" Terry Slack pers.comm. August 2008. "One notable kill" - in 1993 over a 1,000 common murres and other seabirds were killed in fishing nets. Martin Keeley, Friends of Boundary Bay, personal communication. "In 2005.." bird kill records Beached Bird Survey, Bird Studies Canada, pers. comm..

Page 133. "creates a bycatch problem" - bycatch discards can be over 50%: Pitcher & Chuenpagdee 1994. "subject of legal challenges" e.g. the 1973 Calder case recognized existence of Aboriginal title; 1993 Delgamuukw clarified that Aboriginal rights continued to exist. "Boldt Decision" - this was a US federal district court decision, known after the presiding judge. "Lummi fishing fleet rapidly grew" Lummi 2003. "In 1990 the Musqueam successfully challenged" - Regina v. Sparrow: the Sparrow Decision involved a Musqueam fisher using a particular type of net. "The most marketable salmon in North America.." Nelson 2006.

Page 134. "a kidney parasite, *Parvicapsula minibicornis*" Jones 2002. "Pacific Salmonid Enhancement Program" Schubert 1982.

Page 135. *"we had to get people to think of the river..."* Price 1997; Serpentine Enhancement Society: Ibid. "a 1995 study of coho in the Columbia River" Glavin 2000; Lichatowich 2001.

Page 136. "Swaren Singh, Prahim Singh and Oudam Singh" Peter Oldershaw http://members. shaw.ca/j.a.brown/OysterCo.html "sand flats of Boundary Bay are the best environment" Quayle 1988; Shellfish Growers Association http://www.bcsga.ca. "Mud Bay oysters" e.g. Crescent Oyster Co., B.C.Packers. "When the water warms in spring" Quayle 1988

Page 137. "Community Oyster Farm" - http://whatcomshellfish.wsu.edu/Drayton/oysterfarm "Clam gardens" Williams 2006. "B.C. Packers oyster plant" - for more photos see http://members. shaw.ca/j.a.brown/DeltaBCP.html.

Page 138. "poaching is a problem" Seattle Times September 14 1999. "Drug dealers have been known to target crab fishers" Seattle Times, August 10 2003. See also Lummi website.

Page 139. "Unregulated market hunting" Leach 1982.

Page 140. "Pacific halibut and lingcod populations.." Wallace & Dalsgaard 1998. "once abundant rockfish" Bernard Hanby speaking at FBCN/CPAWS Marine Protected Areas Symposium, 13 May 2004, UBC, Vancouver. "Basking sharks" Pacific Wildlife Foundation www.pwlf.org; "offshore Pacific right whales" Webb 1988.

Page 141. "97 whales were harpooned" Merilees 1985. "A few recent sightings.." Mary Taitt pers. comm.; Southern Gulf Islands Atlas 2006, Community Mapping Network. "Populations rebounded" - the years 1973 to 1993 saw an increase in harbour seals: Baird 2001. "20,000 were shot" Lichatowich 2001. "fishers even shot local eagles" Hancock 2004.

The Riches of the Land

Page 144. "Traditional Aboriginal farming practices" Carlson 2001. "Semiahmoo...adopted potato growing" Work 1945; Suttles 1987; Glavin 2000. John Work's journal of 1835 mentions that Mr Yale of the HBC had given *"each of the principal men a little garden"* and was encouraging them to raise potatoes. "The first Hudson's Bay Company farm" Work 1945. *"very well regulated"* with *"an excellent farm....."* Sir George Simpson, 25 November 1841 in Williams 1973.

Page 145. *"the cornerstone of a healthy and prosperous society"* in Koroscil 2000. "was one of the main proponents of the Sumas Lake drainage" for more about this scheme and John Oliver's tenure as Premier, see Cherrington 1992. "The Booth sisters" Ladner 1972s.

Page 146. "To make way for hay fields" - the delta yielded up to 3.5 tons of hay/acre according to Philips 2003. "only horses could be turned loose to feed on the "saltings" Dr. Bert Brink, personal communication, 2006. "Dutch farmers played a key role" Winter 1968; Surrey Archives. In 2004, there were about 70 dairy farms in the Boundary Bay/Drayton Harbor watershed.

Page 147. George Triggs: *"fair day's pay"* in Hastings 1981. "Dolly Wade wrote" Dolly Wade's article in a Crescent Beach Association monthly newsletter, undated, personal collection of Don Munro. Dolly's parents arrived in Crescent Beach in 1917. "Potato culture" information from Delta Museum & Archives special exhibit: "All eyes on the potato!" January 25 2007.

Page 148. *"Our three enemies were drought, frost and the California market."* Dennison undated."marsh known as the Tule" Pt. Roberts Hist. Society members pers. comm. Oct. 2007. "Black Angus cattle" Sally Roberts pers. comm. Oct. 2007.

Page 149. "Tents to Townhouses": Hastings 1981; Philips 2003; Hayes 2005; and community archives.

Page 150. "Delta farmland reached $750/ha ($300 an acre)" Jeffcott 1949; Ladner 1972. *"always been a constant of good times"* Gen Matheson then 76, interviewed by Trudi Beutel in "Summers at bay remembered" in Delta Optimist, July 17 1999. "300 new families…" The Weekly Optimist, Ladner, B.C. May 3 1928. "3,600 cars…" The Weekly Optimist, Ladner, B.C. July 4 1929.

Page 151. "production of sugarbeet seeds" Dr. Bert Brink, pers. comm. 2006. *"Of course Delta needs a bridge or a tunnel"* Delta Optimist, January 3 1946. "The Deas Island tunnel" Hastings 1968; Hayes 2005.

Page 152. "With 120,000 ha ... of arable land within the lower Fraser Valley" Winter 1968.

Page 153. Peat harvesting information: from Western Peat Co. Ltd. 1947 brochure supplied by John Fehr, Sun-gro Peat; *When Burns Bog fed the War Machine* by David W. Robertson, Vancouver Sun, March 5 2005; Biggs 1973; Don de Mille, Burns Bog biologist, at Federation of B.C. Naturalists symposium, May 14 2004, Vancouver.

Page 154. "In 1913, the Forslund brothers" Waite 1977. "The Agricultural Land Reserve" ALC website www.alc.gov.bc.ca; Klohn Leonoff 1992; Township of Langley http://www.tol.bc.ca; B.Smith pers. comm. 2007. Not all ALR land is actively farmed.

Page 155. "Cranberries ... are the province's biggest berry crop" www.agf.gov.bc.ca accessed Dec 2007. "largest raspberry crop in USA" Lynden Museum.

Page 156. "Only about 3,000 people" B.C. Ministry of Agriculture and Fisheries website: www.agf.gov.bc.ca. "1,230 farms in Langley" latest figure, 1991, Township of Langley website: www.tol.bc.ca. There were 500 farms in Surrey in 2007 www.surrey.ca. "important source of employment" Winter 1968. "John Berry hired about twenty men" Waite 1977. "Punjabis prided themselves" Jagpal 1994. *"we got established here with muscles and money"* Ibid.

Page 157. "Green choices": in Metro Vancouver there were 26 organic farms in 2001 and 412 in 2006. www.metro.cupe.ca/updir/metro/SCIBulletin2007-9.pdf *"restore, maintain and enhance"* Fraser Valley Organic Producers Assn. website www.fvopa.ca accessed November 2004.

Page 158. "burrowing owls and horned larks were extirpated" Butler & Campbell 1987. "Yellow-billed cuckoos... western bluebirds" Ibid. The western bluebird died out as a breeding bird in Surrey in 1938, and is now extirpated from the Boundary Bay area. Mountain bluebirds are casual on migration. "Nighthawks used to nest even in downtown Vancouver" Rosemary Taylor pers. comm. 2006. They were common until the 1970s, with a few still seen on migration today. "swallows and swifts are declining" Partners in Flight 2007. "Purple martins" - they were common breeding birds in the 1890s but declined precipitously by the 1940s. They have recently begun to come back at Blackie Spit, where some nest each year. "Bald eagles, at the top of the food chain" Gray & Tuominem 1998. "numbers have rebounded" - the bald eagle was removed from the US endangered species list in June 2007. "blueberry and cranberry producers...." Schreier 1998.

Page 159. "suburban homeowners use, on average, ten times as many…" US National Wildlife Federation quoted on eartheasy website: http://eartheasy.com/grow_lawn_care.htm accessed 2007. "Snow geese ... playing fields": in winter 2007-08 there were 100,000 snow geese in the Fraser delta and there was a shortage of habitat in Richmond, though not in Delta. "At least 45 bird species" Butler 1999. "over 70 species" Markus Merkens, pers. comm. 2006. Delta Farmland and Wildlife Trust: www.deltafarmland.ca; the Trust is a joint initiative of the Delta Farmers Institute and the Boundary Bay Conservation Committee.

Page 160. Greenhouse information from Barry Smith, pers. comm.; Ministry of Agriculture. Delta Corporation acreage numbers are somewhat different. "This far surpassed" www.agf.gov. bc.ca. The gross revenue from Surrey farms in 2004 was $106 million,of which $12 million were from cranberries, $6 million from strawberries. In Langley, farm sales were $203.4 m in 2001, from 40% of the total agricultural land in the Fraser Valley. Township of Langley www.tol.bc.ca.

Page 161. "A seminal study of agriculture" Klohn Leonoff 1992. "BC Ministry of Agriculture" www.agf.gov.bc.ca. "an outbreak of avian flu" - 17 million poultry were culled as a result of the viral infection in the lower Fraser Valley in spring 2004.

Conservation of Boundary Bay

Page 163. "*I believe a chance to see wildlife ....*" Roderick Haig-Brown, conservationist, 1908-1976, *Writings and Reflections* quoted in Centennial Wildlife Society 1987. "Wildlife declines and extinctions" Diamond et al. 1989; Environment Canada 2007. For example: the US Fish and Wildlife Service data show that between 1950 and 1970, 2.39 billion birds died in the USA from human causes and diseases associated with crowding and habitat restriction; in the 1970s, 196 million birds were killed every year, 61% from hunting, and 39% poisoned from lead shot or pollutants, killed by hydro wires or road collisions, or perished in disease outbreaks. About a third of these were on the Pacific Flyway - see Baldassare and Bolen 1994. Environment Canada identifies 30 bird taxa endangered by habitat loss to agriculture in Canada, 21 bird taxa to forestry, 18 to urbanisation, 14 to pollution, 10 to livestock ranching, 10 to harvest, 5 to recreation and tourism and 4 to exotic species interaction - Env. Canada 2007.

Page 164. "A brief catalogue of projects" numerous contemporary media accounts of the time in Surrey Archives and Delta Museum & Archives; also Burns 1997; McKinnon 1996. "Two railway companies envisioned port terminals" in 1904 the Great Northern Railway, and in 1911 the Canadian Northern Pacific Railway.

Page 166. "Number of western sandpipers" - decreases of up to 80% were announced by the Canadian Wildlife Service in summer 2004, compared with numbers prior to 1992. CWS news report, in Delta Optimist July 14 2004; pers.comm. CWS. "Vancouver International Airport" information from YVR website www.yvr.com. "wildlife habitat was lost" $9 million was paid in habitat compensation costs by YVR. $2.5 million of this was entrusted to the Vancouver Foundation as an endowment fund for the Delta Farmland and Wildlife Trust, www.deltafarmland.ca.

Page 167. "475,000 tonnes of waste" Vancouver landfill website www.city.vancouver.bc.ca/engsvcs/solidwaste/landfill/facts.htm accessed November 2007.

Page 168. "The frontier was an unregulated place" for a history of hunting and conservation on the lower Fraser see Leach 1982. "*the people of Blaine...*" Stewart 1959. "*hunted down and destroyed, regardless of cost*" Dr William Hornaday, 1914, in Centennial Wildlife Society 1987. Hornaday also advocated the killing of all birds of prey.

Page 169. "*biggest pile of feathered and furred quarry*" National Audubon Society www.audubon.org/bird/cbc/. "Wood ducks" and "Reifel Bird Sanctuary was set up" in 1964 - see Leach 1982.

Page 171. "Greenbelt' lands" Murray & Taitt 1992. "recommendations of an Environmental Assessment Panel" - EAP Report on the Roberts Bank Port Expansion 1979. "*full expansion of the port would present an unacceptable threat to the Roberts Bank ecosystem*" Len Marchand, Minister of Environment, 1979. "The Birds of the Fraser River delta" Butler and Campbell 1987.

Page 172. "*no comparable sites exist along the Pacific coast between California and Alaska*" Ibid. "Ramsar site, after the Convention of 1971" - The Convention on Wetlands of International Importance especially to waterfowl was held in Ramsar, Iran. Key wetland sites across the world have been protected under this convention to which Canada is a signatory. Boundary Bay far exceeds the criteria yet the B.C. government has not listed it as a Ramsar site. "there would be at least 44 golf courses" for more information on this and following paragraphs, see Murray & Taitt 1992. TDL hearings, see for example Tom Perry, MLA, remarks during discussion in 1989 Legislative Session: recorded in Hansard, Thursday June 22 1989.

Page 173. "study of Boundary Bay" - a series of government studies including the Delta Agricultural Study, Klohn Leonoff 1992; the Delta Rural Land Use Study, Norecol, Dames & Moore 1994; a Burns Bog study, Berris 1993; and a wildlife study, Butler (Ed.) 1992. "biosphere reserve" Murray & Taitt 1992. "Boundary Bay, South Arm Marshes and Sturgeon Bank Wildlife Management Areas" - Roberts Bank was recommended by the B.C. Ministry of Environment but has never been formally designated. "With many hundreds of people" much Burns Bog advocacy was organized by Burns Bog Conservation Society, under Eliza Olson, www.burnsbog.org.

Page 174. "B.C. Breeding bird surveys" from Partners in Flight, pers. comm. Tanya Luszcz, 2007. "one in four birds is at risk" study of the National Audubon Society 2007, www.audubon.org accessed Jan. 23 2008.

Page 175. "How long will black bears continue to live?" This issue was addressed by the Technical Review meetings for the Burns Bog Ecosystem Review. For the population to stay genetically viable, it was thought that at least one outside bear needed to enter the bog once a generation and breed with resident bears: McIntosh and Robertson 1999.

Page 177. "Nature Canada reports" www.naturecanada.ca/climate_change_birds.asp

Page 178. "in 1892, dairy farmers" Philips 2003. "Aquifers" Brookswood aquifer under Langley is one of the largest in the lower Fraser Valley. "Poorly operating septic systems" Scales 1997. "The critical point for marked degradation" Payette 1997.

Page 179. "problems were described in a study" Swain & Holms 1988. "Smog" GVRD 2003.

Page 180. "is now 4 cm higher" MWALP (BC Min. Environment) 2002. "Sea surface temperatures" Ibid. "constant car travel is linked with obesity" Frank et al. 2004

Page 181. "one third of all artificial lighting" Bower 2000. "true darkness is needed for biological activities" Ibid.; Longcore & Rich 2004. "falcons hunting" R.Swanston pers.comm.

Page 182. "death toll from night lighting" Bower 2000. and Fatal Light Awareness Program (FLAP) www.flap.org. "a reduction in some moth populations" Longcore & Rich 2004. "aquatic species are also at risk" Ibid. "Studies are finding..." Bower 2000; Longcore & Rich 2004. "is now being proven" information in this section from Bird 2007.

BIBLIOGRAPHY

Akrigg, G.P.V and Helen B. 1986. *British Columbia Place Names*. Sono Nis Press, Victoria.

Ames, Kenneth, Herbert D.G. Maschner. 1999. *Peoples of the Northwest Coast, their Archaeology and Prehistory*. Thames and Hudson. London.

Appleby, Geraldine. 1961. *Tsawwassen Legends*. Ladner, B.C. Dunning Press.

Arcas Consulting Archaeologists Ltd. 1991. *Archaeological Investigations at Tsawwassen B.C. Vols.I - IV*. Prepared for the Construction Branch, South Coast Region, MOTH, Burnaby, B.C.

Baird, Kim. 1999. *Main Table Notes*. Tsawwassen Treaty Negotiations. Tsawwassen First Nation. Delta, B.C.

Baird, Robin W. 2001. *Status of Harbour Seals, Phoca vitulina, in Canada*. Can. Field-Naturalist 115(4): 663-675.

Baldassare, Guy, A. and Eric G. Bolen. 1994. *Waterfowl Ecology and Management*. John Wiley and Sons Inc. New York.

Bawlf, Samuel. 2003. *The Secret Voyage of Sir Francis Drake*. Douglas and McIntyre. Vancouver, Toronto. 400 pp.

BC Ministry of Water, Land and Air Protection. 2002. *Indicators of Climate Change for British Columbia*. 48 pp.

Berris, Catherine. 1993. *Burns Bog analysis*. Boundary Bay Studies. B.C. Ministry of Environment, Lands and Parks, Victoria. 60pp

Bierwert, Crisca. 1999. *Brushed by Cedar, Living by the River, Coast Salish Figures of Power*. University of Arizona Press, Tucson, 314 pp.

Biggs, Wayne G. 1973. *An ecological and land use study of Burns Bog, Delta. B.C.* MSc Thesis UBC. Vancouver, B.C.

Bird, William. 2007. *Natural Thinking. Investigating the links between the natural environment, biodiversity and mental health*. Royal Society for the Protection of Birds, UK.

Blunt, David J., Don J. Easterbrook, Nathaniel W. Rutter. 1987. *Chronology of Pleistocene Sediments, Puget Lowland*. In Schuster, J. Eric. 1987. (Ed.) *Selected Papers on the Geology of Washington*. WA Division of Geological and Earth Resources. Bulletin 77. WA State Dept. of Natural Resources. pp 321 - 353.

Boas, Franz. 2002. *Indian Myths and Legends from the North Pacific Coast of America*. 1895 Edition translated by Dietrich Bertz. Edited and annotated by Randy Bouchard and Dorothy Kennedy. Indian Language Project. Talon Books, B.C.

Bocking, Richard C. 1997. *Mighty River. A Portrait of the Fraser*. Douglas & McIntyre. Vancouver, B.C. University of Washington Press. Seattle, WA. 294 pp.

Boehm, S.G. 1973. *Cultural and non-cultural variation in the artifact and faunal samples from the St. Mungo Cannery Site, B.C. DgRr 2*. MA Thesis. Dept. of Anthropology, UVIC, Victoria.

Bower, Joe. 2000. *The Dark Side of Light* Audubon Magazine. http://magazine.audubon.org/darksideoflight.html

Boxberger, Daniel L. 2000. *To Fish in Common - Ethnohistory of Lummi Indian Salmon Fishing.* University of Washington. Seattle. WA.

Boyd, Robert. 1999. *The Coming of the Spirit of Pestilence.* UBC Press, Vancouver and Toronto. Univ. of WA Press, Seattle and London.

Brown, Jack. *Surrey's History*; a personal website. www.members.shaw.ca/j.a.brown/surrey.html.

Bunyan, D.E. 1978. *Pursuing the Past. A General Account of British Columbia Archaeology.* UBC Museum of Anthropology. Number 4. Vancouver,

Burns, Bill. 1997. *Discover Burns Bog.* Hurricane Press. Vancouver, B.C.

Butler, Robert, Wayne Campbell. 1987. *The Birds of the Fraser River delta: populations, ecology and international significance.* Occasional Paper No. 65. CWS, Environment Canada

Butler, R.W. (Ed.) 1992. *Abundance, distribution and conservation of birds in the vicinity of Boundary Bay, British Columbia.* Canadian Wildlife Service, Pacific and Yukon Region, Delta, B.C. Technical Report Series No. 155, and Wildlife Working Report No. WR-52. 134pp.

Butler, R.W. 1999. *Winter abundance and distribution of shorebirds and songbirds on farmlands on the Fraser River delta, British Columbia, 1989-1991.* The Canadian Field Naturalist, Vol. 113, Number 3: 390-396. The Ottawa Field Naturalists Club, Ottawa

Calvert Gay. 1970. *The St. Mungo Cannery Site: a preliminary report.* In. Carlson, Roy L. (Ed.) 1970. *Archaeology in British Columbia; New Discoveries.* B.C. Studies Spec. Issue 6&7. pp54-94.

Carlson, Keith, Thor. (Ed.) 2001. *A Stò:lo Coast Salish Historical Atlas.* Douglas & McIntyre, Vancouver. Univ. of Washington Press, Seattle and Stò:lo Heritage Trust, Chilliwack.

Carlson, Roy. (Ed.) 1983. *Indian Art. Traditions of the Northwest Coast.* Archaeological Press. Simon Fraser University, Burnaby. B.C.

Carlson, Roy L. & Luke Dalla Bona. 1996. *Early Human Occupation in British Columbia.* UBC Press. Vancouver.

Centennial Wildlife Society (Ed.). 1987. *Our Wildlife Heritage. 100 Years of Wildlife Management.* Published by The Centennial Wildlife Society of British Columbia. 192 pp.

Cherrington, John A. 1992. *The Fraser Valley: a History.* Harbour Publishing. Madeira Park, B.C. 391 pp.

Church, Michael, and Wendy Hales. 2007. *The Tidal Marshes of Fraser Delta.* Discovery. Vol.36 No.1. 28-33. VNHS

Clague, J.J., J.L. Luternauer, S.E. Pullan and J.A. Hunter. 1991. *Postglacial deltaic sediments, south Fraser River delta, B.C.* Can. Journ. Earth Sci.28: 1386 1393.

Clague, J.J, J.L. Luternauer, and D.C. Mosher. (Eds.) 1998. *Geology and Natural Hazards of the Fraser River Delta, British Columbia.* Geol. Surv. of Canada Bull. 525. Natural Resources Canada.

Clague John J. and Bob Turner. 2003. *Vancouver, City on the Edge.* Tricouni Press. Vancouver. B.C. 191 pp.

Clark, Richard. E. 1980. *Point Roberts USA ~ The history of a Canadian Enclave.* Textype Publishing, Bellingham WA.

Community Mapping Network, Friends of Semiahmoo Bay Society. 2007. *Georgia Basin Habitat Atlas: Boundary Bay.* www.georgiabasin.net; printed version available from Friends of Sem. Bay. Soc. www.birdsonthebay.com

Crawford, Michael, H. 1998. *The Origins of Native Americans: Evidence from Anthropological Genetics.* Cambridge University Press, UK.

Crockford, Susan, J. 2005. *Native dog types in North America before arrival of European dogs.* In Proceedings of 30th World Congress Small Animal Veterinary Association May 11-14 www.vin.com/proceedings.

Cutter, Donald C. 1991. *Malaspina and Galiano: Spanish Voyages to the Northwest Coast 1791 and 1792.* Douglas & McIntyre, Vancouver and Toronto, University of Washington Press, Seattle.

David, Andrew. 1992. *Vancouver's Artists.* In Proceedings of the Vancouver Conference on Exploration and Discovery. Simon Fraser University. History Department.

Dennison, W.J. undated. *Memoirs of a Ladner Lad, 1920s to 1990s.* Unpublished memoir held at Delta Museum & Archives.

Diamond, Antony, Rudolf Shreiber, W. Cronkite, Roger Tory Peterson. 1989. *Save the Birds.* Pro Natur and International Council for Bird Preservation.

Drake, Sir Francis. 1628. *The World Encompassed.* Hakluyt Society edition, with appendices and introduction by W.S.W. Vaux. 1854. Reprinted by Burt Franklin New York. New York. 1975.

Egan, Brian (Ed.). 1999. *Helping the Land Heal ~ Ecological Restoration in B.C.* Proceedings. B.C. Environmental Network Educ. Foundation, Victoria.

Environment Canada 2007. *Species at Risk.* www.speciesatrisk.gc.ca

Espinosa y Tello, trans. Cecil Jane. 1930. *A Spanish Voyage to Vancouver and the Northwest Coast of America, 1792.* London, UK. 1971 Edition AMS Press, New York, USA.

Fladmark, Knut. 1986. *British Columbia Prehistory.* Archaeological Survey of Canada; Nat. Mus. of Man, Ottawa.

Flannery, T. 2001. *The Eternal Frontier - An Ecological History of North America and its Peoples.* Atlantic Monthly Press. NY.

Flannery, T. 2005. *The Weather Makers.* Text Publishing, Melbourne, Australia.

Finlay, Victoria. 2002. *Colour - Travels through the paintbox.* Sceptre. Hodder and Stoughton, London.

Fisheries and Oceans Canada. 2001. *Fish Stocks of the Pacific Coast.*

Frank, Lawrence D., Martin A. Andresen, Thomas L. Schmid. 2004. *Obesity relationships with community design, physical activity and time spent in cars.* Am J Prev Med 2004 27(2):87-96

Gibbon, Dr John Murray. 1951. *The Romance of the Canadian Canoe.* Ryerson Press. Toronto. 145pp.

Glavin, Terry. 1994. *A Ghost in the Water.* New Star books. Transmontanus. Vancouver.

Glavin, Terry. 2000. *The Last Great Sea. A voyage through the human and natural history of the North Pacific Ocean.* GreyStone Books. Douglas & McIntyre.

Goetzmann, William H. and Glyndwr Williams. 1992. *The Atlas of North American Exploration.* Swanston Publishing Ltd. Prentice Hall.

Gray, Colin & Taina Tuominem (Eds.). 1998. *Health of the Fraser River Aquatic Ecosystem. Vol. 1 & 2.* Research synthesis Fraser River Action Plan. Env. Canada.

Green Timbers Historical Society. Undated. *The History of Green Timbers.* Unpublished manuscript.

GVRD. 2003. *Forecast & Backcast of the 2000 Emission Inventory for the Lower Fraser Valley Airshed 1985-2025.* Greater Vancouver Regional District Policy & Planning Department. http://www.gvrd.bc.ca/air/pdfs/ 2000EmissionInventoryForecast.pdf

Hales, Wendy. 2002. *The impact of human activity on deltaic sedimentation, marshes of the Fraser River Delta, British Columbia.* In *The Changing Face of the Lower Fraser River Estuary* Symposium Proceedings. Fraser Basin Council. New Westminster. FRC, FBC, FREMP, GVRD.

Halliday, Jan and Gail Chehak. 2000. *Native Peoples of the Northwest - A Travelers Guide to Land, Art and Culture.* In cooperation with the affiliated tribes of NW Indians. Sasquatch Books. Seattle. 319pp.

Ham, Leonard. 1982. *Seasonality, shell midden layers and Coast Salish subsistence activities at the Crescent Beach site, DgRr1.* PhD thesis. Dept. of Anthropology, UBC, Vancouver. B.C.

Hancock, D. 2004. *The Bald Eagle of Alaska, B.C. and Washington.* Hancock House Publ. Surrey, B.C./Blaine WA.

Harris, Cole. 1994. *Voices of Disaster. Smallpox around the Strait of Georgia in 1782.* Ethnohistory 41: 591-626.

Harris, Cole, 1998. *Social Power & Cultural Change in Pre-colonial BC.* B.C. Studies. 115/116 pp 45-82.

Hastings, Margaret Lang. 1981. *Along the Way…. An account of pioneering White Rock and surrounding district in British Columbia.* City of White Rock. B.C.

Hastings, W.W. 1968. *Boundary Bay - another White Rock?* Enterprise: A Magazine for Credit Union members. Vol. 28.8. Sept. 1968. p6-9.

Hayes, Derek. 1999. *Historical Atlas of British Columbia and the Pacific Northwest.* Cavendish Books, Vancouver, B.C.

Hayes, Derek. 2001. *Historical Atlas of the North Pacific Ocean.* North Pacific Marine Science Organisation. Douglas & McIntyre. Vancouver. Toronto.

Hayes, Derek. 2005. *Historical Atlas of Vancouver and the Lower Fraser Valley.* Douglas & McIntyre, Vancouver.

Hebda, Richard Joseph. 1977. *The Palaeoecology of a raised bog and associated deltaic sedments of the Fraser River delta.* PhD thesis. University of British Columbia.

Hebda, R. 1991. *Burns Bog: vegetation and future.* Discovery. 20:1, 13-16.

Henderson, D. 2000. *Salvage Operation Angers Tribe, Concerns White House.* In *Archaeology,* A publication of the Archaeological Institute of America. Vol. 53. Number 3 May/June.

Hrutfiord, Jan. 2002. *Back in Time.* All Point Bulletin. March 2; pp 4-5.

Hutcherson, W. 1982. *Landing at Ladner.* Carlton Press. New York.

Hutcherson, E. 2002. *Looking back at a town called Ladner.* Trafford Publishing.

Jagpal, Sarjeet Singh. 1994. *Becoming Canadians. Pioneer Sikhs in their own words.* Harbour Publishing, Madeira Park, B.C. 166 pp.

Jeffcott, Percival, R. 1949. *Nooksack Tales and Trails. Historical Stories of Whatcom Co.* WA. Sedro-Woolley Courier Times. Ferndale. WA. 136 pp.

Jenness, Diamond. 1955. *The Faith of a Coast Salish Indian.* Anthropology in British Columbia. Memoir No.3. B.C. Provincial Museum. Victoria. 92pp.

Jones, Simon. 2002. *Changes & consequences of late-run sockeye salmon migration behaviour* In *The Changing Face of the Lower Fraser River Estuary* Symposium Proceedings. 18 April 2002. Fraser Basin Council. New Westminster. Fraser River Coalition, FBC, FREMP & GVRD. pp 43-44.

Jordan, Terry G. 1982. *Division of the Land.* In *This Remarkable Continent - An Atlas of US and Canadian Society and Cultures.* (Ed.) John F. Rooney. Published for the Society for the North American Cultural Survey. Texas A& M University Press. pp 54 -69.

Justice, Clive, L. 2000. *Mr. Menzies Garden Legacy, Plant Collecting on the Northwest Coast.* Cavendish Books, Vancouver, B.C.

Kelsey, Harry. 1998. *Sir Francis Drake: The Queen's Pirate.* Yale University Press.

Kendrick, John. Undated. *Jose Maria Narvaez.(1765- 1840). Biographical notes for Heritage Day.* White Rock Historical Society. White Rock Museum and Archives collection.

Kendrick, John. 1999. *Alejandro Malaspina: Portrait of a Visionary.* McGill & Queen's University Press, Montreal, Kingston, London and Ithaca.

Kidd, Thomas. 1973. *A History of Richmond Municipality.* Richmond.

Klohn Leonoff Ltd. 1992. *Delta Agricultural Study.* B.C. Ministry of Agriculture, Fisheries and Food, Agriculture Canada, B.C. Agricultural Land Commission, Delta Farmers Institute and the Corporation of Delta,

Knudsen, E. Eric. 2000. *Sustainable Fisheries Management: Pacific Salmon.* CRC Press.

Koroscil, Paul M. 2000. *British Columbia: Settlement History.* Simon Fraser University, British Columbia.

Kuhnlein, Harriet, V. and Nancy J. Turner. 1991. *Traditional plant foods of Canadian Indigenous peoples: Nutrition, Botany and Use, Food and Nutrition.* In History and Anthropology Vol.8. Gordon & Breach Science Publ. Philadelphia.

Ladner, Leon J. 1972. *The Ladners of Ladner by covered wagon to the welfare state.* Mitchell Press. Vancouver

Ladner, Thomas E. 1979. *Above the Sand Heads. A Vivid Account of Life on the Delta of the Fraser River. 1868-1900.* Edna G. Ladner. Burnaby. B.C.

Lamb, W. Kaye, Editor. 1984. *The Voyage of George Vancouver 1791-1795.* Volume 11. The Hakluyt Society, London.

Leach, Barry. 1982. *Waterfowl on a Pacific Estuary. B.C.* Provincial Museum. Special Publication #5. Victoria. B.C.

Levanthes, Louise. 1994. *When China ruled the Seas.* Simon & Shuster. NY.

Lichatowich, Jim. 2001. *Salmon without Rivers.* Island Press, Washington D.C.

Longcore, Travis and Catherine Rich. 2004. *Ecological Light Pollution.* Front Ecol Environ 2 (4): 191-198.

Lord, John, Keast. 1866. *The Naturalist in Vancouver Island and British Columbia.* Two volumes. Richard Bentley, London.

Lummi Nation. 2004. www.lummi-nsn.org

Luternauer J.L., D.C. Mosher, J.J. Clague, R.J. Atkins. 1998. *Sedimentary Environments of the Fraser Delta.* In Clague, J.J, J.L. Luternauer and D.C. Mosher. (Eds.) 1998. *Geology and Natural Hazards of the Fraser River Delta, British Columbia.* Geol. Survey of Can.Bull. 525, Nat. Res. Canada. pp 27 -39.

Luxton, Donald and Associates. 1998. *Delta's Rural Heritage: An Inventory.* Corporation of Delta, B.C. 122 pp.

MacDonald, Jake. 2004. *More than Zero.* Western Living Summer 2004. p 39-43.

MacLachlan, Morag. (Ed.) 1998. *The Fort Langley Journals. 1827-1830.* UBC Press. Vancouver, B.C. 280 pp.

Manley, W.F. 2002. *Postglacial Flooding of the Bering Land Bridge: A Geospatial Animation.* INSTAAR, Uni. of Colorado. http://instaar.colorado.edu/QGISL/bering_land_bridge/

Mathewes, Rolf W. 1991. *Connections between palaeoenvironments and palaeoethnobotany in coastal British Columbia.* In *New light on early farming - recent developments in palaeoethnobotany.* Ch.29. Edinburgh University Press, Scotland.

Mathewes R.W. & Heusser. 1981. *A 12,000 year palynological record of temperature and precipitation trends in SW B.C.* Can. J. Bot. 59 pp 707 -710.

Mathewes, Rolf W. and John J. Clague. 1994. *Detection of large prehistoric earthquakes in the Pacific Northwest by microfossil analysis.* Science. Vol. 264. 29 April. pp 688 -691.

Matson, R.G. 1976. *The Glenrose Cannery Site.* Mercury Series. Archeological Survey of Canada. Paper number 52. National Museum of Man. Ottawa.

Matson R.G, Deanna Ludowicz and William Boyd. 1980. *Excavations at Beach Grove in 1980.* Report to Heritage Conservation Branch. Victoria. B.C.

Matson R.G, and Gary Coupland.1995. *The Prehistory of the Northwest Coast.* Academic Press. California. 364 pp.

Matson, R.G, Gary Coupland and Quentin Mackie (Eds.). 2003. *Emerging from the Mist. Studies in Northwest Coast Culture History.* UBC Press. Vancouver.

Matson, R.G. 1996. *The Old Cordilleran Component at the Glenrose Cannery Site* In Carlson, Roy & L. Dalla Bona (Eds.) *Early Human Occupation in B.C.* UBC Press pp 111-122.

Maud, Ralph. 1978. *The Salish People, The local contribution of Charles Hill-Tout,* Volume 111: *The Mainland Halkomelem.* Talon Books, Vancouver. 165 pp

McIntosh, K.A. and I. Robertson. 1999. *Status of Black Bears. In Summary Report: Technical Review Meetings in Support of the Burns Bog Ecosystem Review.* Sims, Richard, Jeff Matheson and Sergei Yazvenko. 2000. Environmental Assessment Office. Victoria.

McKinnon, Stan, 1996. *History of the Nicomekl and Serpentine Rivers* In Proceeding of Serpentine River Stakeholders Workshop, April 4/5 1997. Surrey, B.C..

McLay, Eric, Kelly Bannister, Lea Joe, Brian Thom, George Nicholas. 2004. *Respecting the Ancestors.* Report of the Hul'qumi'num Heritage Law Case Study. Hul'qumi'num Treaty Group.

Menzies, Archibald. 1792. Letters to Sir Joseph Banks. Series 16. www.sl.nsw.gov. au/banks/series

Merk, Frederick. (Ed.) 1968. *Fur Trade and Empire. George Simpson's Journal. 1824 -25.* Rev. Ed. The Belknap Press of Harvard Uni. Cambridge, Mass.

Merilees, William. 1985. *Humpbacks in Our Strait.* Waters. Vol. 8. Journal of Vancouver Aquarium. Vancouver, B.C.

Mitchell, Leslie S. 1996. *The Archaeology of the Dead at Boundary Bay, B.C. A history and critical analysis.* SFU graduate thesis: www.sfu.ca/archaeology/ dept/gradstu/these/masters/mitchell.htm.

Mozino, Jose Mariano Mozino Suarez de Figueroa. 1970. *Noticias de Nutka: an account of Nootka Sound in 1792.* Transl., Ed. Iris Higbie Wilson. McClelland & Steweart. Toronto. 142 pp.

Murray, Anne & Mary Taitt. (Eds.) 1992. *Ours to Preserve.* Boundary Bay Biosphere Reserve. Boundary Bay Conservation Committee. 49pp.

Murray, Anne. 2004. *Wildlife Inventories in the Boundary Bay Watershed.* In

Proceedings of the 2003 Georgia Basin/ Puget Sound Research Conference, Puget Sound Action Team/Env. Canada.

MWALP. 2002. *Environmental Trends in British Columbia 2002.* Min. Water, Land and Air Protection, Victoria. B.C.

Naish, John M. 1992. *Health of Vancouver and his men.* In Chapter 11. Proceedings of the Vancouver Conference on Exploration and Discovery. Simon Fraser University History Department.

Naish, John M. 1996. *The Interwoven Lives of George Vancouver, Archibald Menzies, Joseph Whidbey and Peter Puget. Exploring the Pacific Northwest Coast.* Canadian Studies Vol. 17. The Edwin Mellen Press. Lewiston, Lampeter.

Nelson, Stuart. 2006. Revised edition. *Fraser River Sockeye Salmon Benchmark Study. A Business Perspective on Fraser Sockeye.* Nelson Bros. Fisheries Ltd. for AAC, CAFI Program. www.ats.agr.cg.ca/ Can/4228_e.htm - accessed 2007.

Newcombe, C.F. (Ed.) 1923. *Menzies' Journal of Vancouver's Voyage April - October 1792.* Archives of B.C. Memoir, Number 5, Victoria, B.C. Canada

Norecol, Dames and Moore. 1994. *Our Legacy for Future Generations.* Vol. 1 and 2. Delta Rural Land Use Study, prepared for The Corporation of Delta.

North, M.E.A. M.W. Dunn & J.M. Teversham. 1979. *Vegetation of the Southwestern Fraser Lowland 1858-1880.* Lands Directorate, Vancouver

Page, Nick. 1999. *Intervention vs. Non-intervention in Restoration Planning: Implications for Beach Grove Lagoon.* Coast River Environmental Services Ltd. Vancouver. In Egan 1999.

Parsons, Marlene R. 1981 *Fraser River Estuary - Heritage Resource Inventory.* Heritage Conservation Branch. B.C. Ministry. Victoria. B.C.

Payette, Krista.1997. *Urban Salmon Habitat Program, Stewardship Advisory Committee report.* In Proceedings of Serpentine River Stakeholders Workshop. Surrey, B.C. pp 23-27.

Pepin, Maureen L. 1998. *Roads and other Place Names in Langley, B.C.* Langley Centennial Mus. & National Exhibition Centre. Township of Langley. 103 pp.

Philips, Terrence. 2003. *Harvesting the Fraser ~ A History of Early Delta.* Second Edition. Delta Museum & Archives.

Pielou E.C. 1991. *After the Ice Age - The Return of Life to Glaciated North America.* University of Chicago Press. Chicago and London.

Pitcher, Tony J. & Rattana (Ying) Chuenpagdee. (Eds.) 1994. *Bycatches in fisheries & their impact on the ecosystem.* Workshop proceedings. UBC Fisheries Centre. Accessed from www.fisheries.ubc. ca/publications. April 2004.

Pojar, Jim and Andy MacKinnon. (Eds.) 1994. *Plants of Coastal British Columbia, including Washington, Oregon and Alaska.* B.C. Forest Service, Research Program. Lone Pine Publishing. Vancouver, B.C., Edmonton, Alberta and Redmond, Washington. 526 pp.

Price, L. 1997. *Serpentine Enhancement Society report.* In Proceedings of Serpentine River Stakeholders' Workshop. Surrey, B.C.

Quayle, D.B. 1988. *Pacific Oyster Culture in British Columbia.* Can.Bull.Fisheries and Aquatic Sciences 218. DFO Canada.

Rogers, G.C. 1998. *Earthquakes* In Clague et al. 1998.

Scales, Peter. 1997. *Langley Environmental Partners Society report.* In. Proceedings of the Serpentine River Stakeholders' workshop. Surrey, B.C. pp 64 -69.

Schreier, Hans, K.J. Hall, S.J. Brown, B. Wernick, C. Berka, W. Belzer and K. Pettit. 1998. *Agriculture: an important non-point source of pollution.* In Gray and Tuominen, 1998.

Schubert, N.D. 1982. *A Biophysical Survey of 30 Lower Fraser Valley Streams.* Report of Fisheries and Aquatic Sciences. DFO. New Westminster.

Scouler, John. 1905. *Journal of a Voyage to NW America.* Quarterly of the Oregon Historical Society, Vol. 6.

Sewid-Smith, Daisy. 1999. *Incorporating Traditional Knowledge, Interests and Values of Aboriginal Peoples into Restoration of Natural Systems Projects.* In Egan 1999. SHS. 1995. *Looking back at Surrey.* Surrey Historical Society, Vol. 1. 92pp.

Smith, Marian. 1950. *The Nooksack, the Chilliwack and the Middle Fraser.* Pacific Northwest Quarterly 41(4):330-341.

Smith, Risa and Jenny Fraser. 2002. *Indicators of Climate Change for British Columbia 2002.* B.C. Ministry of Water, Land and Air Protection, Victoria. B.C.

Sobel, Dava. 1996. *Longitude.* Fourth Estate. London. pp 184.

Stein, Julie K. 1992. *Deciphering a Shell Midden.* Academic Press, Seattle, WA.

Stewart, Hilary. 1977. *Indian Fishing. Early Methods on the Northwest Coast.* J.J. Douglas, Ltd. Vancouver.

Stewart, Hilary. 1984. *Cedar: Tree of Life to the NW Indians.* Douglas & McIntyre, Vancouver, Seattle.

Stewart, Hilary. 1996. *Stone, bone, antler and shell: Artifacts of the Northwest Coast.* Douglas & McIntyre, Vancouver. University of Washington Press, Seattle.

Stewart, Margaret. 1959. *How They Lived.* Surrey Museum Press.

Stilson, M.Leland. 2003. *A Field Guide to Washington State Archaeology.* Dept. of Archaeology and Historic Preservation. www.dahp.wa.gov

Suttles, Wayne. 1951. *Economic Life of the Coast Salish of Haro and Rosario Straits.* PhD.Thesis. University of Washington.

Suttles, W. 1955. *Katzie Ethnographic Notes.* Anthropology in B.C. Memoir No. 2. B.C. Prov. Mus. Victoria. 31pp.

Suttles, Wayne. 1987. *Coast Salish Essays.* Talonbooks. University of Washington Press, Seattle and London, 320 pp.

Suttles, Wayne (Ed.). 1990. *Handbook of North American Indians, Northwest Coast. Vol. 7.* Smithsonian Institute. Washington. 796 pp.

Swain, L.G. and G.B. Holms. 1988. *Fraser - Delta Area, Boundary Bay and its tributaries, water quality assessment and objectives.* Water Resource Report. Water Man. Branch, B.C. MOE.

Szychter, Gwen. 1996. *Ladner's Landing of Yesteryear.* 1997. *Beyond Ladner's Landing.* 1998. *Across the Bridge from Ladner's Landing.* 1999. *Port Guichon. Forgotten neighbour of Ladner's Landing.* 2007. *Chewassin, Tsawwassen, or Chiltinm ~ The land facing the sea.* All published by Gwen Szychter.

Tanner, Adrienne. 2002. Globe and Mail Newspaper, Sat. July 20.

Taylor, Gordon D. 1958. *Delta's Century of Progress.* Delta Cent. Com. Delta.

Thom, Brian. 1997. *The Whalen Farm Site.* www.home.istar.ca/~bthom/whalen-new.htm

Thomson, R.E. 1981. *Oceanography of the British Columbia coast.* Fisheries and Oceans Canada, Ottawa.

Thor, Jonas. 2002. *Icelanders in North America. The First Settlers.* University of Manitoba Press. Winnipeg. pp 306.

Thrift, Henry T. 1929. *Fortified Indian Encampment here when settlers first came.* Surrey Gazette. October 17. Reprinted In *Reminiscences of Henry T. Thrift.* Undated document. Surrey Museum Press.

Thrift, Henry T. 1931. *Reminiscences of events of an historical character occurring in the experience of Henry T. Thrift of White Rock, B.C.* Surrey Museum Press.

Thrower, Norman J.W. (Ed.) 1984. *Sir Francis Drake and the Famous Voyage. 1577 - 1580.* University of California, Los Angeles Press.

Todd, Kim. 2002. *Tinkering with Eden. A Natural History of Exotic Species in America.* W.W. Norton & Company. New York. London. pp 302.

Treleaven, G. Fern. 1992. *The Surrey Story.* Surrey Historical Society

Tsawwassen First Nation. 2004. *Land Facing the Sea, A Fact Book.* 26 pp.

Turner, Nancy and M. Bell. 1971. *Ethnobotany of Coast Salish Indians of Vancouver Island.* Economic Botany 25:63-104.

Turner, Nancy J. 2001. *Plant Technology of First Peoples in British Columbia*. Royal British Columbia Museum Handbook; UBC Press, Vancouver.

Vaughan, Thomas. 1977. *Voyages of Enlightenment, Malaspina on the Northwest Coast, 1791/92*. With E.A.P. Crownhart-Vaughan, M.P. de Iglesias. Oregon Historical Society. Portland.

Wagner, Henry R. 1933. *Spanish Explorations in the Strait of Juan de Fuca*. Santa Ana Fine Arts Press, USA. Reprinted 1971. HMS Press. NY. NY.

Waite, Donald E. 1977. *The Langley Story*. Don Waite Publishing. Maple Ridge, B.C. 280 pp.

Wallace, Scott and Johane Dalsgaard. 1998. *Back to the Future: a comparison of ecosystem structure of the Strait of Georgia 100 years ago and present day*. UBC. Vancouver, B.C. http:// mehp. vetmed.ucdavis.edu/pdfs/MPApdfs/Wallace Dalsgaard1998.pdf accessed 21/08/03

Ward, Peggy. 1980. *Explore the Fraser Estuary*. Environment Canada

Webb, Robert Lloyd, 1988. *On the Northwest, Commercial Whaling in the Pacific Northwest 1790-1967*, UBC Press, Vancouver, B.C. 425 pp.

Welsh, Don. 2004. *Steatite Plaque and a Carved Tooth from Semiahmoo Spit*. The Midden. Vol.36. No. 1&2. pp 14-18

Whatcom County. 1992. *Nomination form for National Register of Historic Places* submitted by Lummi Cultural Department, accepted by Washington State Advisory Council. In *Draft Environmental Impact Statement*. The Resort at Lily Point. Whatcom Co. Dept. of Public Works. Bellingham. WA.

Whiteside, Richard V. 1974. *The Surrey Pioneers*. Printed by Evergreen Press.

Williams, Glyndwr. (Ed.) 1973. *Simpson's Letters to London. 1841-42*. Hudson's Bay Record Soc. Vol. XXIX.

Williams, Glyn. 2002. *Voyages of Delusion. The Search for the Northwest Passage in the Age of Reason*. Harper Collins, London. 467pp.

Williams Judith. 2006. *Clam Gardens*. Transmontanus 15. New Star Books. Vancouver, B.C.

Winter, George R. 1968. In *Lower Fraser Valley, Evolution of a Cultural Landscape*. (Ed.) Alfred H. Siemens. Dept of Geog. UBC Tantalus Research Ltd.

Work, John. 1824. *John Work's Journal. Nov. - Dec. 1824*. Washington Historical Quarterly 111. (1912) accessed on line at http://www.xmission.com/~drudy/mtman/html/jwork/work02.html

Work, John. 1945. *The Journal of John Work*, Jan.- Oct. 1835. Archives of British Columbia. Memoir No.X. Victoria. B.C.

Wright, J.V. 1995. *A History of the Native People of Canada. Volume 1 (10,000 - 1,000 B.C.)* Mercury Series, Archeological Survey of Canada, Paper 152; Can. Mus. of Civilization. 564 pp.

Yates, Steve. 1992. *Orcas, Eagles and Kings, The Natural History of Puget Sound and Georgia Strait*. Primavera Press, Florida and Washington, 236 pp

Yesaki, Mitsuo. 2001. *Charcoal Production for the Salmon Canning Industry in B.C.* B.C. Historical News. Vol. 34. No. 2. www.goldseal.com

Yesaki, Mitsuo, Harold Steves and Kathy Steves. 2005. *Steveston Cannery Row*. Peninsula Publishing. BC

PHOTO CREDITS

The photographs and maps in this book, unless listed below, are the work of David Blevins; please see the website **www.blevinsphoto.com** for more information and to order David's prints. All archival photos have been given a sepia colour tone but are otherwise unaltered.

Page i. View of Mud Bay ca.1965; from the collection of Surrey Archives: 82.001-12.

Page iv-v. Fishing Boats on the Fraser; courtesy of Delta Museum & Archives: 1983-214-2.

Page 8. Chipped basalt core from Tsawwassen site. DgRs-2:73691. photo courtesy of the UBC Laboratory of Archaeology, Vancouver, B.C. with permission of Semiahmoo First Nation and Tsawwassen First Nation.

Page 10. Anthropomorphic carving on antler from Glenrose Cannery. DgRr-6:2687. Photo courtesy of the UBC Laboratory of Archaeology with permission of Semiahmoo First Nation, Tsawwassen First Nation, Stò:lo Nation, Stò:lo Tribal Council and Musqueam First Nation.

Page 16. Petroglyph, Kwomais Point: photograph courtesy of Don Welsh.

Page 22. Whalen Farm excavation, view to east trench, 1949. Photo courtesy of the UBC Laboratory of Archaeology, DfRs1 Whalen Farm. Photograph by Charles E. Borden.

Page 25. Unilateral barbed antler point from Beach Grove, DgRs-1:617, side 2. Photo: UBC Laboratory of Archaeology with permission of Semiahmoo First Nation and Tsawwassen First Nation.

Page 26. Coast Salish clam basket, 978.049.453; photograph White Rock Museum & Archives. Coast Salish Woman, Caw Wacham, 1847, sketched by Paul Kane (1810-1871). Reproduced from Wikimedia Commons.

Page 29. Carved stone head from Beach Grove, with permission of Semiahmoo First Nation and Tsawwassen First Nation; Stanley Triggs photo; Vancouver Public Library 85724A 1962 CD 256.

Page 32. Hand maul, Tsawwassen, DgRs-2:43. Photo: UBC Laboratory of Archaeology with permission of Semiahmoo First Nation and Tsawwassen First Nation.

Page 33. Coast Salish wool dogs; drawing by Cameron J. Pye from a forensic reconstruction. Courtesy Susan Crockford, Pacific Identifications Inc. Victoria, B.C.

Page 34. Net anchor/sinker stone, perforated and ground granitic cobble. DgRs-2:74267, side 2. Photo: UBC Laboratory of Archaeology with permission of Semiahmoo First Nation and Tsawwassen First Nation.

Page 36. Chipped basalt projectile point from St. Mungo site, DgRr-2:585. Photo : UBC Laboratory of Archaeology, with permission of Semiahmoo First Nation, Tsawwassen First Nation, Stò:lo Nation, Stò:lo Tribal Council, Musqueam First Nation and the Vancouver Museum.

Page 38. Antler haft incised with zoomorphic design, Tsawwassen. DgRs-2: 73946. Photo courtesy of the UBC Lab. of Archaeology with permission from Semiahmoo FN and Tsawwassen FN.

Page 41. Ground slate, corner notched biface. DgRr-1:8199, Crescent Beach. Photo: UBC Laboratory of Archaeology with permission from Semiahmoo First Nation and Tsawwassen First Nation.

Page 52. "Goodfellow's" fishing camp, ca.1895. Courtesy Delta Museum and Archives 1970-1-120.

Page 53. Reef net drawing by Hilary Stewart from Indian Fishing 1977, page 94; reproduced with permission of the artist.

Page 57. Fraser River canoes, 1887, photograph by William Notman Studio. Vancouver Public Library 30829 CD 151

Page 60. "*A Chart shewing part of the coast of Northwest America.*" Plate 5 of *Atlas for a Voyage of Discovery to the North Pacific Ocean and Around the World.* George Vancouver 1798. Public Record Office CI 700 B.C. 1.

Page 64. Title page of the 1760 edition of *Systema Naturae* by Carl Linnaeus, Wikimedia Commons.

Page 67. "*Carta que comprehende los interiers y veril de la costa desde los 48 de Latitud N..*" Juan Carrasco, 1791. Museo Naval, Madrid.

Page 68. *Portrait of a chief of Puerto del Descanso* (Descano Bay, Gabriola). 1792 watercolour [José Cardero]. Museo de America, Madrid. Bausa no. 55.

Page 70. *Native canoes approaching the Sutil and Mexicana.* 1792 watercolour by Jose Cardero. Museo Naval, Madrid. 1792

Page 78. The flood down Delta Street, Ladner, January 1895. Courtesy of Delta Museum & Archives 1991-22-18.

Page 82. Fort Langley, 1862. Image C-09126, courtesy of Royal B.C. Museum, B.C. Archives.

Page 84. Richard Moody's 1858 sketch map of the Fraser delta. Vancouver Centennial Bibliography.

Page 86. Fort Langley 1862. Image A-04313 courtesy of Royal B.C. Museum, B.C. Archives.

Page 87. Cariboo gold miners' cabin, 1904. Image A-06021 Royal B.C. Museum, B.C. Archives.

Page 88. White Rock, 1917. Image C-02779; Royal B.C. Museum, B.C. Archives.

Page 90. Steveston, 1890. Bailey Bros. photo. City of Vancouver Archives, Out P671.

Page 91. Mrs M.D. Beveridge and children clearing land at 182 St Surrey, 1918. From the collection of Surrey Archives, 1.1.54

Page 93. Point Roberts' lighthouse, ca.1912-1915. Courtesy of Point Roberts Historical Society

Page 96. Burning of Chinatown, Ladner, July 6 1929. Courtesy of Delta Museum & Archives 1981-36-93.

Page 97. After the Steveston fire, 1918. City of Richmond Archive # 1977 11 2

Page 101. Based on a map by North, Dunn & Teversham 1979, with permission.

Page 102. Photograph: Anne Murray.

Page 103. Break in the dike at Ladner, 1899. Image E-03692, courtesy of Royal B.C. Museum, B.C. Archives.

Page 107. Dredge, 'Beaver No. 2' ca.1908. Richmond City Archives # 1978 14 5

Page 111. Giant stump at Point Roberts; photograph courtesy Point Roberts Historical Society.

Page 113. Herbert Gilley logging crew; from the collection of Surrey Archives 121.90

Page 118. Fishing on the Fraser; courtesy of Delta Museum & Archives 1983 -214-4

Page 122. A.P.A Cannery in Point Roberts; courtesy of Delta Museum & Archives 1970 - 1- 383

Page 123. Fishing scene on the Fraser; courtesy of Delta Museum & Archives 1970-1-268

Page 125. Anglo-B.C. Packing Co. receiving salmon, Garry Point Cannery Fraser River 189-. Bailey Bros. photo, City of Vancouver Archives: Indust P4

Page 126. Fish trap off Point Roberts; courtesy of Delta Museum & Archives 1983-214-13.

Page 130. Fraser River sturgeon, 1912; Delta Museum & Archives 1970-1-580

Page 136. B.C. Packers oyster plant, 112th St. Delta. Photo John Christopherson; courtesy Delta Museum & Archives 1994-006-12

Page 137. B.C. Packers oyster plant, Boundary Bay. Photo John Christopherson. courtesy Delta Museum & Archives 1994 - 6 -10

Page 138. Crabber, 1965; from the collection of Surrey Archives 82.001/16

Page 139. True Oliver, Boundary Bay; courtesy Delta Museum & Archives 1983-211-3

Page 145. Haying in Delta Image A-03954; Royal B.C. Museum, B.C. Archives.

Page 146. Plowing scene, Ladner ca.190-? Courtesy of Delta Museum & Archives 1980-52-55.

Page 147. Surrey landscape, ca.194-. Image I- 27865, courtesy Royal B.C. Museum, B.C. Archives.

Page 149. T.E. Ladner and family, Beach Grove, ca.1903; courtesy of Delta Museum & Archives 1983-263-19(CP)

Page 150. Semiahmoo Park, ca.1950s; from the collection of Surrey Archives #150.01

Pages 151 & 152. Construction of Highway 99, Boundary Bay, 1960; courtesy of Delta Museum & Archives 2003-2-5 and 2003-2-2

Page 153. Peat harvest, Burns Bog, 1947. Image I-27875, courtesy of Royal B.C. Museum, B.C. Archives.

Page 170. Wrecked autos at high tide, 1965; from collection of Surrey Archives 82.001/40

Page 196. Gulf Road, Point Roberts; photograph courtesy Point Roberts Historical Society.

Page 230. Photograph of Anne Murray by Sheena Studios, Surrey; www.sheena.ca

INDEX

Index of selected topics; illustrations shown in **bold**.

Adams River *120, 127*

Agricultural Land Reserve *143, 154–**157**, 172*

Airfields *166*

Air quality *160, 171,* **179**, *180*

Albion Box *132*

Alien plants (invasives) *95, 117, 176, 177*

Alien species *95, 136, 163, 165*
- bullfrog, American *176*
- pigeon, rock *174, 176,* **177**
- rats *174, 176*
- starling, European *159, 174, 176*

American Border Peak **4**

Amphibians *174, 175, 176*

Annacis Island *36, 178*

Annieville *123*

Anthropomorphic figure **10**

Aquifers *178*

Archaeology *8–10,* **22**–*41, 185, 190*

Art and Culture *37–***39**

B.C. Heritage Cons. Act *40*

Back-up lands *165*

Baker, Lt. Joseph *72, 73*

Ballast water *165, 176*

Beach Grove *6, 17,* **25**, *26,* **29**, *31,* **102**, *106, 108,* **139**, *140, 149, 150*

Bee farming (Apiculture) *146,* **154**

Bellingham *87, 191*

Bellingham Bay *66, 70, 72, 86, 94*

Benson, Henry *103, 178*

Bering, Vitus *61, 64, 65*

Bering Strait *9, 10, 61*

Berry, John *156*

Beveridge family **91**

Biodiversity *161, 163, 171, 174, 175, 182*

Birch Bay (Tsau'wuch) *14, 30, 34-36, 49, 50, 54-55, 66, 70, 72, 75, 77, 93, 94, 114, 147*

Birch Point *20, 36, 54*

Birds *31, 36, 37, 85, 99, 106, 107, 128, 132, 139, 143, 158-160, 174-177*
- bluebird, western *107, 158*
- bittern, American **175**
- condor, Californian *17, 85*
- crows *16, 24, 31, 36, 95, 174*
- dunlin **17**, *106*
- eagle, bald *16, 17, 31, 120, 126, 141, 158,* **159**, *175, 191*
- falcons *106, 181*
- geese *36, 46, 54, 106, 139, 143, 159, 169,* **173**
- grouse, ruffed *99, 106*
- gulls *16, 24, 31, 36, 69, 159, 166*
- hawks *95, 143, 158, 160*
- heron, great blue *36, 107, 132, 143, 160*
- junco, dark-eyed **85**
- mallard **31**, *107, 139*
- merganser, hooded **171**
- nighthawk, common *95, 107, 158*
- owls *31, 95, 143, 158, 160, 181*
- pigeon, rock *174, 176,* **177**
- seabirds *128, 132*
- shorebirds **17**, *104, 106, 143, 158, 165, 166, 172, 175*
- swans *36, 106, 139, 143, 168*
- tern, Caspian *166, 177*
- thrush, Swainson's *37*
- wigeon, American **58**, *139, 159*

Blackie Spit *6, 91, 136, 158, 170*

Blaine *6, 34, 35, 55, 82, 86, 88, 92–94, 106,* **108**, *113, 123, 127, 137, 138, 165, 168, 178, 179,* **187**, *195*

Blaine Harbor *96, 108*

Bodega y Quadra, Don *62, 65*

Boldt Decision *133*

Booth family *91, 145*

Borden, Carl *22, 34*

Borden number *40*

Bose, Harry *112*

Boundary Bay exploration *62, 66–77, 80–82*

Boundary Bay formation *20–21*

Boundary Bay middens *29–36*

Boundary Bay Reg. Park *29, 102, 106, 173, 193*

Boundary Bay views *11, 12,* **20-21**, **42**, **52**, **58**, **149**, **164**, **181**

Boundary Commission *52, 83, 85*

Boundary Marker One **83**, *187*

Britannia Heritage Shipyard **184**, **186**

Broughton, Lt. William *71–72*

Brownsville *94, 194*

Brunswick Point *58, 103, 106*

Burgess, Jim *129*

Burney, James *62*

Burns Bog *viii-ix, 20-21, 97, 110, 115, 116, 151,* **153**, *164, 166,* **167**, *173, 175, 180*

Cain Creek **108**

California Creek *20, 54, 92, 106, 108, 110, 137*

Campbell, Archibald *83, 85*

Campbell Valley *154, 194*

Canada Land Inventory *154*

Canadian Pacific Rwy. *94, 147*

Canneries *36, 54, 93, 119, 120,*

123–126, 164, 195
- Alaska Packers Association
52, 122, 126, 187, 196
- George and Barker 196
- Garry Point 125
- Gulf of Georgia 185, 186
- Wadham's 126
Canoe Pass 87, 103, 106, 132
Captain John of Soowhalie
130
Cardero, Jose 68, 69, 70
Caw Wacham 26
c'a:yum 33
Cedar See Trees
Centennial Beach 170
Chanique 51, 110
Chantrell Creek 110
Chapea, Chief 50
Charles, Pierre 110
Charles Culture 28, 36
Cheam 57, 134
Chelhtenem 14-15, 35, 52-54,
74-75, 122
Chilliwack 85
Chilukthan Slough 77, 91, 95,
103, 151, 192
Chinatowns 96, 124
Chinook Jargon 56, 86
Chow Wing 112
Christmas Bird Count 169
Chung Mor Ping (Chung
Chuck) 148
Clam beds/gardens 15, 45,
137
Clam digging 46, 52, 87, 137,
147
Clams 10, 23, 29, 32, 33, 37, 54,
58, 136–138, 176
Climate 6, 16
- change 3, 4, 18, 134, 143,
161, 163, 177, 180
Cloverdale 38, 92, 97, 104, 113,
117, 147
Coast Meridian 89, 94
Coast Salish 1, 3, 11–19,
26–52, 53-57, 58, 59, 61,
68, 70, 75, 77, 80–82,
85–88, 96, 109–110, 115,
116, 120–122, 124, 125,
130, 131, 137, 139, 185,
188–191
Colebrook 112

Colebrook Dyking Dist. 104
Columbia River 80, 81, 125,
130, 135
Community growth 149–152
Conde de Revillagigedo 68
Cook, Captain James 62, 64,
65, 72, 75
Copper 25, 44, 64, 68
Cowichan First Nation 16,
43, 50, 56, 59, 80, 121,
192
Crabbing 138, 147
Crabs, Dungeness 54, 127,
138, 147
Crescent Beach 6, 29, 38, 41,
55, 86, 110, 113, 120,
138, 147, 192
Crescent Slough 90, 148
Crops 102, 117, 143–145, 148,
151, 154–155, 159, 160
- berries 143, 153, 155, 158
- cabbage 146
- corn 144
- hay 51, 102, 110, 145
- potatoes 102, 142–144,
147–148, 152, 155
Cultus Lake 15, 120, 127
c'usna7um 36
Custer Prairie 92, 196

Dakota Creek 20, 54, 92, 106,
108, 137
Deas Island 90, 151, 152, 170,
193
De Hezeta, Bruno 62
Delta 7, 10, 13, 20, 28, 31, 34,
36, 58, 89, 91, 94, 97,
101, 103, 115, 123,
143–146, 150–156,
160, 165–167, 171–173,
178, 182, 185, 187, 188,
191–193
Delta Farmland & Wildlife
Trust 159-161, 173
Delta Nature Reserve 115
Derby Reach 84
Development proposals
164–167, 172–173
Diseases 134, 161, 177
- scurvy 65
Dogs, Coast Salish 10, 16, 27,
31, 33, 34, 46

Dominion Fishery Act 122
Douglas, David 81
Douglas, Governor James 63,
73, 87
Douglas Treaty 56
Drake, Sir Francis 61, 62, 65
Drayton Harbor 6, 34, 43,
54, 67, 87, 92, 108, 113,
137, 146, 165, 179, 187,
195, 196
Dyking, ditching & drainage
94, 100, 102, 103–107,
108, 112, 145–147, 158,
170, 178

East Delta 102, 145, 146, 178
East Wenatchee 9
Eburne Mound 36, 59
Ecological impacts 95,
99–117, 158–159
Eelgrass 128, 166
Elgin 91, 94, 112, 113, 147
Elgin Heritage Park 105, 193
Eliza, Don Francisco 66, 67
English Bluff 13, 32, 33
Epidemics 48–51
Ethnographers 13, 17
Expeditions 61–77, 80–82
Explorers, European 27, 37,
39, 45, 99
Explorers, Russian 39, 61,
64, 66

Fannin, John 96, 168
Farming 143–148, 155–157,
158-161 See also
Greenhouses
- Aboriginal 45, 46, 54, 116,
144
- cattle 100, 109, 143–145, 148
- dairy 102, 146–148, 158
- organic 157–158
Finn Slough 106
Fires 7, 82, 96–97, 150
First Ancestors 13–17
First Salmon Ceremony 19,
53
Fish 16, 31, 32, 69, 107, 108,
118, 119–141, 127,
131, 132, 133, 174. See
also Salmon

- eulachon *10, 46, 85, 119, 122, 128, 132*
- flounder, starry *3, 16, 127, 132*
- halibut, Pacific *119, 123, 127, 132, 133, 140*
- herring, Pacific *31, 32, 119, 125, 127–128, 186*
- lingcod *119, 126, 140*
- rockfish *132, 140*
- shark, basking *140*
- smelt, surf *119, 127*
- sturgeon, white *10, 16, 47, 52, 57, 58, 75, 77, 119, 121, 122, 128, **129**–130, 132, 133, 174*
Fisheries
- Aboriginal *119, 133, 134*
- Aboriginal Fisheries Strategy *133*
- commercial *119, 122–134*
Fishing *123, **125**, **129**, **131**–133*
- fish traps *125–**126***
- gill nets *122, 127, 131–133*
- seining *131, 133*
- weirs *43, 122, 125*
- *See also* Reef Nets
Floodplain *6, 20, 146, 160, 165*
Floods *102, 103, 105*
Forests *7, 18-19, 45–46, **73**, 80, **109**, 112. See also* Trees
Forestry *96, 110, **111**–**113**, 114, 147, 170*
Fort Langley *50, 56, 81, **82**–**86**, 87, 109, 122, 130, 185–186, 189*
Fraser, Simon *80*
Fraser Canyon *6, 10, 57, 87, 114, 120, 125*
Fraser delta surveys *101*
Fraser salmon runs ***118**, 120–123, 125, 134. See also* Salmon
Fraser Valley *3, 4, 86, 107, 114, 117, 130, 139, 143, 146, 148, 157, 158, 161, 179*
Fur trade *45, 61, 64, 100*

Gabriola Island *71*
Galiano, Dionisio *63, 64, 69–73, 76, 77*
Garry Point Park *186*
Gilley, Herbert ***113***

Glenrose Cannery *7, **10**, 16, 26, 28, 36, 37, 120*
Gold rush *87, 88, 103, 145*
Golf courses *105, 108, 143, 157, 160, 172*
Goodfellow's Camp ***52***
Grauer Beach *150*
Great Fraser Midden *36*
Great Northern Railway *94, 149*
Greenbelt Lands *171*
Greenhouses *155, **160**, 181*
Green Timbers Urban Forest *94, 114, 173*
Green Zone, Metro Vancouver *172*
Groundwater *21, 146, 178*
Gulf Islands *43, 46, 54, 56, 58, 83, 116, 124*

Habitats *84-85, 99-**101**, 102-117, 143, 163-167. See also* Forests, Plants
Haig-Brown, Roderick *163*
Halkomelem *11, 13, 37, 43, 49, 52, 56, 57, 59, 80, 81, 110, 121*
Hall's Prairie *87, 92, 94, 109, 110, 117, 147*
Hanwell, Captain Henry *81*
Harris, John *51, 92*
Harrison, John *63*
Harrison River *120*
Hatzic Rock *7, 57, 187*
Hawkins, John S. *83*
Hazelmere *113*
Health *27, 143, 158, 170, 180, 182*
Heritage buildings *147, 154, 185, 192–196*
- Burrvilla ***193***
- Eric & Sarah Anderson Cabin ***189***
Hernandez, Juan *62*
Hill-Tout, Charles *29*
Hooker, Sir William *81*
Horses *45, 81, 93, 112, 143, 145, 146, 147, 151, 155*
Huckaway's Island *105*
Hudson's Bay Company *50, 54, 56, 57, 80-**82**, 109,*

122, 123, 144, 186, 194
Hunting
- Aboriginal *3, 8, 23, 25–26, 33–34, 46, 51, 55–57, 67, 85, 89*
- bounty *109, 141*
- elk *17, 46, 55, 67, 110*
- fur trade *100*
- waterfowl ***139**–140, 168–**169***
Hyland Creek *179*

Ice Ages *2–4, 6, 62, 107, 119*
Icelanders *93, 124*
Insects
- Butterfly, satyr comma ***174***
- Mosquitoes *46, 107*
Irrigation ***98, 161***
Jefferson, President Thomas *88*
Joe, Chief Harry *13*
Juan de Fuca Strait *4, 48, 60, 62, 63, 66, 69, 70, 72*

Kane, Paul *26*
Katzie *13, 43, 57*
Kennerly, C.B.R *52*
Kensington Prairie *92, 117*
Kikayt *58*
Kinley, George *52, 110*
Kirkland, Leila May *103*
Kuno, Gihei *124*
Kwantlen First Nation *57, 58, 130*
Kwo'tsesleq *86*
Kwomais Point *12, **16**, 38, 55, 67, 70, 86*

Ladner *29, **78**, 94–**96**, **103**, 106, 132, 148, 151, 169, 171, 188, 191, 192*
Ladner's Landing *91, 94, 103, 123, 147, 149*
Ladner, Thomas and William *91, 145, **149**, 178*
Ladner Marsh *45, 151*
Lake Terrell *55, 67, 110*
Lake Whatcom *55*
Langley *5, 6, 56, 57, **80**, 89, 97, 110, 144, 146, 147, 154, 155, 156, 178*
Langley Prairie ***80***

Langley Railway Belt *89, 94*
Lichens *37, 113, 115*
Lighthouse Park *15*
Light pollution *160, 181, 182*
Lily Point *14-**15**, 20,* **35***, 52-54,*
    74-75, **122***, 126, 196*
Lind, James *65*
Linnaeus, Carl **64**
Little Campbell River *5, 43,*
    54, 95, 106, 107, 110,
    113, 122, **135***, 137,* **155***,*
    168
Llama farm **155**
Locarno Beach Culture *28, 36*
Logging *See* Forestry
London Farm *185, 186, 195*
Lord, John Keast *85, 130*
Lulu Island *58, 103, 109, 178*
Lummi *15, 19, 34, 35, 41, 43,*
    49, 51–56, 59, 66, 85–86,
    110, 121, 122, 133

Malaspina, Alejandro *64, 69*
Mammals, land *48, 52, 61, 64,*
    174, 175, 176
 - bear *7, 17, 34,* **47***, 79, 109,*
    120, 191
 - beaver, American *10, 55, 56,*
    80, 100, 106, 107, 159,
    174, 191
 - cougar *79, 109, 168*
 - coyote *95, 109, 155*
 - deer, Columbian black-tailed
    3, 7, 10, 17, 34, 46, 52,
    56, 58, 110, 181
 - elk, Roosevelt *3, 10, 17,*
    31, 32, 34, 52, 55, 67,
    80, 110
 - marten *17, 34*
 - mountain goat *27, 46*
 - muskrat *106, 107*
 - prehistoric *3, 8, 9*
 - wolverine *17*
 - wolves *17, 79, 109, 168*
Mammals, marine
 - otter, river **183**
 - porpoises *3, 17, 34, 141, 165*
 - seal, harbour *10, 17, 31–34,*
    52, 64, 106, 126, 141
 - sea lions *34, 126, 128*
 - whales *3, 17, 69, 77, 128,*
    140–141, *165, 174*
Maple Beach *33, 52, 71, 149*

Maps *5,* **14, 30, 60, 67, 84, 101**
Marpole Culture *28, 32, 36*
Massey, George *151, 152, 164*
McKenna-McBride
    Commission *13*
McKinnon, Allan *104, 112*
McLean brothers *103*
McMillan, James *80, 81, 82*
McNally Creek *110*
Menzies, Archibald *49, 65, 66,*
    72, 81, 114
Metro Vancouver *152, 166,*
    178, 180
Middens *10,* **22***, 23–36, 40–41,*
    44, 54
Migratory Bird Act *140*
Miners **87, 88***, 145*
Moody, Colonel Richard
    84, 85
Mountains
 - Cascade *4,*
 - Coast *6*
 - North Shore *4, 178, 179*
 - Olympic *4*
Mount Baker *55, 59, 70, 81, 94,*
    142, *143*
Mount Mazama *18*
Mozino, Mariano *65*
Mud Bay ***i***, *20, 85, 91, 94,* **95***,*
    102 - 104, 136, 145, 152,
    169, 173
Murray Creek *154*
Museums *168, 185, 188,*
    189–191
Musqueam First Nation *36,*
    43, 56–59, 77, 80, 121,
    130, 133, 134

Nanaimo *56, 59, 147*
Narvaez Gervete, Jose Maria
    44, 54, 64, 66–69, 77
National Historic Sites
    186–188
Nature Legacy program *173*
Navigation *63–64*
New Westminster *6,* **57***, 90,*
    94, 97, 103, 114, 147
Nicomekl River *20, 77,* **80***,*
    90, 95, 100, 104, 105,
    107, 110, 112, 113, 122,
    135, 146, 147, 156, 179,

193, 194
Nooksack River *50, 110*
Nooksack Indian Tribe *41,*
    43, 51, 56, 59, 81, 85
North Delta *10, 20, 36, 94, 99,*
    108, 167, **178**
Northern Straits Salish *11,*
    13, 34–35, 43, 54, 56,
    86, 121
nuwnuwuluch *34*

Ocean Park *12, 16, 38,* **42***, 55,*
    75, 110
Old Cordilleran Culture *10,*
    28
Oliver, John *103, 145*
Olson, Sarah **93**
Orcas Island *15*
Oregon Treaty *82, 83*
Oxen *144, 145*
Oyster growers **136, 137**

P'qals *15*
Pacific Flyway *165, 168*
Pacific Salmon Commission/
    Treaty *133*
Pacific Salmonid Enh.Prog.
    134
Panorama Ridge *3, 102, 112*
Pantoja y Arriaga, Juan *53,*
    66, 68
Parks *2, 6, 38, 89, 95, 106, 108,*
    111, 114, 115, 170, 173,
    175, 187, **192-193**
Pauquachin *56*
Peace Arch *83, 165, 188*
Peat *7, 97, 104*
 - harvesting *151,* **153***, 167*
Pesticides *141, 152, 157, 158,*
    171
Petroglyphs **16***, 35, 38, 191,*
    192
Pierre, Peter (Old Pierre)
    13–16, 43, 58
Pioneer names *90, 90–93, 92,*
    94, 111, 112, 145, 147,
    148, 154, 156, 166, 178
Pioneer settlement *79, 83, 85,*
    89–97, 144, 147
Pitt Lake *15, 45*
Plants *50, 75, 81, 99, 100*

- ferns *7, 18, **47**, 109, 113, 144*
- grasses *7, 16, 21, 145, 146*
- marsh *24, 25, 44, 46, 57, 75, 100–101, 115, 117, 166*
  - asparagus, sea *76*
  - wapato *45, 46, 51, 144*
- mosses *113, 115*
  - sphagnum *21, 153*
  - shrubs *18, 21, 34, 37, 73, 105, 177*
  - blueberry, bog *46, **47**, 116, 177*
  - bog-rosemary *21*
  - cranberry, bog *46, 105, 115, 144, 177*
  - juniper, Rocky Mountain *50, 75*
  - Labrador tea *21, 115*
  - Oregon grape *37, 73*
- wildflowers *34, 65, 73–75, 81, 114–117, 147*
  - camas, blue *45–46, 54, 116, 144*
  - gumweed, entire-leaved ***105**, 114, **115***
  - lily, pink fawn *114, **116***
  - orchids *114, 147*
Plover Ferry *185, **187***
Poaching *130, 138*
Point Elliott Treaty *55, 59, 85*
Point Roberts *3, 4, 13–16, 20, 29, 33, 35, 44, 50–56, 67–71, 75–77, 80–84, 87, **93**, 106, 110, **111**, 119–122, 124, **126**, 132, 148, 151, 165, 168, 178, 187, **196***
Polk, President James *82, 83*
Portage Park ***80***
Pre-emption *89–90, 100*
Prehistory *8–10*
Prevost, Captain James *84*
Puerto del Descanso, Chief of *68*
Puget, Lt. Peter *49, 53, 71–77*
Puget Sound *3, 4, 9, 10, 29, 45, 48, 50, 61, 72, 77, 86, 92, 133*

Radiocarbon dating *24*
Railways *89, 93–95, 113, 147, 165*

Ramsar Site *172*
Red ochre *27, 37*
Red tide *177*
Redwood Park *2, 6*
Reef nets *35, 43, **53**, 120–122*
Reifel Bird Sanctuary *169*
Richards, Captain George *84*
Richmond *59, 106, **107**. 109, 115, 128, 166, 172, 178*
  See also Steveston
Richmond Nature Park *115*
Roads *84, 89, 93–95, 103, 114, 147, **151–152**, 165, 167*
- Gulf Road *148, **196***
- Old Yale Road *94, 114, 194*
Robert's Town *87*
Roberts Bank *58, 76, 87, 103, 130, 151, **162***
- B.C. Ferry Terminal *32, 165*
- Port *95, **162**, 165–166, 171*
Roosevelt, Theodore *93, 168*
Royal Engineers *84, 85, 89, 101*

s7iluch *34*
Saanich First Nation *43, 50–53, 56*
Salamanca, Secundino *69*
Salishan languages *11, 43, 56, 69, 86*
Salmon *19, 46, **47**, 52, 54, 55, 58, **81**, 107, **118**, 120, **121**–127, 132–135, 168*
- coastal cutthroat trout *121, 168*
- steelhead *121, 127*
Salmon hatcheries *134*
Salmon River *100*
Samish *43*
Sand Heads *84*
Sands, Doug *138*
San Juan Islands *11, 31, 43, 49, 56, 66, 83*
Saturna Island *66*
Scouler, Dr John *50, 81*
Scowlitz First Nation *57*
sc'ultunum *4-**15**, **35**, 52-54, 74-75, **122***
Sea Island *166*
Seasons, Coast Salish *46–**47***
Semiahmoo Bay *15, 38, 54, 75,*

*77, 80, 83, 89, 90, **108**, 116, 123, 149, 151, 165*
Semiahmoo First Nation *6, 15, 34, 38, 43, 50, 51, 52, 54, 55, 56, 85, 86, 109, 144*
Semiahmoo Park *150, 185, 195*
Semiahmoo Spit *6, 34, 41, 54, **55**, 66, 87, 92, 187, **195***
Sencot'en Alliance *56*
Serpentine Fen Wildlife Area *169*
Serpentine River *i, 7, 20, 86, 95, 104, 105, 107, 112, 135, 137, 146, 154, 156, 169, 178, 179*
Shannon, William *110*
Shellfish *16, 24, 25, 29, 32, 33, 36, 44, 52, 55, 122, 127, 133, 176, 179. See also* clams, crabs
- cockle, heart *4, 23, 29, 33, 54, 137*
- mussels, bay *10, 16, 23, 29, 32, 33, 44, 54, 55*
- oyster, Pacific *104, 127, **136–137***
Shellfisheries *136–138*
Ships
- Beaver *50*
- Cadboro *54, 81*
- Chatham *71, 72*
- Discovery *72–73*
- Mexicana *69, **70**, 71*
- Recovery *87*
- Santa Saturnina *66–69*
- Sutil *69, **70**, 71*
- William and Ann *81*
Si'ke village *34*
Simpson, Sir George *54, 144*
Singh, Swaren, Prahim and Oudam *136*
Skagit River *50*
Skaiametl *57, 58*
Sloughs *95, 107, 108, 132, 147, 178*
Smallpox *48 - 51*
Smekwats, Smakw'ets *15*
Smith, Harlan *32, 35*
South Arm Marshes *103, 173*
South Surrey *6, 20, 84, 94,**105**, 110,**112**, **142**, 188, 193*

Species at Risk *117, 174*
Spetifore Farm *148, 172,* **173**
Splockton, Joe *49*
Springs, artesian *108, 146,* **178**
Squid, opal *127*
Sqwema:yes. *See* Kwomais
St. Mungo Cannery *10,* **36**, *37*
Steller, Georg Wilhelm *65*
Stevens, Homer *124*
Steves, William Herbert *123*
Steveston *90,* **97**, *106, 124,* **125**
Stewart, Margaret *99*
stl'elup *13, 32,33, 49, 52*
Stò:lo *13, 17, 40, 41, 43, 49, 51, 80, 86, 109, 122, 130, 134, 144, 187*
Stò:lo Nation Society *57*
Stò:lo Tribal Council *57*
Stone tools *8, 25–27,* **32**, **34**, **36**, **41**
Strait of Georgia **iv-v**, *3, 4, 6, 8–11, 13, 19, 20, 28, 33, 39, 43–45, 48, 49, 51, 61–67, 70, 72, 81, 84, 87, 103, 121, 127, 132, 140,* **141**, *168, 180*
Sturgeon Bank *77, 81, 106, 128, 166*
Sullivan *113*
Sumas Lake *130, 145*
Sunnyside Acres *114, 173*
Surrey *2, 7,* **89**, **91**, *99, 106, 107, 111,* **113**, *114, 145, 146,* **147**, *151, 154, 156, 165, 172, 173, 175, 178,* **188-189**
Surrey Dyking District *104*
Survey grid *88,* **89**
suw'q'weqsun **36**
Swaneset *13*
Swenson family *103*

Telegraph Line *94*
Th'emqellem *86*
Thunderbird *17*
Tl'ektines *58, 59*
Tongue Spit **55**, *92*
Totem poles *185,* **191–192**
Trails
  - Anderson Brigade *87*
  - Boundary Bay dyke *106*

  - Fort Langley *94*
  - Grease *128*
  - Semiahmoo *6, 94, 96*
  - Semiahmoo Heritage *193*
  - Whatcom *87, 94*
Transformers *13–14, 16*
Treaty of Washington *82*
Trees *3, 6, 37, 45,* **73**, *75, 92, 96, 100, 105, 106, 109, 177, 192*
  - cedar, yellow *18*
  - Douglas-fir *3,* **7**, *16, 45,* **73**, *94, 99, 109, 110,* **112**
  - hemlock, western *18, 37,* **65**, *73, 109*
  - maple, bigleaf **46**
  - redcedar, western *6,* **18–19**, **39**, *99, 109, 192*
  - redwood, dawn **2**
  - willow *46, 80, 81, 100, 105, 106*
Triggs, George *147*
True Oliver *139*
Tsartlip *56*
Tsawwassen *6, 29, 77, 108, 110, 148,* **173**, *179, 191*
Tsawwassen Beach *33, 151*
Tsawwassen First Nation *3, 13, 15, 19, 27, 32–34, 43, 52, 56–58, 80, 103, 115, 121, 134, 191, 192*
Tsawwassen First Nation Reserve *58, 103, 151*
Tsawwassen Forest *34*
Tsawwassen village site (Stl'alep) **8**, *13,* **32**, *33,* **34**, **38**, *49, 52*
Tselhtenem *14-***15**, **35**, *52-54,* **74-75**, **122**
Tseycum *56*
Tsi'lich *34*
ttunuxun **31**
Tynehead *113, 173*
Tynehead hatchery *135*

Valdes y Flores Bazan, Cayetano *69–73, 76*
Vancouver *59, 66, 77, 94, 108, 126*
Vancouver, Captain George *49, 60, 65, 70, 71, 72, 77, 114, 128*

Vancouver Island *7, 11, 43, 44, 48, 54, 57, 58, 59, 62, 66, 77, 82, 85, 87, 121*
Vancouver landfill *167*
Vernaci, Juan *69*
Victoria *56, 77, 87, 117, 147*

Wade, Dolly *147*
Wade, Ted *116*
Waldron Island *54*
Waller, John *125*
Waller, Kate *93*
Waterfowl *6, 17, 24,* **31**, *32, 46, 54,* **58**, *100, 104, 107, 139, 140, 143, 159, 160, 166, 168-***171**, *172,* **173**
Water quality *100, 106, 107, 134, 146, 171, 178*
Weather *62,* **63**, *64, 70–***71**, **78**, **102, 103**
Westham Island *106, 132, 140, 148, 154*
Westman, Eythor and Hannes *127*
Wetlands **i**, *46, 99, 100, 117, 140, 168, 175, 176*
Whalen, Michael *33*
Whalen Farm **22**, *25, 33, 34, 188*
Whatcom (old Bellingham) *87*
Whatcom County *155, 173, 178*
Whidbey, Joseph *45, 70, 72*
Whidbey Island *77*
White Rock *6, 20, 38, 54, 84, 85,* **88**, *106, 112, 96, 107, 109, 112, 113, 149, 178, 179, 189, 192, 195*
Wildlife Management Areas *173*
Woodward, Elizabeth *145*
Woodward's Landing *151, 186*
Work, John *82, 96, 100, 110*
Xa:ls (Khaals) *14, 15*
Xa:ytem *7, 57*
xwut *37*

Yale *7, 87, 89, 94, 114*
Yukulta (Lekwiltok) *55, 86*

About the author:

Anne Murray has had a life-long interest in birds, nature, history and different cultures. Born and educated in England, where she received her BSc (physics and geology), she has taught mathematics and science to every age group. She has lived in Papua New Guinea, Thailand, Alberta, and for the last two decades, in Tsawwassen, British Columbia.

Anne Murray

Anne volunteers with a number of nature organizations, including Nature Canada, B.C. Nature, and the Delta Farmland and Wildlife Trust. Among other awards for her conservation work, she was a recipient of the Queen's Golden Jubilee medal. Anne is married, with three daughters.

David Blevins

About the photographer:

David Blevins is a forest ecologist and award winning nature photographer. His photographs have been published internationally in magazines, calendars, and books. His academic and photography work both come from an appreciation of the natural patterns that can be found in what at first seems to be a chaotic world. Wildlife behavior, the organization of a plant, or the natural patterns of landscapes are all sources of inspiration for his photography.

David created most of the photographs for this book while completing a Ph.D. in forest ecology at the University of British Columbia, Canada. He now lives in North Carolina where he continues to work on both science and photography projects. More of his photography can be viewed on his website **www.blevinsphoto.com**.

THE INTERNATIONAL MARINE SAILBOAT LIBRARY

SAILBOAT HULL & DECK REPAIR

DON CASEY

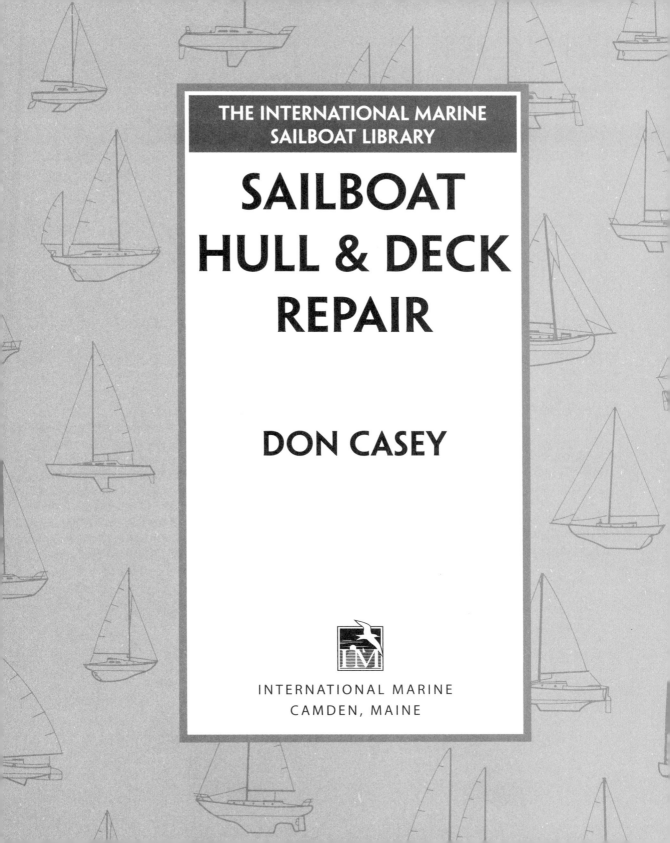

INTERNATIONAL MARINE

CAMDEN, MAINE

CONTENTS

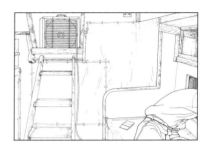

Introduction   **4**

Leaks   **6**

Choosing a Sealant   8

Rebedding Deck Hardware   10

Preparing a Cored Deck for New
   Hardware   14

Sealing Chainplates   15

Sealing Portholes—A Temporary
   Solution   17

Rebedding Deadlights   18

Replacing Portlights   21

Mast Boots   24

Hull-to-Deck Joint   24

Centerboard Trunks   27

Through-Hull Fittings   27

Pressurizing to Find Leaks   29

Restoring the Gloss   **30**

Buffing   32

Sanding   34

Scratch Repair   36

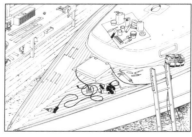

Deck Repairs   **40**

Stress Cracks   42

Voids   45

Crazing (Alligatoring)   47

Renewing Nonskid   51

Teak Decks   55

Laminate Repair   **64**

Understanding Polyester Resin   66

Grinding Is Essential   69

The Basics of Fiberglass Lay-up   70

When to Use Epoxy   74

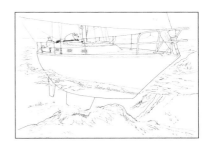

Core Problems   78

Delamination   80

Wet Core   82

Damaged Core   85

Reinstalling the Skin   87

Strengthening   89

Stiffening a Skin   90

Hull Repairs   92

Gouges   94

Blisters   98

Impact Damage   104

Keel and Rudder Damage   114

Weeping Keel   116

Keel/Centerboard Pivot Problems   119

Hull Damage Around Fins and Skegs   120

Damaged Rudder   121

Blade/Shaft Movement   123

External Ballast   124

Index   126

Copyright Information   134

INTRODUCTION:

BEAUTY IS MORE THAN SKIN DEEP

Fiberglass.

Legendary yacht designer L. Francis Herreshoff has—somewhat inelegantly, it seems to me—called this versatile material "frozen snot." Maybe so, but how many of Herreshoff's beautiful wooden boats have ended up as lobster condominiums or fuel for a boat shed stove, while snot-built boats of the same age, no matter how undeserving of immortality, continue to ply the world's oceans, bays, and estuaries? Wooden boats regularly die early deaths of natural causes; fiberglass boats must be assassinated.

Don't get me wrong; I love wooden boats, and in particular I love Herreshoff's wooden boats. There is something magical about taking straight lumber and manipulating it into the flowing contours of a boat. The craftsmanship of the builder is obvious: planks steamed to linguini and worried into compound curves, knees cut from a natural crook to harness the tree's full strength, precise dovetail joints wedding shelf and beam. Such craft is far less evident in a hull formed by painting a boat-shaped mold with a thick layer of fiber and sticky glop. That the dried glop pops out of the mold with the same graceful curves yields no redemption.

But it isn't redemption that is called for; it's perspective. If we open the leaded glass doors in the galley of one of Mr. Herreshoff's classic wooden yachts, we are likely to encounter fine china. Why not wooden plates, woven bowls? Because china dishes—molded bone-reinforced clay—are infinitely more serviceable.

Likewise molded glass-reinforced plastic boats.

Fiberglass is malleable, durable, and easy to maintain. These characteristics, widely known, have made fiberglass the overwhelming material of choice for boat construction for more than three decades. If you want a boat to display, wood has much to recommend it, but for a boat to *use*, fiberglass is hard to beat.

A lesser-known virtue of fiberglass is that it is easy to repair. A fiberglass hull's seamless nature leads many boatowners to conclude that repair must be difficult. Any assertion to the contrary too often elicits raised eyebrows. In the pages that follow, we hope to quell the skeptics with astoundingly clear explanations, but the only way you can fully purge yourself of any nagging doubts is to buy a can of gelcoat paste or a bit of glass cloth and resin and give it a whirl. You'll wonder what you were worried about.

While this book is confined to hull and deck repairs to fiberglass boats, it is not limited to fiberglass repairs. Fiberglass boats are not *all* fiberglass. Decks, for example, may be cored with plywood, balsa, or foam, railed with aluminum, covered with teak, outfitted with bronze, interrupted with acrylic, penetrated with stainless steel, and booted with rubber. Virtually all of these components require regular maintenance and occasional repair, and they must be assembled properly and carefully if the boat is to be dry.

Watertight joints are our first order of business. Boatowners today don't need even a passing acquaintance with oakum caulking and firming irons; molded hulls are completely seamless, and rare is the fiberglass *hull* that leaks, no matter how old. Deck leaks are, unfortunately, another matter. The dirty little secret of fiberglass boats is that most are only slightly more watertight than a colander. Spray? Rain? Wash-down water? A significant amount of all three finds its way below.

Deck leaks don't just wet the contents of lockers, drip on bunks, and trickle across soles; they destroy wood core, corrode chainplates, and delaminate bulkheads. Identifying and eliminating leaks is essential. This book details the most effective technique for sealing joints and bedding hardware, and it provides specific sealant recommendations for various uses. It instructs you in portlight replacement, hull-to-deck joints, and centerboard trunk repairs. It also shows you how to test your work and how to locate pesky leaks.

Often all that is wrong with a fiberglass hull is a chalky surface or a few scratches. Restoring the gloss can be the easiest of repairs to fiberglass; it is where we begin our exposition of this material.

The ravages of time affect decks more than hulls. An older fiberglass deck is likely to be webbed with hairline cracks, even pocked with open voids, and may have stress cracks radiating from corners or from beneath hardware. Fortunately there are easy ways to repair these blemishes. Step-by-step instructions for restoring the deck to perfection are provided.

Deck repairs are complicated by the necessity of providing effective nonskid surfaces. Owners of boats with molded-in nonskid will find the included instructions for renewing those surfaces useful. Those with planked decks will be more interested in the section detailing the care and repair of teak overlay.

Eventually, of course, a hull-and-deck-repair book for fiberglass boats must come around to repairs requiring fiberglass lay-up, but not without first providing clear and concise descriptions of the various materials to be used. When should you use polyester resin and when epoxy? What is vinylester? Cloth, mat, or roving? You will find answers to these questions and more in Chapter 4.

Armed with an understanding of the materials involved and guided by clear illustrations, you are ready to take on more complicated repairs. Chapter 5 shows you how to repair deck delamination and how to replace spongy core. Chapter 6 focuses on hull repairs—dealing with gouges, repairing blisters, and reconstructing after impact damage. A quick look in Chapter 7 at repairing common rudder and keel problems, and you will have taken the cannon.

When all boats were built of wood, a truly professional repair required the skills of someone with years of experience. Not so with fiberglass. Pay attention and give it a try, and you will discover that there is virtually no repair to a fiberglass hull or deck that a motivated owner can't do as well (if not as quickly) as a pro.

Frozen snot, indeed!

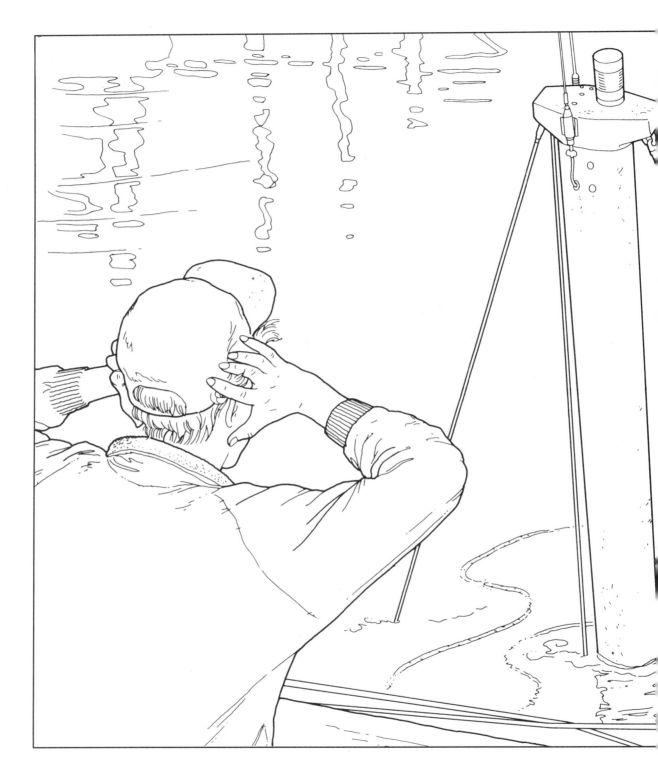

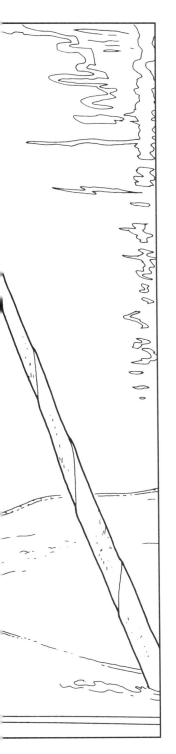

LEAKS

Leaks are insidious. A tiny leak, left unattended for months or years, can easily result in damage that will cost thousands of dollars to have repaired, or take innumerable hours if you make the repairs yourself.

There are the obvious things: ruined interior varnish below leaking ports, mildewed upholstery from trickles from the hull-to-deck joint, a punky cabin sole from "mysterious" rainwater intrusion.

As serious as these are, they're small potatoes. The biggest risk from leaks is to the deck core, and you may not see any evidence of a leak until major damage is already done.

The decks of most fiberglass boats are made up of a plywood or balsa core sandwiched between two skins of fiberglass. (Closed-cell foam, more resistant to saturation but no less susceptible to delamination, is found in relatively few production boats.) If water penetrates the fiberglass skin and gets into the core, the result is likely to be failure of the bond between the core and the skin(s). This core delamination weakens the deck. Delamination is accelerated if the boat is subjected to temperatures that cause the trapped water to freeze and expand.

The water entering a cored deck cannot get back out; the flow is one way, like filling a jug. Balsa cores become saturated and mushy. Plywood soon rots. In both cases, the only solution is cutting away the fiberglass skin and replacing the core. After you do this job once, knowing full well that it could have been prevented with four-bits' worth of caulk and an hour's worth of effort, you will become religious about maintaining a watertight seal around any hole in the deck.

You walk into a marine store and there they are, dozens of different cartridges and tubes standing on shelves, stacked in bins, and hanging in blister cards. Geez, how many different kinds of marine sealants can there be?

Three. That's it. Three. Understand these three and you have the selection process whipped.

POLYSULFIDE

Polysulfide is the Swiss Army knife in marine sealants; you can use it for almost everything. Often called Thiokol (a trademark for the polymer that is the main ingredient of all polysulfide sealants regardless of manufacturer), polysulfide is a synthetic rubber with excellent adhesive characteristics. As a bedding compound it allows for movements associated with stress and temperature change, yet maintains the integrity of the seal by gripping tenaciously to both surfaces. Polysulfide is also an excellent caulking compound since it can be sanded after it cures and it takes paint well.

Use polysulfide for everything except plastic. Polysulfide bonds as well to plastic surfaces as to any other, but the solvents in the sealant attack some plastics, causing them to harden and split. Specifically, don't use polysulfide to bed plastic portlights, either acrylic (Plexiglas) or polycarbonate (Lexan). Don't use it to bed plastic deck fittings (including portlight frames); plastic marine fittings are generally either ABS or PVC, and polysulfide will attack both. Any plastic fitting made of epoxy, nylon, or Delrin—such as quality plastic through-hull fittings—may be safely bedded with polysulfide.

The black caulking between the planks of a teak deck is polysulfide. For this application, a two-part polysulfide gives the best results. Because polysulfide adheres well to teak (a special primer improves adhesion), and because it is unaffected by harsh teak cleaners, it is also the best choice for bedding teak rails and trim.

Polysulfides are the slowest curing of the three types of sealant, often taking a week or more to reach full cure.

POLYURETHANE

Polyurethane is the bulldog of marine sealants—once it gets a grip, it doesn't turn loose. Polyurethane is such a tenacious adhesive that its bond should be thought of as permanent; if there is any likelihood that you will want to separate the two parts later, don't use polyurethane to seal them.

Use polyurethane anywhere you want a permanent joint. This is the best sealant for the hull-to-deck joint. It is also a good choice for through-hull fittings and for toerails and rubrails, but not if they are raw teak because some teak cleaners soften it. Like polysulfide, polyurethane should not be used on most plastics—acrylic, polycarbonate, PVC, or ABS.

The cure time for polyurethane is generally shorter than polysulfide, but still may be up to a week.

SILICONE

Silicone can seem like the snake oil of the marine sealant trio. A bead of this modern miracle is too often expected to cure any and every leak. And it

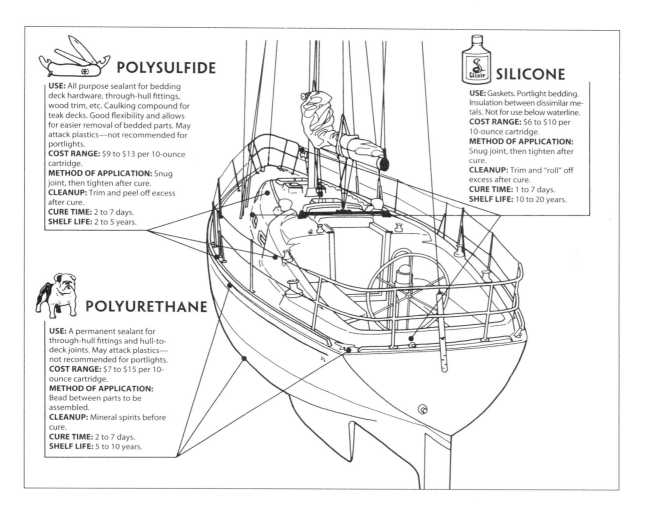

POLYSULFIDE

USE: All purpose sealant for bedding deck hardware, through-hull fittings, wood trim, etc. Caulking compound for teak decks. Good flexibility and allows for easier removal of bedded parts. May attack plastics—not recommended for portlights.
COST RANGE: $9 to $13 per 10-ounce cartridge.
METHOD OF APPLICATION: Snug joint, then tighten after cure.
CLEANUP: Trim and peel off excess after cure.
CURE TIME: 2 to 7 days.
SHELF LIFE: 2 to 5 years.

SILICONE

USE: Gaskets. Portlight bedding. Insulation between dissimilar metals. Not for use below waterline.
COST RANGE: $6 to $10 per 10-ounce cartridge.
METHOD OF APPLICATION: Snug joint, then tighten after cure.
CLEANUP: Trim and "roll" off excess after cure.
CURE TIME: 1 to 7 days.
SHELF LIFE: 10 to 20 years.

POLYURETHANE

USE: A permanent sealant for through-hull fittings and hull-to-deck joints. May attack plastics—not recommended for portlights.
COST RANGE: $7 to $15 per 10-ounce cartridge.
METHOD OF APPLICATION: Bead between parts to be assembled.
CLEANUP: Mineral spirits before cure.
CURE TIME: 2 to 7 days.
SHELF LIFE: 5 to 10 years.

does—for about as long as it used to take the magic elixir salesman to slip out of town. Then the bead releases its grip, and what started out as a tube full of promise ends up as a dangling rubber worm. All is not lost—with a hook and the right wrist action, you can at least catch dinner.

Silicone sealant is a gasket material—period. If you think of silicone's adhesive abilities as temporary at best, you will find it is the best product for a number of sealing requirements. It is the only one of the marine sealant trio than can be safely used to bed plastic. It is an excellent insulator between dissimilar metals—use it when mounting stainless hardware to an aluminum spar. It is the perfect gasket material between components that must be periodically dismantled—beneath hatch slides, for example.

Silicone retains its resilience for decades and is unaffected by most chemicals, but it should not be used below the waterline. Because it depends upon mechanical compression to maintain its seal, silicone is not the best choice for sealing hardware on a cored deck. Exposed silicone is a magnet for dirt but repels paint like an opposite pole, so never fillet with silicone, and don't use this sealant on any surface you plan to paint.

Silicone sealants typically set in a few minutes and usually reach full cure in less than 24 hours.

A USEFUL HYBRID

THERE IS A BIG ADVANTAGE TO USING A SEALANT with good adhesive properties. An adhesive sealant maintains its seal even when stresses pull or pry the bedded components apart, the sealant stretching and compressing like the bellows joining the two sides of an accordion.

This accordion effect would be especially useful for plastic portlight installations where the portlights are not bolted in place but rather clamped between an inner and outer frame. As the cabin sides expand and contract with temperature changes or flex with rigging stresses, the space between the frames varies.

Applied properly (see "Rebedding Deadlights"), silicone sealant can accommodate these variations, but it is not easy to set the portlight in a uniform thickness of silicone. Although silicone has amazing elasticity, its lack of adhesion means it must always be under pressure to maintain a watertight seal. If the gasket formed by the cured silicone is thin anywhere around the portlight, the seal is sure to fail, probably sooner rather than later.

Either polysulfide or polyurethane would provide a more dependable seal, but polysulfide is certain to attack the plastic, and polyurethane prohibits any future disassembly. Fortunately a chemist somewhere, one who undoubtedly owns a boat and tried to bed plastic portlights, cooked up a new goo that is part silicone and part polyurethane. Marketed by BoatLife as Life Seal, this is a more durable sealant than silicone for portlights and other plastic fittings.

REBEDDING DECK HARDWARE

Fiberglass boats are notorious leakers. Wood is, to a degree, self-sealing; a leak swells the wood, pinching off further leakage. Not so with fiberglass. Once the seal between the fiberglass and the hardware is broken, it will leak unabated until you reseal it. The seal can be broken by stress, by deterioration, or by temperature changes. Wrenching the top of a lifeline stanchion can break the seal at the base. Sunlight and chemicals erode sealants. In cold weather the deck may literally contract away from the hardware.

Every seal on the deck (and hull) of a fiberglass boat should be carefully examined at least annually, and at any sign of failure, the joint should be opened, cleaned, and resealed. This modest investment is guaranteed to return greater relative dividends than even your most profitable stock fund.

1 Gather all the necessary materials. If you are using a cartridge—economical if you have quite a bit of rebedding to do—you need a caulking gun. Have masking tape and adhesive cleaner on hand to control squeeze-out.

2 Remove the fitting. This is usually the hardest part of the job, either because access to the fasteners is difficult to gain or because the bolts are frozen—or both. Access sometimes requires removing headliners or cabinetry, but don't try to avoid this by simply running a bead of sealant around the fitting. If you do that, eventually you will still be removing the fitting, only this time in preparation for major deck repair.

For access to the fasteners securing wooden components, the bungs hiding the bolt heads will have to be removed. This can be accomplished by drilling a small hole in the center of the bung and threading a screw into it; when the point of the screw finds the screw head below the bung, continuing to turn the screwdriver will lift the bung. Extracting bungs this way can sometimes damage the bung hole. A safer method is to drill the bung with a bit slightly smaller than the diameter of the bung, then carefully remove the remaining ring of material with a small chisel.

If the fitting was installed with polyurethane, removing the fasteners may have little effect. Trying to pry the fitting loose is likely to result in damage to the deck and the fitting. Heating the fitting (especially metal fittings) or the deck can coax the polyurethane to release its grip.

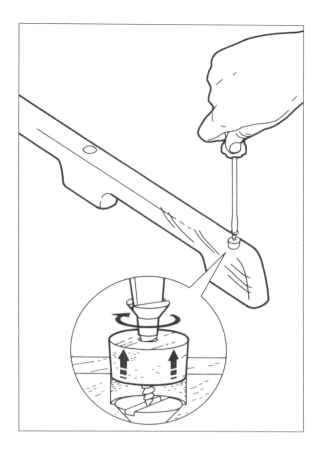

HEADLINERS

HEADLINERS ARE AS VARIED AS BOATS. If the headliner is fiberglass, you likely cannot remove it (without removing the deck). Occasionally manufacturers bolt hardware to the deck before installing it over the headliner. You will have to cut or drill the headliner beneath the fasteners to gain access. Reinstall the hardware with longer bolts through spacers and a backing plate that covers the cutout.

When the headliner is made up of panels, it is usually captured by trim pieces screwed in place. Panels may also attach with Velcro.

Sewn headliners are typically stapled to wooden strips across the overhead. You can't see the staples because they are through the excess material on the back side at the seams. You gain access by removing the trim piece at the forward or aft end of the liner and pulling the liner loose at that end. Work the staples out with a flat screwdriver at the seams until you uncover the desired area. Be sure you use Monel staples when you replace the liner. For more on headliners, see *Canvaswork and Sail Repair* in this series.

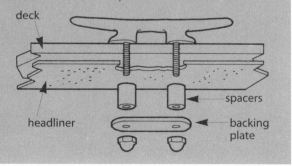

deck

spacers

headliner

backing plate

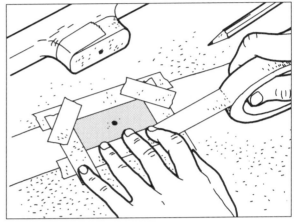

3 Clean off the old bedding. Every trace of the old sealant must be removed. Use a blade, sandpaper, or a wire brush as required, and clean both the deck and the fitting with acetone.

4 Mask adjacent areas. Cleaning up the squeeze-out with solvent takes twice as long as masking and is ten times more messy. Dry-fit the part and trace around it with a pencil. This is the time to strengthen the mounting location if required (see "Deck Repairs"). Mask the deck ⅛ inch outside the pencil line and mask the edge of the fitting.

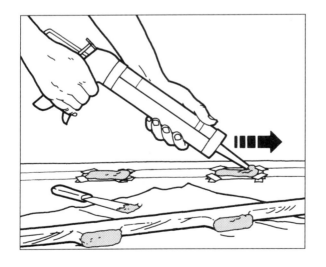

5 Coat both surfaces with sealant. Cut the tip of the tube or cartridge at a 45° angle—close to the tip for a thin bead, farther back for a thicker bead. (Cartridges have an inner seal you will have to puncture with an ice pick.) Apply the sealant with a forward motion, pushing the bead in front of the nozzle. Coat both surfaces to make sure there will not be any gaps in the bond; use a putty knife to spread the sealant evenly, like buttering bread. Before inserting the mounting bolts—not screws—run a ring of sealant around each just below the head. NEVER apply sealant around the fasteners on the underside of the deck; if the seal with the outer skin breaks, you want the water to pass into the cabin where it will be noticed.

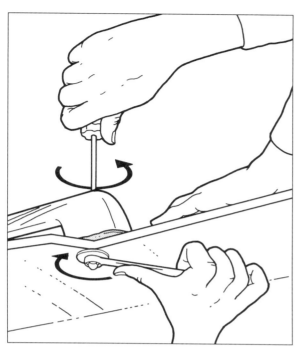

6 Assemble the parts and "snug" the fasteners enough to squeeze sealant out all the way around.

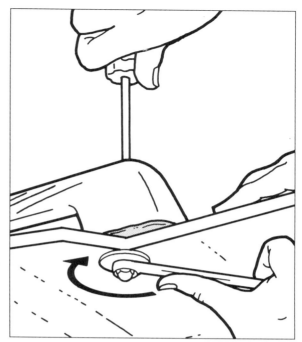

7 Wait until the sealant partially cures—30 minutes for silicone, 24 hours for polysulfide or polyurethane—then fully tighten the bolts by turning the nuts only to prevent breaking the seal around the shank of the bolt. If the fitting is attached with screws, withdraw them one at a time, run a bead of sealant around the shank beneath the head, reinstall each in turn, then drive them all home evenly.

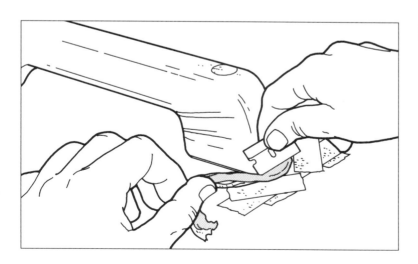

8 Trim away the excess squeeze-out by running a razor blade around the fitting, then peeling away the masking tape. Never leave a fillet around the edge; silicone attracts dirt, polyurethane yellows, and polysulfide weakens in the sun, so you want the least amount of sealant visible—only the thin edge beneath the fitting. Install new bungs, matching color and grain and setting them with varnish.

PREPARING A CORED DECK FOR NEW HARDWARE

As good as marine sealants are, you should never depend on them to keep water out of the core of a deck or hull. Anytime you drill or cut a hole in the deck, seal the exposed core with epoxy before mounting any hardware. If you are rebedding old hardware for the first time, be certain that the core has been properly sealed, or follow this procedure before reinstalling the fitting.

1 Drill all fastener holes oversize. A large hole—for a through-hull fitting, for example—doesn't need to be cut oversize.

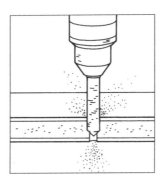

2 Remove all the core within 1/2 inch of the hole. You can do this easily with a bent nail chucked into a power drill. Vacuum the pulverized core from the cavity; whatever you can't remove will act as a filler.

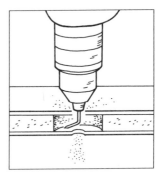

3 Fill the cavity with epoxy. The most secure way is a two-step process. First seal the bottom hole with duct tape, then pour catalyzed epoxy into the top hole. When the cavity is full, puncture the tape and let the epoxy run out back into your glue container. Filling the cavity with unthickened epoxy allows the epoxy to better penetrate the edge of the core. Retape the bottom hole. If there are several mounting holes, fill each and drain in turn until all have been treated and all bottom holes resealed.

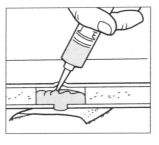

The second step is to thicken the epoxy (the same mix you have already poured through the holes) with colloidal silica to a mayonnaise consistency. Now fill each cavity level with the deck and allow the epoxy to cure fully.

4 Redrill the mounting holes through the cured epoxy. Sand and clean the area that will be under the fitting. Now you are ready to bed the new hardware as detailed in the previous section.

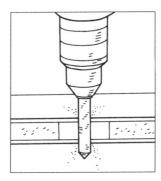

SEALING CHAINPLATES

When water finds its way below, very often the culprit is a leaking chainplate. Chainplates' propensity to leak is understandable; they are bedded under moderate fixed stress, but once under sail the windward chainplates are alternately yanked and eased while the leeward chainplates are virtually released. This tries the grip of any sealant. They are also stressed in unfair directions by poor sheet leads, shroud encounters with the dock, and by the use of shrouds for body support or as handholds for coming aboard. When the seal fails, rain and spray gathered by the attached shroud or stay runs down the wire and across the turnbuckle directly to the chainplate.

As annoying and potentially damaging as a leak into the cabin is, the larger risk is often from chainplates that appear to be watertight. The danger is usually not to the deck; most manufacturers know enough not to have chainplate openings located in a cored section of the deck (but you should check yours). It is the rig that is at risk. If the seal at the deck breaks, water penetrates, but additional sealant lower on the chainplate stops the leak before it enters the cabin. This results in the chainplate sitting in a ring of water. Despite the corrosion resistance of stainless steel, this situation will, over time, almost certainly result in chainplate failure. Because the erosion is hidden by the deck and/or sealant, the only way to detect this problem—short of catastrophic failure—is to pull the chainplate and examine it. If you have never fully examined your chainplates, or if it has been a few years, you are strongly urged to pull them before you rebed them.

1 Remove the trim plate if there is one. This can usually be taped up out of the way, but rebedding is much easier if you disconnect the shroud or stay by slackening the turnbuckle and pulling the pin. Disconnect only one shroud at a time. Before releasing a stay always set up a halyard to support the mast.

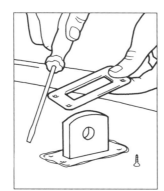

2 Pull the mounting bolts below deck and extract the chainplate. It is only necessary to remove the chainplate if you want to check it for signs of corrosion. If it doesn't come out easily, pass a long, round screwdriver shaft through the pinhole and support the end on a wooden block while lifting on the handle.

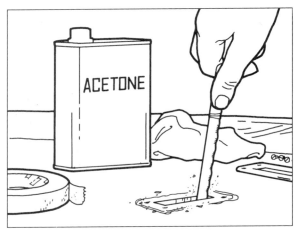

3 Dig all of the old caulk from the hole. A piece of hacksaw blade can be useful for this, but be careful not to enlarge the hole through the deck; the tighter the chainplate fits, the less it will move, and the longer your bedding job will last. Clean the deck, trim plate, and chainplate of old bedding. Examine the chainplate in the caulk area carefully; any pitting, cracks, or brown discoloration indicate replacement. Wipe down the deck, trim plate, chainplate, and the inside of the hole with acetone.

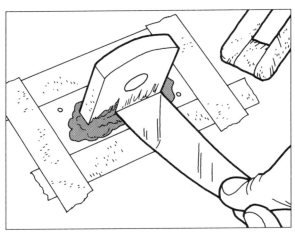

4 Reinstall the chainplate if you removed it. Dry-fit the trim plate and trace around it with a pencil. Mask the deck outside of the pencil line, the chainplate above the trim plate, and the top surface of the trim plate. Push a generous bead of polysulfide sealant into the space between the chainplate and the deck all the way around the chainplate. Use the flat of a flexible putty knife to force sealant into the crack. Butter the deck and the bottom of the trim plate with sealant.

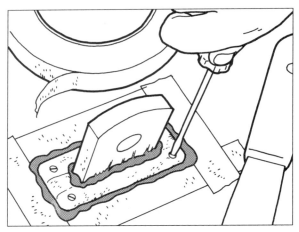

5 Fit the trim plate in position and install its fasteners. Because the trim plate screws are generally quite small, there is little to be gained by two-stage tightening, so tighten these screws fully. Sealant should squeeze out of the slot and all around the plate.

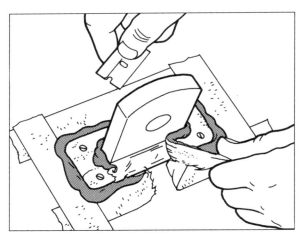

6 When the sealant is sufficiently cured, trace around the trim plate and the chainplate with a razor blade and remove the masking.

SEALING PORTHOLES—A TEMPORARY SOLUTION

When a porthole develops a leak, what you should do is rebed it properly. But maybe you're 500 miles offshore, and removing a porthole doesn't seem like a very good idea. Or maybe your end-of-season haulout is only three weeks away. Or maybe you just don't have the time or the inclination to get involved in such a job at the moment. You could just let the sucker leak, but a more sensible solution is a temporary repair.

1 Wipe the frame, cabin side, and portlight thoroughly with *alcohol* to remove any oil or grease. *Never use acetone or other strong solvents on plastic portlights or hatches.*

2 Mask both the portlight and the cabin side about 1/8 inch from the frame.

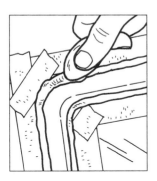

3 Push a thin bead of silicone sealant into the corners the frame forms with the portlight and with the cabin side. Drag a fingertip or a plastic spoon through the bead to form a concave fillet all the way around both edges of the frame.

4 Give the silicone about 30 minutes to dry, then peel the masking tape away slowly. The silicone will seal the port for a few weeks or months (depending on conditions). When it is time to do the job properly, lift one edge of the fillet with a blade and the silicone should pull away in a single strip; any residue can be "rolled" off with a thumb.

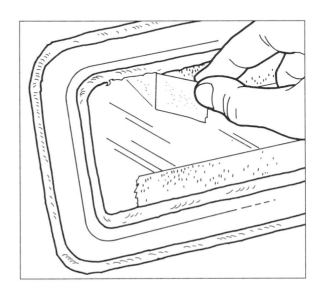

REBEDDING DEADLIGHTS

Opening portholes are rebedded like any other piece of hardware—by removing them, cleaning away all old caulk, buttering the cabin side and the outside flange with fresh sealant, snugging the fasteners, then tightening fully after the sealant has cured. Use polysulfide if the frames are metal, silicone (or a silicone hybrid) if they are plastic.

Getting a watertight seal around a fixed port is a bit more exacting.

1 Dismantle the deadlight. The deadlight pane is typically captured between inner and outer trim rings bolted together, either by through-bolts or by machine screws that thread into sockets on the outer ring.

NOTE: You may find a rubber gasket between the inner frame and the plastic pane. This is not a seal; it's a spacer. Boat manufacturers often installed acrylic windows thinner than the space between the trim rings, making up the difference with a rubber gasket. If you aren't replacing the pane, you will need this gasket. If it has hardened or deteriorated, cut a new one. Don't use soft gasket material; this will allow the pane to be pushed away from the outer frame, probably resulting in seal failure. Eliminating the gasket altogether by installing thicker portlights is the best plan (see next section).

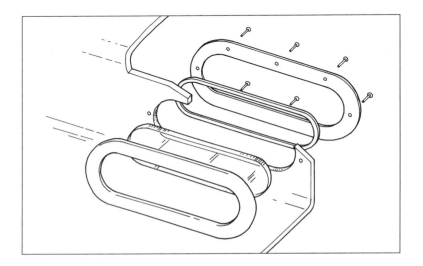

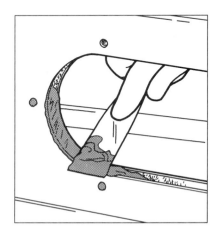

2 Remove all traces of old sealant and wipe all the surfaces with rubbing (isopropyl) alcohol. Check the edges of the cutout in the cabin side. If there is exposed core, dig it out beyond the screw holes and fill the cavity with epoxy thickened with colloidal silica. After the epoxy cures, redrill the fastener holes.

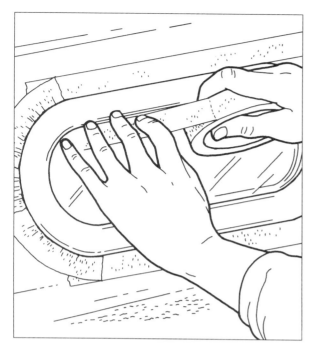

3 Reassemble the cleaned parts and mask both the portlight and the cabin side at the edge of the outer frame. Disassemble.

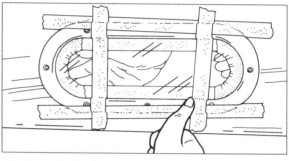

4 Tape the inside ring in place, then position the pane—with the gasket, if fitted—in position inside the cutout. A couple of strips of tape across the pane and frame inside the cabin will hold the pane in place.

5 Back up on deck, fill the space between the plastic and the cabin side with silicone-based adhesive sealant (Life Seal or equivalent). Continue applying sealant to the pane and cabin side until the unmasked edges of both are coated. Distribute the bedding with a putty knife.

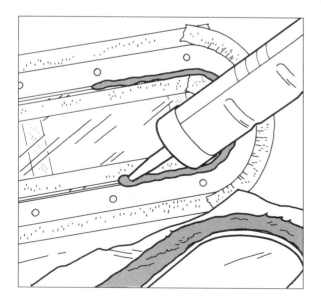

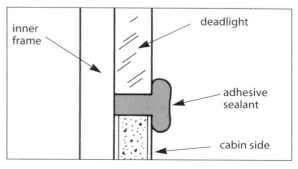

inner frame

deadlight

adhesive sealant

cabin side

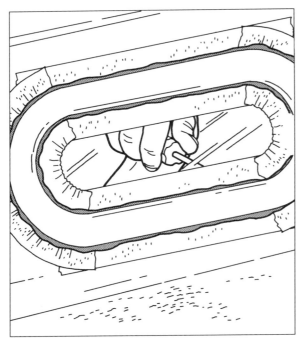

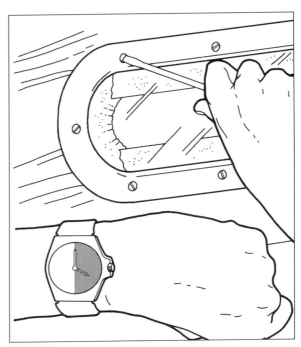

6 Butter the underside of the outside ring and install it. Snug the mounting bolts or machine screws until sealant squeezes from under every edge of the ring. If the rings are through-bolted, don't forget to put a bead of sealant under the head of each fastener.

7 Let the silicone cure for 30 minutes, then tighten the screws.

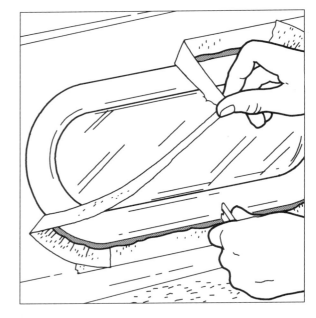

8 Run a new razor blade around both edges of the outside frame, then peel away the masking tape.

REPLACING PORTLIGHTS

Old acrylic portlights get scratched, cloudy, and crazed, but because this happens gradually we often fail to notice until they are almost opaque. Replacing portlights is easy and inexpensive. You are sure to be amazed at the difference it will make in both the look of the boat and the clarity of the view.

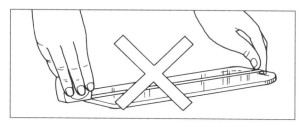

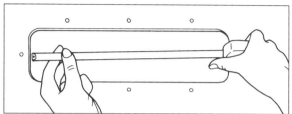

1 Measure the opening. Don't blindly copy the old pane. Mount the inner and outer frames without a pane and measure the space between them. This is the correct thickness for your new plastic—perhaps a few thousandths thinner to allow for expansion. Using a rubber gasket to fill the space is a poor compromise.

Also look at how the old portlight fits the opening. Often the corner radius of the opening and that of the pane are completely different, resulting in excessive gap at the corners. The new plastic should fit the opening (or the frame if it has a capturing flange) with an even gap of about 1/8 inch all the way around. If the old pane isn't a good fit, cut a stiff paper pattern.

2 Take the measurements and patterns to the plastics supplier and have the pieces cut, or buy a sheet of plastic the proper thickness and cut them yourself. Both acrylic and polycarbonate can be cut and drilled with standard woodworking tools, but the edges must be well supported to prevent chipping. Special plastics blades will give the best results. Scrape away any slag with the back (smooth edge) of a hacksaw blade, then file or sand out any chips to eliminate any points that might lead to cracking. Leave the protective film on the plastic while fabricating and mounting.

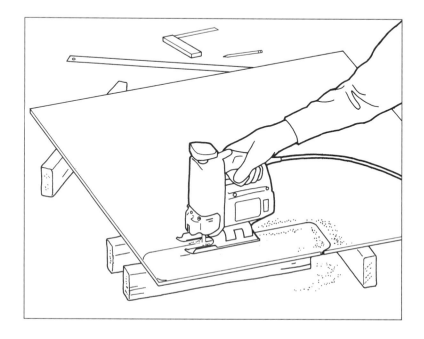

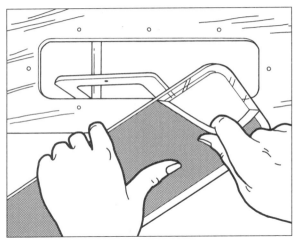

3 Dry-fit the new portlight and trace around the inside of both frames with the corner of a razor blade to cut the protective film. Dismantle the assembly and peel the film from the edges of the plastic.

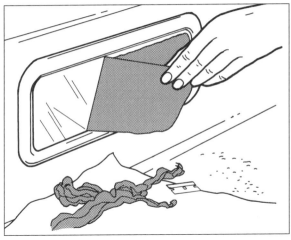

4 Bed the new pane as detailed in the previous section. When the sealant is cured, trace the frame with a blade and peel away the protective film.

ACRYLIC OR POLYCARBONATE?

WHEN CHOOSING THE MATERIAL FOR REPLACEMENT PORTLIGHTS, you have two choices. Acrylic—sold under such brand names as Plexiglas, Lucite, and Acrylite—is the plastic you are most likely to be replacing. Manufacturers use it because it is adequate and relatively cheap. Those may be good enough reasons for you to chose acrylic as well. Acrylic can be brittle and has historically exhibited a tendency to craze. Crazing is less of a problem with today's formulations, but stressed acrylic still cracks. However, for spans typical of boat portlights, acrylic of appropriate thickness is unlikely to break even in extreme conditions.

Polycarbonate—most familiar as Lexan—is not just a better acrylic. It is an entirely different thermoplastic and has an impact resistance roughly 20 times greater than acrylic. Polycarbonate's remarkable strength makes it the undisputed best choice for the wide spans of hatches and oversize windows, but it is probably overkill for most portlight installations. Polycarbonate is softer than acrylic and thus easier to scratch, and it tends to darken with age.

With the cost of polycarbonate about 2½ times that of acrylic, there is little reason to spend the extra money for polycarbonate portlights unless you are heading out to the high latitudes. Acrylic provides adequate strength for all but the most extreme conditions, is more scratch-resistant, and will remain bright longer. Of course if you would feel more secure with bulletproof portlights, that's what you should install.

Acrylics and polycarbonates are both available with coatings to make them more scuff-resistant, but in the marine environment these coatings invariably peel off like a bad sunburn. Opt for the less expensive basic untreated plastic; lost luster can be easily restored with a quality plastic polish.

Both plastics are easy to fabricate with common woodworking tools. Polycarbonate shows less tendency to chip, but more tendency to heat up in the cut and bind the blade. Lubricate the blade with beeswax or bar soap, and use a moderate blade speed.

DEAD FRAMES

THREADED SOCKETS, ESPECIALLY IN ALUMINUM frames, are prone to corrode and may strip when you back out the mounting screws. If that happens, you can still reuse the frames by drilling through the

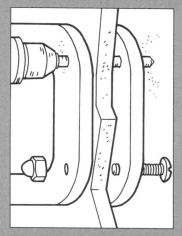

outside frame (down the center of the stripped sockets), countersinking the new holes, and through-bolting the frames with barrel screws or with oval-head screws and cap nuts.

Sometimes aluminum frames are so corroded underneath that they simply disintegrate when dismantled; if that happens, you can buy new frames from the original manufacturer, special order frames through Bomar, Vetus (W. H. Denouden), or Hood Yacht Systems, or have them fabricated by a local machine shop.

Deadlights are sometimes installed without a frame, captured in a rubber gasket like an old automobile windshield. The gasket may have a metal or plastic insert to "lock" it in position. Rubber-mounted deadlights are more often found on small boats and are probably safe enough for a boat used inshore. You work a new pane into the gasket with the aid of a soap solution and a rounded prying tool—like putting a tire on a rim—but by

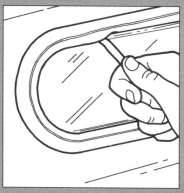

the time it is time to replace the pane, the rubber gasket is usually too brittle to stretch over the plastic without tearing. You may be able to find replacement gaskets, but if the boat will be operated outside of protected waters, where the punch of a boarding wave could push the portlight out of the gasket, consider reinstalling the pane with a rigid mounting system.

If you don't object to the change in look, one of the most secure methods of installing a deadlight is to simply cut the plastic pane an inch or so larger than the opening all around and

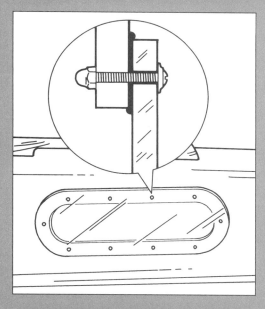

through-bolt the pane directly to the cabin side—well-bedded, of course. Drill the mounting holes a drill size or two oversize to allow the plastic to expand and contract. Space the holes at roughly 12 times the thickness of the pane. Don't countersink the holes; the wedge effect of countersunk fasteners will eventually crack the plastic. Instead, use panhead bolts (preferably with oversize heads) or use finishing washers under oval-head bolts. Hide the raw edges of the hole in the cabin side with trim.

MAST BOOTS

Leaks around a keel-stepped mast indicate boot failure. The time to install a new mast boot (or coat) is when the mast is being stepped. Universal molded replacement boots are available, but a cheaper, more versatile, and more durable alternative is a section of inner tube; your nearest tire dealer probably has a bin full of discards that would provide a suitable section.

Slip the boot up the mast inside out and upside down. Once the mast is stepped and the rubber chocks are in place, slide the boot down and clamp the lower end to the mast with a boot clamp (a BIG hose clamp). Now turn the top of the boot down over the clamp—like rolling down a sweatsock—and stretch it over the deck flange. A clamp around the flange completes the seal. If the mast has an extruded sailtrack, before installing the boot, fill the track in the clamp area with polysulfide sealant or epoxy putty and let the compound dry.

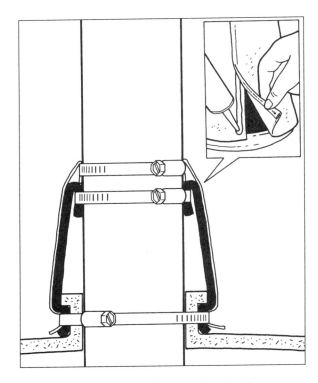

Extend the life of the boot by protecting it with a canvas coat. Cut the canvas so the edges overlap and "button" them with a bead of polyurethane after the coat is installed. The canvas should be captured under its own clamp on the mast, but it can share the flange clamp with the boot.

HULL-TO-DECK JOINT

In older fiberglass boats, deck joints are often an annoying source of leaks. Rare is the old fiberglass boat that doesn't exhibit a bead of silicone somewhere along the edge of the caprail or toerail, placed there in some past ill-fated attempt to stop the intrusion.

Better fastening techniques and better joint compounds have improved hull-to-deck joints, and these improvements can be applied to older boats.

EVALUATING THE JOINT

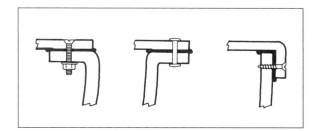

1 Most hull-to-deck joints fall into one of three categories: inboard flange, outboard flange, or shoebox. The best joints are fiberglassed together into a single strong and leak-free unit, but few boats are built this way. Most are joined mechanically with rivets, screws, or bolts, and depend on sealant to keep water out.

2 Gaining clear access to the deck joint almost always involves removing the rail, which requires bedding anyway. The outboard flange is generally the easiest joint to reseal. Cabin joinery can make access to the other types impossible without virtually dismantling the interior. In such cases, a compromise repair may be the best alternative.

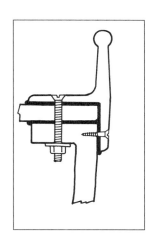

REBEDDING

How big the job of repairing a leaking hull-to-deck joint is depends almost entirely on how much dismantling is required to get to the joint fasteners. The good news is that if you do it right, you are unlikely to need to do it again for at least 20 years. Don't cut corners.

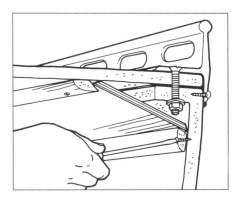

1 Remove the rail. Metal rails may be through-bolted (best), screwed in place (a distant second), or sometimes riveted (bad). Wood rails will be bolted or screwed; drill out the bungs to get at the heads of the fasteners. The nuts are often accessible behind easily removed panels in the cabin.

2 Remove mechanical fasteners holding the joint together. Don't be surprised if these are widely spaced; manufacturers often installed only a sufficient quantity to hold the flange together until the rail was installed, depending on the rail fasteners to do double duty. Grind the heads off any rivets and punch them out.

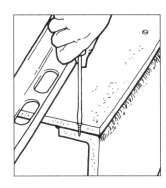

3 Reef out old bedding compound. Early fiberglass boats were bedded with an oil-based mastic that eventually dries out and shrinks. For a secure seal, all the old caulking must be removed. Clean the joint thoroughly with acetone, using a sharpened putty knife to separate the flanges as much as possible without damaging the laminate.

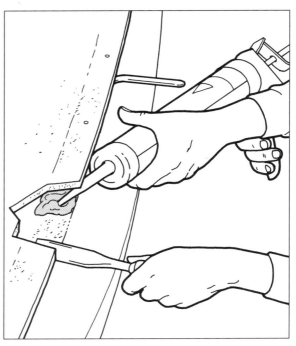

FIBERGLASSING

THE BEST SOLUTION FOR A LEAKING HULL-to-deck joint is to join the two parts permanently with fiberglass lay-up. This can be done either inside or outside, depending on access and the design of the joint. Instructions for laying up fiberglass are provided in "Laminate Repair."

4 With the flange pried open, fill the gap with polyurethane sealant (3M-5200 or equivalent). For a permanent seal, it is imperative to have a continuous line of sealant that passes both outboard and inboard of the fastener holes.

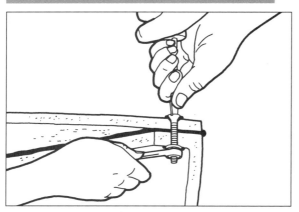

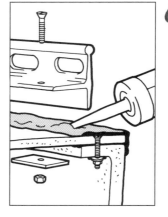

6 Reinstall the rail, bedding it as detailed earlier. Be sure to use backing plates or at least oversize shoulder washers under nuts holding slotted rail or T-track.

5 Refasten the flange. Through-bolt if possible. If not, use self-tapping screws, not rivets. Generously bed the bolts or screws in sealant.

CENTERBOARD TRUNKS

Wooden centerboard trunks are notorious leakers, but fiberglass trunks seldom leak except around the pivot pin. This is easily avoided by making the pin part of an internal frame that slips inside the case, but more often the pivot pin in a true centerboard boat—one without an external keel—passes through the trunk from inside the boat, sealed on either side by rubber grommets. When the grommets harden, a leak occurs. Tightening the nut is not the solution and may distort the case enough to jam the centerboard. Stop the leak by replacing the grommets.

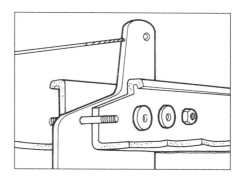

Pulling the pivot pin with the boat afloat will admit a significant flow. This is a job better done with the boat out of the water. Be sure the centerboard is well supported when you pull the pivot pin or the board will drop out of the trunk.

In a keel/centerboard boat, the pivot pin usually passes through the stub keel. A leak here cannot be repaired effectively with grommets or sealant. Details for repairing the pivot pin of a keel/centerboard boat are provided in "Keel and Rudder Damage."

THROUGH-HULL FITTINGS

Through-hull fittings last a very long time, but occasionally require replacement. Modifications to the boat may necessitate an additional through-hull, but the prudent skipper will minimize the number of holes through the hull of his boat by tying into existing through-hulls whenever practical.

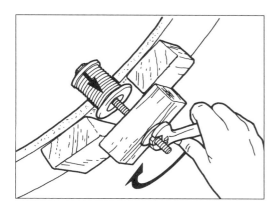

1 Removal of the old through-hull is easy with the aid of a long bolt and a washer large enough to sit on the through-hull. With the retaining nut or seacock removed, pass the bolt through the through-hull and through a wooden block outside the hull. Block the ends of the wood clear of the hull and tighten the nut on the bolt.

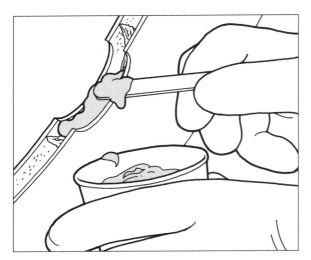

2 The hull must be solid around a through-hull. If the hull is cored, hollow an area around the hole at least as large as the flange of the seacock and fill the hollow with epoxy thickened to mayonnaise consistency with colloidal silica.

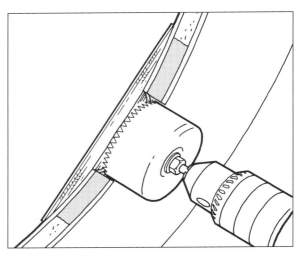

3 Fitting a flat seacock flange to the concave inner surface of a hull requires a contoured backer block. A backer block also spreads the load on the hull. The easiest backer block is a ring cut from plywood and shaped to the hull with a sander, but if you use plywood, saturate it with epoxy before you install it (cured), or the slightest seacock leak will quickly destroy it. Laminating incrementally larger circles of fiberglass to the hull to build up a flat island is a better approach (see "Laminate Repair" for laminating instructions). Capturing the flange bolts under the laminated pad to eliminate bolt holes through the hull makes a very nice installation. Drill the center hole in the pad (from outside) after the laminates cure.

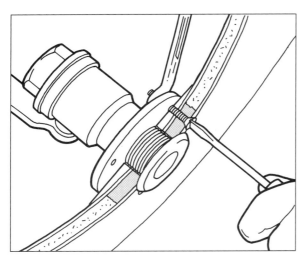

4 Assemble the through-hull and the seacock. If the through-hull bottoms out into the seacock, you need additional pad thickness. If you have not glassed in studs, drill and countersink mounting holes through the hull. Bolt the seacock in place; never install a seacock by simply threading it onto the through-hull. Bed the mounting bolts well.

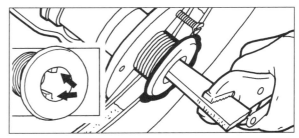

5 Remove the through-hull, turning it with a file, steel plate, or hardwood wedge against the internal ears. Butter the hole, the edge, and the through-hull threads and flange with polyurethane sealant. Reinstall the through-hull and tighten; caulk should squeeze out all around the outside flange. Clean away the excess and use some of it to fair the heads of the flange bolts.

PRESSURIZING TO FIND LEAKS

Some leaks into the cabin are obvious, but most aren't. Water may leak through the deck, then travel along the top of a headliner 10 feet or more before finding an exit and dripping out. The traditional way of finding leaks is to flood the deck, moving the hose incrementally "up" the deck until the drip appears. This method often fails. Here is a method that requires a bit more effort, but it will locate every leak.

 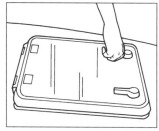

1 Shut all seacocks and close all hatches.

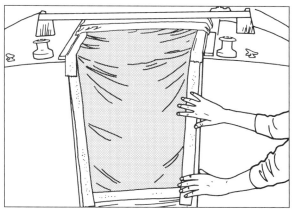

2 Use duct tape to seal all openings you don't expect to be airtight, i.e., ventilators, cockpit hatches, hawsepipes, etc. Seal the companionway with plastic sheeting (a garbage bag will be adequate) edge-taped over the hatch and the dropboards.

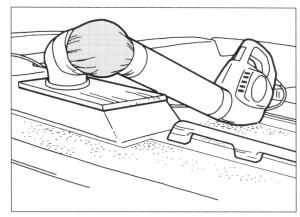

3 Insert the nozzle of a small electric leaf blower into an open ventilator or deck plate and seal it with tape. A shop-vac with the hose on the "blow" side will also serve.

With the blower running (give it five minutes to pressurize all the internal spaces), sponge soapy water over all ports, hatches, and hardware. Anywhere you see bubbles, you have a deck leak.

After you rebed the identified fitting, you can pressure-test again to confirm that the leak is

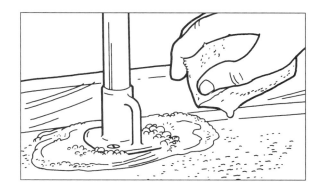

resolved, but don't leave the tape in place for more than a few hours—never overnight—or you will have great difficulty removing it.

RESTORING THE GLOSS

Production fiberglass boats are built by laying multiple laminates into a boat-shaped mold. The interior of the mold is polished mirror-bright and coated with a releasing agent (wax); then the first layer, called gelcoat, is sprayed onto the mold surface. The initial layer of fiberglass is applied to the "back" side of the gelcoat, and additional layers are added until the builder achieves the desired thickness.

This is opposite of the way most other products are manufactured, where the last step in production is to spray on the finish—presumably the reason it's called the finish. Gelcoat is the start.

Gelcoat also differs from paint in other important ways. The bond between paint and the underlying surface is mechanical—that Passion Fruit Crimson enamel on your old Roadmaster is hanging on (or not) by gripping microscopic scratches put there by sanding or chemically etching the metal. Between gelcoat and the underlying laminates, the bond is chemical; the resin saturating the first layer of glass material combines with the exposed surface of the gelcoat to form a single mass—not unlike pouring warm gelatin over cold. This is called chemical cross-linking, and it occurs because gelcoat resin and the polyester resin used to saturate the layers of fiberglass material are the same basic product. Gelcoat is essentially pigmented polyester resin.

Gelcoat resin has poor flow characteristics. Good paints are self-leveling—like water—drying to a smooth, glossy finish, but gelcoat resin behaves more like plaster, taking on the texture of the application tool. It can be thinned and sprayed to get a reasonably smooth finish, but the "wet-look" gloss characteristic of new fiberglass boats is due entirely to the highly polished interior surface of the mold.

Gelcoat is much thicker than a paint finish. For example, the dry film thickness (DFT) of a typical polyurethane finish (Awlgrip) is 1.5 to 2 mils (0.0015 to 0.002 inch) thick. The thickness of the gelcoat layer of a boat just popped from the mold is 20 mils, give or take 3 or 4 mils. In other words, the paint on a painted surface is typically thinner than a single page of this book, while a layer of gelcoat will normally be about 10 pages thick.

A well-applied gelcoat (like everything else, there are quality differences between manufacturers) will generally last 10 years with minimal or no care. Protected with an annual coat of wax and compounded in later years, gelcoat can maintain its gloss for 20 years or more. The longevity of gelcoat is due primarily to its thickness. When the surface dulls and chalks, the "dead" layer can be abraded off and the fresh surface underneath polished to restore the gloss.

Thickness can also be the enemy. If the builder applies the gelcoat too thickly—often done with the best intentions on early fiberglass boats—it eventually cracks like dried mud. A faulty resin formulation can also cause cracking and crazing.

Except for color matching, gelcoat repairs are easy and straightforward.

BUFFING

The most common surface malady of fiberglass boats is a dull finish. This is brought on almost entirely by exposure and can be delayed significantly by regularly waxing the gelcoat. When unprotected gelcoat becomes dull and porous, perhaps even chalky, waxing will no longer restore the gloss. The damaged surface must be removed by buffing the gelcoat with rubbing compound.

START WITH A CLEAN SURFACE

1 Wash. Scrub the surface thoroughly with a solution of 1 cup of detergent per gallon of water; choose a liquid detergent, such as Wisk. To make the solution even more effective, fortify it with trisodium phosphate (TSP), available at any hardware store. If the surface shows any signs of mildew, add a cup of chlorine bleach to the mix. Rinse the surface thoroughly and let it dry.

2 Degrease. Soap solutions may fail to remove oil or grease from the porous gelcoat. To degrease the surface, sweep it with an MEK-soaked rag. (Acetone can also be used, but the slower-evaporating MEK holds contaminants in suspension longer.) Protect your hands with rubber gloves and turn the rag often, changing it when a clean area is no longer available.

3 Dewax. Rubbing compound works like very fine sandpaper, and wax on the surface can cause uneven cutting. In addition, if the surface has silicone on it (nine boats out of ten do), the compound drags the silicone into the bottom of microscopic scratches, which will cause you grief if you ever paint the hull. Wipe the hull with rags soaked in toluene or a proprietary dewax solvent. Wipe in a single direction, usually diagonally downward toward the waterline.

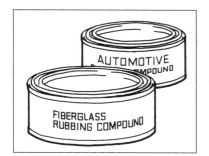

CHOOSE THE RIGHT COMPOUND

Gelcoat is much softer than paint and requires a gentler rubbing compound. Select a compound formulated for fiberglass. If the gelcoat is in especially bad shape, the heavier abrasion of an automotive compound can provide faster surface removal, but it must be used with caution to avoid cutting through.

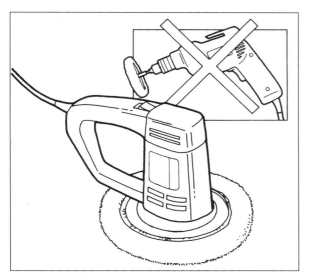

PLUG IN

Rubbing compound can be buffed out by hand if the area is small, but hand-buffing an entire boat is not recommended. An orbital polisher is far cheaper than an artificial elbow. Don't try chucking a buffing disk into your electric drill; it will eat right through the gelcoat, or you'll burn up the drill running it slow.

THE RIGHT PRESSURE

How much of the surface the compound removes relates directly to how much pressure you apply. Since you always want to remove as little gelcoat as necessary, never use any more pressure than is required. You will have to experiment with how much that is.

Whether you are compounding a small repair by hand or an entire hull with the aid of a machine, the process is the same. Working a small area at a time, apply the compound to the surface by hand, then buff it with a circular motion. Use heavier pressure initially, then progressively reduce the pressure until the surface becomes glassy.

If the gelcoat shows swirl marks, buff them out with a very fine finishing compound.

SANDING

Sometimes the dead layer of old gelcoat is so deep that removing it with rubbing compound becomes interminable. In that case, the process can be accelerated by sanding the surface first. This only works if the gelcoat is thick; if you sand through the gelcoat, it is too thin to restore and you will have to paint the surface to restore its gloss.

HIGH SPEED AND HIGH RISK

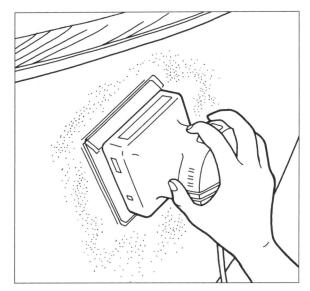

1 The safest way to sand gelcoat is by hand, but you can slash the time required to remove the dead surface layer by using a power sander. You will need a ¼-sheet finishing sander—called a palm sander. Load it with 120-grit aluminum-oxide paper (it's brown). It is a good idea to start in an inconspicuous spot to make sure your gelcoat is thick enough to take this treatment. Keep in mind that the sander is working at about 200 orbits per second, so keep it moving and don't sand any area more than a few seconds. Apply only as much pressure as needed to maintain contact. This first pass removes most of the material; if the gelcoat doesn't get transparent, good results from the remaining steps are likely.

2 Don't let the sander run over any high spots, ridges, or corners, or it will cut through the gelcoat regardless of how thick it is. Change paper when the amount of sanding dust diminishes.

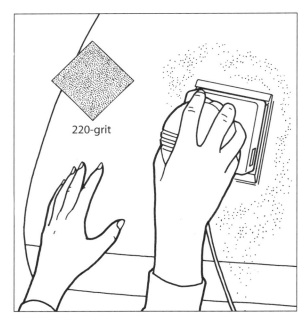

220-grit

3 When you have run the sander over the entire area, change to 220-grit paper and do it again.

WET SAND

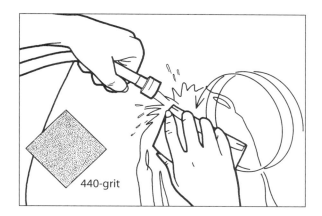

440-grit

1 Remove the scratch marks power sanding left behind by wet sanding the surface with 400-grit wet-or-dry (silicone carbide) sandpaper. Hand sand with a circular motion, keeping a trickle of water running on the sanding area.

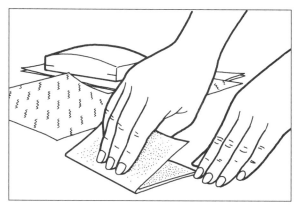

2 To ensure a uniform surface, backing sandpaper with a rubber or wooden block is usually a good idea, but when the grit is very fine—320 or higher—you will get the same results and perhaps better control from finger-backed sanding. Fold the sandpaper as shown to keep the paper from sanding itself and to provide three fresh faces from each piece of paper.

3 Wear cloth garden gloves—the kind with the hard dots—to save the tips of your fingers.

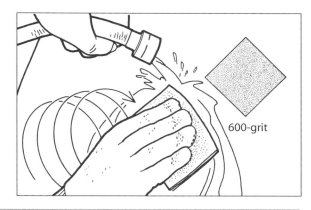

600-grit

4 Make a final pass with 600-grit wet-or-dry paper and the surface should be ready to buff to a like-new gloss.

WHAT?

THE HIGH SPEEDS OF PALM SANDERS—about 14,000 rpm—can result in an ear-damaging shriek. **Earplugs** are available from any drugstore for about a buck; buy a pair and use them. Not only will they save your hearing, but by eliminating the fatigue that accompanies such an assault on the senses, they actually make this job much easier.

SCRATCH REPAIR

Scratches are less visible on gelcoat than on paint since they don't cut through to some different color base. If the surrounding gelcoat is in good condition, always make surface damage repairs with gelcoat rather than paint. Even though the gelcoat application may initially be rough, it can be sanded smooth and polished to blend imperceptibly with the rest of the hull. For dealing with deeper gouges that also damage the underlying laminates, see "Hull Repairs."

OPENING A SCRATCH FOR REPAIR

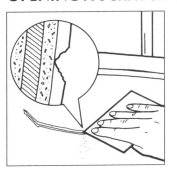

Never try to repair a scratch by simply painting over it with gelcoat. Gelcoat resin is too thin to fill the scratch, and if the resin is thickened to a paste, the paste bridges the scratch rather than filling it. To get a permanent repair, draw the corner of a scraper or screwdriver down the scratch to open it and put a chamfer on both sides.

GELCOAT CHOICES

You will find gelcoat available as both a resin and in a thicker putty form called paste. Paste is what you want for scratch repair. Kits containing a small amount of gelcoat paste and hardener along with a selection of pigments can be purchased for less than $20.

WHAT COLOR IS WHITE?

THE HARDEST PART OF A REPAIR TO THE SURFACE of a fiberglass boat is matching the color. Even professionals that do gelcoat repairs daily have difficulty getting a perfect match. This is one of the few places that may call for conditioning yourself to be happy with a self-assessment of "not bad."

You can purchase gelcoat as unpigmented resin, in a kit with a half-dozen different colors of inorganic pigments, or in "factory" colors for the most popular boats. Because pigments fade, if a boat has seen a few years in the sun, even factory colors won't match exactly.

For small repairs to a white boat, a kit with pigments should serve; getting close is much easier with white, and once the repair is buffed out to a gloss, small shading differences will be unnoticeable.

For colored hulls and larger repairs, getting an adequate match is more difficult. It essentially requires tinting an ounce of gelcoat with one drop of pigment at a time and touching the

resulting mix to the hull until you get a match. Keep track of the number of drops of each tint per ounce to reach the right color. Guys, get your wives or girlfriends to help you with this part; men are eight times more likely to have defective color vision—a minus that becomes a plus if your repair is slightly off (you won't notice).

For additional assistance in matching colors, see *Sailboat Refinishing* in this series.

CATALYZING

The hardener for gelcoat is the same as for any polyester resin—methyl ethyl ketone peroxide, or MEKP. Gelcoat resin usually requires 1 to 2 percent of hardener by volume (follow the manufacturer's instructions). As a general rule, four drops of hardener will catalyze 1 ounce of resin at 1 percent. The mix shouldn't kick (start to harden) in less than 30 minutes. Hardening in about two hours is probably ideal. *Always err on the side of too little hardener.* Also be certain to stir in the hardener thoroughly; if you fail to catalyze every bit of the resin, parts of the repair will be undercured.

SPREADING GELCOAT PASTE

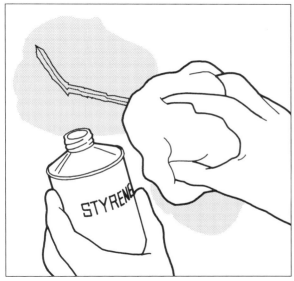

1 Original gelcoat is chemically bonded to the underlying laminates, but this molecular bond applies only to lay-up; the bond between a long-cured hull and an application of fresh gelcoat over a ding or scratch is strictly mechanical—just like paint. Wiping the scratch with styrene just prior to coating *can* partially reactivate the old gelcoat and result in some chemical crosslinking, but as a practical matter this step is usually omitted.

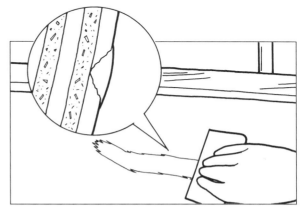

2 Apply gelcoat paste like any other putty; a plastic spreader works best. Let the putty bulge a little behind the spreader; polyester resin shrinks slightly as it cures, and you're going to sand the patch anyway. Just don't let it bulge too much or you'll make extra work for yourself.

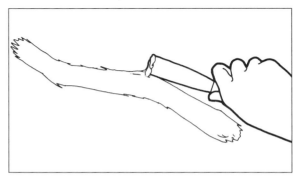

3 Scrape up any excess beyond the patch area.

COVERING THE REPAIR

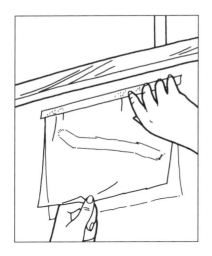 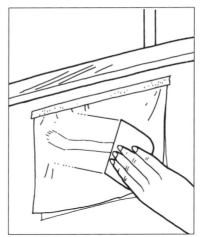 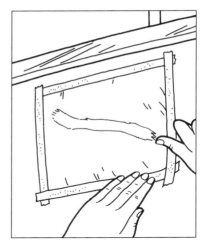

Gelcoat will not fully cure in air. Large repairs require a coating of polyvinyl alcohol (PVA) to seal the surface (see "Laminate Repair"), but to seal a scratch repair, cover it with a sheet of plastic. A section of kitchen "zipper" bag works especially well because it tends to remain smooth and the gelcoat will not adhere to it. Tape one edge of the plastic to the surface just beyond the repair, then smooth the plastic onto the gelcoat and tape down the remaining sides.

SANDING AND POLISHING GELCOAT REPAIRS

After 24 hours, peel away the plastic. The amount of sanding required will depend on how smoothly you applied the gelcoat.

1 For a scratch repair, a 5-inch length of 1 x 2 makes a convenient sanding block. Wrap the block with 120- or 150-grit paper, and use the narrow side to confine your sanding to the new gelcoat. Use short strokes, taking care that the paper is sanding only the patch and not the surrounding surface. Never do this initial sanding without a block backing the paper.

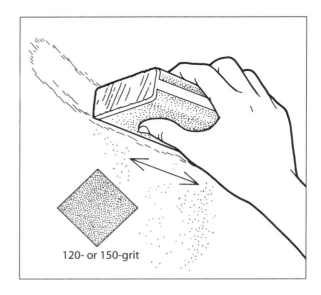

120- or 150-grit

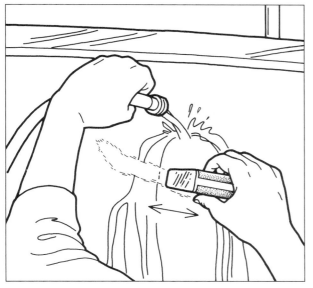

2 When the new gelcoat is flush, put 220-grit wet-or-dry paper on your block and wet sand the repair, feathering it into the old gelcoat until you can detect no ridge with your fingertips.

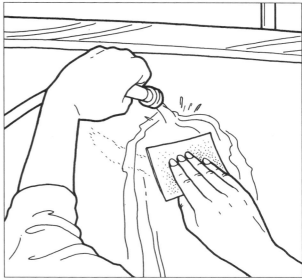

3 Switch to 400-grit wet-or-dry, abandoning the block, and wet sand the surface until it has a uniform appearance. Follow this with 600-grit wet-or dry.

4 Dry the area and use rubbing compound to give the gelcoat a high gloss. On small repairs, you can buff the gelcoat up to a gloss by hand. Give the repair area a fresh coat of wax. If your color match is reasonably good, the repair will be virtually undetectable.

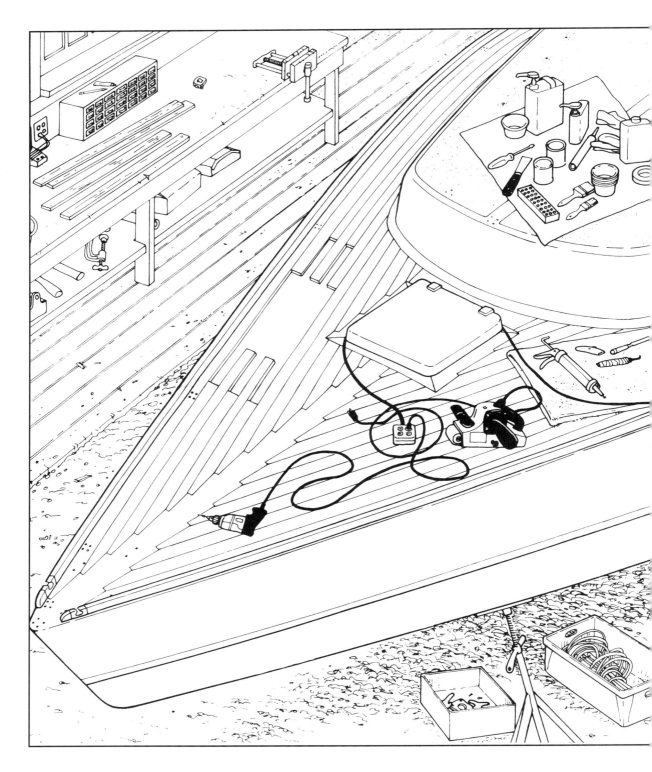

DECK REPAIRS

A fiberglass boat is typically molded in two sections: the hull and the deck. Most of the furniture and machinery is installed inside the open hull before the deck goes on—like filling a box before putting on the lid.

If you stay off the rocks and don't smash into the dock, the hull has a pretty good life—coddled by the water and always half in the shade. The deck, on the other hand, is born to a life of abuse. It sits out in the sun like a piece of Nevada desert. It is assaulted by rain, pollution, and foot. It is eviscerated by openings, pierced by hardware, pried by cleat and stanchion.

You might think that to stand up to such treatment, decks are as strongly built as the hull they cover. You'd be wrong. Weight carried low in a boat has little detrimental impact—a builder can make the hull as thick as he feels like—but weight carried high reduces stability. A deck must first be light; strength is defined by "strong enough." As a result, the need for deck repairs is far more common than the need for repairs to the hull.

Deck repairs can also be more complicated (but not necessarily "harder"). While the surface of a hull is flat or uniformly curved and relatively featureless, a deck is a landscape of corners, angles, curvatures, and textures. Damage often extends under deck-mounted hardware. Backside access may be inhibited by a molded headliner. And to provide stiffness without weight, deck construction generally involves a core.

In this chapter we will confine repairs to surface damage. This is hardly a constraint; most deck problems are limited to the deck's top surface.

RECOGNIZING STRESS CRACKS

Stress cracks are easy to identify by their shape. Typically the cracks run parallel or fan out in starburst pattern. You will see parallel cracks in molded corners, such as around the perimeter of the cockpit sole or where the deck intersects the cabin sides. These suggest weakness in the corner. Parallel cracks also show up on either side of bulkheads or other stiffening components attached to the inside surface of the hull or deck. The concentration of flexing stresses at such "hard spots" causes the gelcoat, and sometimes the underlying laminate, to crack.

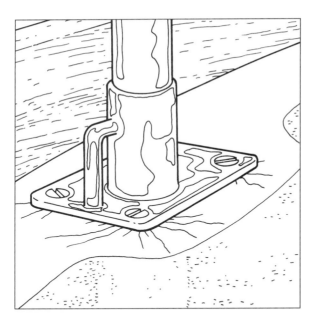

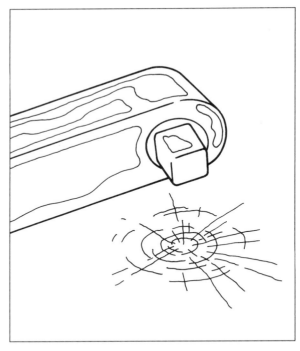

Starburst cracks are also caused by flexing, but in this case the movement centers at a point rather than along an edge. The most common starburst cracking extends from beneath stanchion mounts, brought about by falling against lifelines or by pulling oneself aboard with the top of the stanchion, which literally levers the deck up around the socket mounting holes.

Another cause of starburst cracking is point impact, such as dropping an anchor or a heavy winch handle on deck. (Exterior impact may instead result in concentric cracks—like the pattern of a target.)

ELIMINATING THE CAUSE

Backing plates.

Starburst cracking can usually be stopped by installing generous backing plates on the underside of the deck beneath the offending hardware to spread the load. Wooden plates are the easiest to fabricate, but stainless steel or bronze are better because of their resistance to crushing. Bevel the edges of the backing plate to avoid causing a hard spot. Polished stainless steel plates with threaded holes make for an attractive installation.

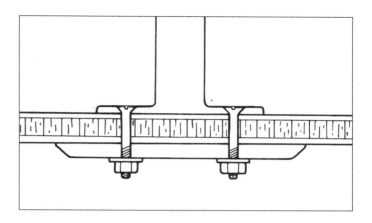

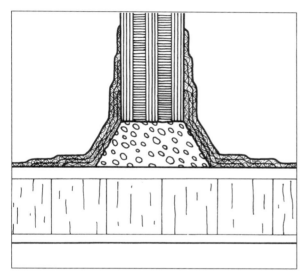

Hard spots.

Hard spots are more common on the hull than the deck, and usually appear where bulkheads attach. Stress cracks around hard spots are likely to return unless you eliminate the hard spot. This typically involves detaching the offending fixture, shaving some material from the edge, then reattaching it mounted on a foam spacer. Realistically, the work required may exceed the benefit, but anytime a bulkhead is detached or a new bulkhead is installed, it should always be mounted with a foam spacer.

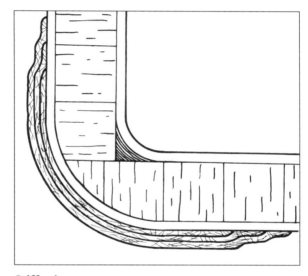

Stiffening.

Stress cracks related to general laminate weakness, such as those that too often appear around cockpit soles, can be prevented by stiffening the area with additional laminates. Laminating instructions are provided in "Laminate Repair." In this case you are trying to add stiffness, not strength, which translates into laminate thickness; use fiberglass mat to quickly build additional thickness.

REPAIRING THE CRACKS

Cracks in the deck typically affect only the gelcoat layer, and perhaps the first layer of mat beneath the gelcoat. Repairs are identical to scratch repair detailed in the previous chapter, except that you may need to remove deck hardware to get full access to the damage. Occasionally flexing has been so severe that stress cracks extend into the woven fabric of the laminate. When this is the case, the strength of the laminate is compromised, and the area must be ground out and relaminated to restore it. Detailed instructions for this type of repair are found in "Hull Repairs."

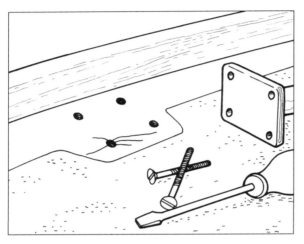

1 Gain access to the entire length of the crack.

2 Open the crack with the corner of a cabinet scraper.

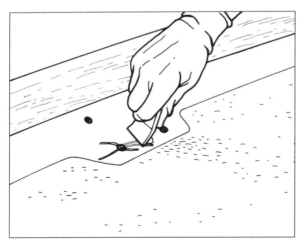

3 Fill it with gelcoat paste.

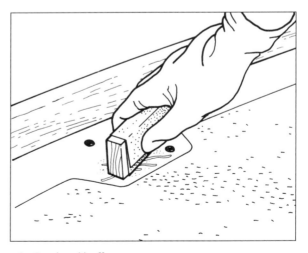

4 Sand and buff.

VOIDS

Voids are thankfully rare in the flat expanses of hull lay-up, but all too common in fiberglass decks. Voids occur when the first layer of cloth is not compressed against the gelcoat (or when a subsequent laminate is not compressed against the previous one). They are often as much a consequence of design as of workmanship. While crisp angles and corners may look stylish, they are more difficult to mold with glass fabric. The fabric resists being forced into a tight corner and after saturation may take a more natural shape, pulling away from the gelcoated mold. The result is a void—a pocket of air beneath the thin gelcoat, perhaps "bird caged" with a few random strands of glass. The first time pressure is applied, the gelcoat breaks away like an eggshell, revealing the crater beneath.

Deck voids are a cosmetic problem and easily repaired.

1 Break away the cracked gelcoat to fully expose the void.

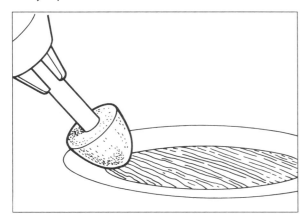

2 Use a rotary grinding point chucked in your drill to grind the interior surface of the cavity. Chamfer the gelcoat all around the void.

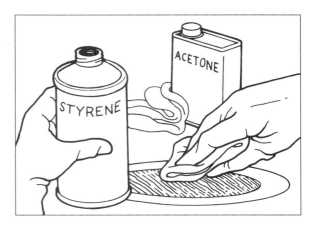

3 Clean the cavity with acetone. For a better bond, wipe the cavity with styrene.

4 Fill the cavity to the bottom of the gelcoat with a putty made from polyester resin and chopped glass. Be sure you use laminating resin, not finishing resin. Epoxy is not recommended because you are going to finish the repair with a layer of gelcoat, and gelcoat does not adhere as well to epoxy as to polyester.

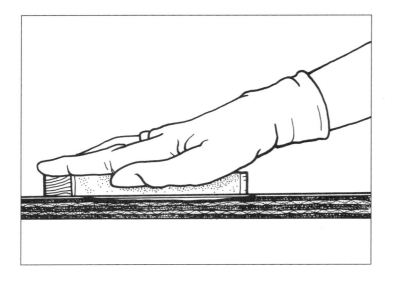

5 When the patch hardens, fill the remaining depression with gelcoat, overfilling slightly. Roll a piece of plastic into the repair and seal the edges with tape.

6 When the gelcoat cures, sand it flush with the surrounding surface and buff it with rubbing compound to restore the gloss.

CRAZING (ALLIGATORING)

Crazing, sometimes called alligatoring, is a random pattern of cracks that, at its worst, can cover the entire surface of a fiberglass boat—both deck and hull. There are two primary causes: flexing and excessively thick gelcoat. If flexing is the culprit, the crazing will be localized. For a repair to be successful, stiffening must be added to the deck in the area where the crazing has occurred.

Fortunately the more common cause of crazing is gelcoat thickness (or occasionally gelcoat formulation). As the hull heats and cools, it expands and contracts. A thin layer of gelcoat accommodates these changes, but thick gelcoat, not reinforced like the underlying laminates, tends to crack. In this case, the crazing is likely to be extensive. That's the bad news; the good news is that the repair doesn't require any structural alterations.

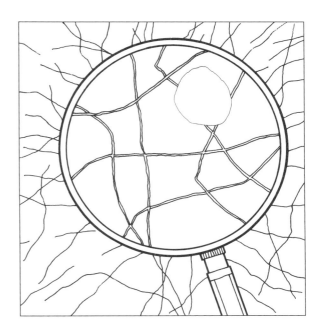

LOCALIZED CRAZING

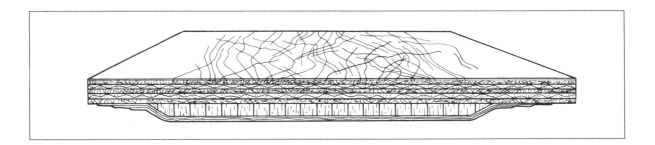

1 Stiffen the crazed area. See "Core Problems" for alternatives and step-by-step instructions.

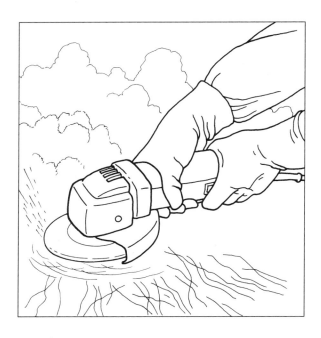

2 Trace each crack with the corner of a cabinet scraper, or if the pattern is too fine, grind the area with a 36-grit sanding disk. Stop when the disk begins to break through the gelcoat; don't grind all the gelcoat away.

3 Paint the cracks or ground area with color-matched gelcoat paste. Seal the surface to let the gelcoat cure.

4 Fair the new gelcoat by block-sanding, then buff to a gloss.

WIDESPREAD CRAZING

Sanding and polishing surface-applied gelcoat is worthwhile when the new gelcoat area is relatively small and the rest of the gelcoat is in good condition, but when the majority of the original gelcoat is damaged, the labor intensive nature of gelcoat application suggests a different approach. The best alternative is painting the entire deck with a two-part polyurethane paint.

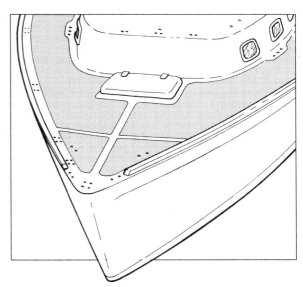

1 Remove as much deck hardware and trim as possible. The quality of your refinishing job is directly related to how much hardware you remove—how unobstructed the deck is when you apply the paint.

2 Clean, degrease, and dewax all the smooth surfaces of the deck. (Nonskid surfaces are restored in a separate process.)

3 Sand the gelcoat thoroughly with 120-grit sandpaper and wipe it dust-free with solvent.

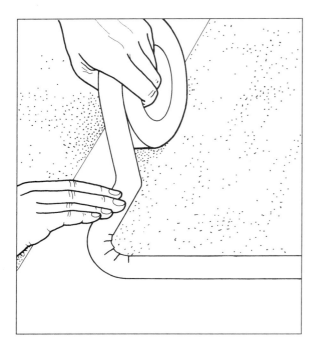

4 Mask nonskid surfaces and any hardware you have elected not to remove.

5 Paint the sanded gelcoat with a high-build epoxy primer. Apply the primer with a foam roller. Two coats are generally necessary to fill all crazing and porosity; machine sand each coat with 120-grit paper.

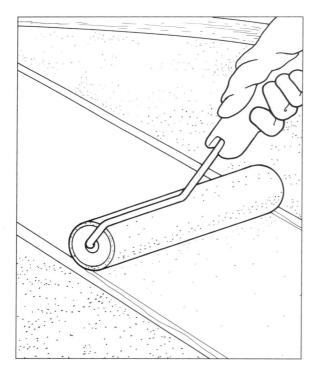

6 Paint the primed surfaces with two coats of two-part polyurethane, following the manufacturer's instructions for rolling and/or brushing the paint. For complete instructions on repainting decks—and all other boat surfaces—see *Sailboat Refinishing* in this book series.

RENEWING NONSKID

If you paint the smooth surfaces of the deck, you will probably want to refinish the nonskid surfaces as well. Painting nonskid surfaces tends to reduce their effectiveness. You can easily offset this by adding grit to the paint.

Always refinish the textured sections of the deck after the smooth portion. There are two reasons for this order. First, the nonskid surface is almost always a darker color than the smooth surfaces, and it is easier to cover a lighter color with a darker one than the other way around. Second, if the final masking is done on the textured surface, it will be hard to get a sharp line between the two. Prepare textured surfaces for refinishing before painting the other parts of the deck.

ENCAPSULATED GRIT

1 Scrub the nonskid thoroughly with a stiff brush, then use terry cloth—sections of old bath towels—to dewax the surface. The rough surface of the terry cloth penetrates the craggy nonskid.

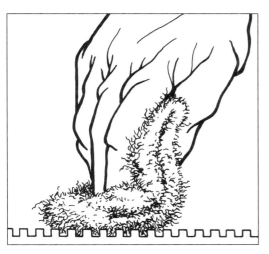

2 You can't sand the bottom surfaces of the nonskid, but abrade it with coarse bronze wool, using short, quick strokes. Fortunately most of the stress on the new paint will be on the top surface, which you can sand with 120-grit paper. Flood the surface and brush-scrub it again, then let it dry.

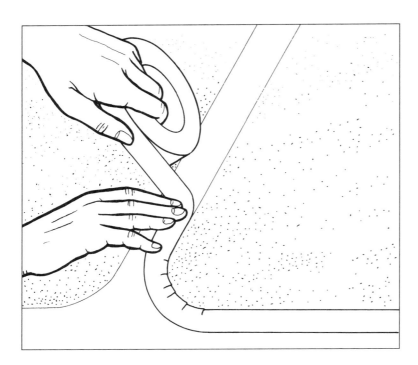

3 After the smooth surfaces are painted and dry, mask them at the mold line of the nonskid.

4 Mix a nonskid paint additive into the paint and roll it on with a medium-nap roller. (There is never any reason to "tip out" the paint on a nonskid surface.) This is the easy way to introduce grit into the paint, but because the additive—usually polymer beads—tends to settle to the bottom of the paint tray, dispersion of the grit on the painted surface can be irregular.

For a more aesthetically pleasing result, first coat the nonskid area with an epoxy primer and cover the wet epoxy with grit sifted from your fingers or a large shaker. When the epoxy kicks, gently sweep off the grit that didn't adhere (you can use it on another nonskid area), and encapsulate the grit that remains with two rolled-on coats of paint.

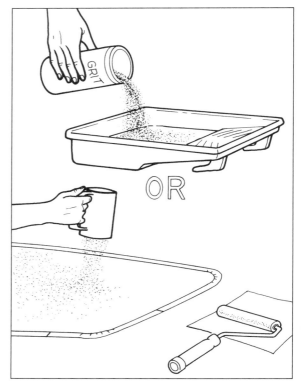

RUBBERIZED OVERLAY

For the best footing, you may want to consider a rubberized nonskid overlay, such as Treadmaster M or Vetus deck covering, also good choices for completely hiding old, worn-out nonskid textures. For overlay application, carry the paint ½ inch into the nonskid area when you paint the deck.

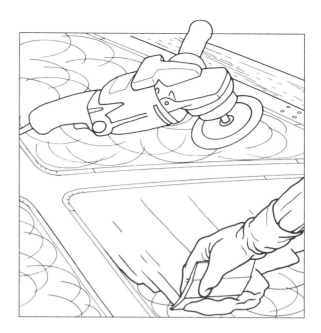

Preparing the surface.
Eliminate all molded texture. Most of it can be quickly taken off with a disk sander and a 36-grit disk. (A belt sander can also be used.) Be careful not to let the sander get outside the textured area. It is neither necessary nor desirable to *grind* away all the pattern. *Fill* the remaining depressions with epoxy putty. When the epoxy cures, sand the surface to fair it and prepare it for the adhesive.

Cutting patterns.

1 Make a pattern from kraft paper for each of the nonskid panels. Cut the paper oversize, then place it on deck to trace the exact outline. Tape across holes cut in the center of the paper to hold it in place. Use a flexible batten to draw curved edges, a can lid for uniform corners. For appearance and drainage, leave at least 1 inch between adjacent panels, at least twice that between the nonskid and rails, coamings, or cabin sides. Write "TOP" on the pattern to avoid confusion when you cut the overlay, and draw a line on it parallel to the centerline of the boat, with an arrow toward the bow.

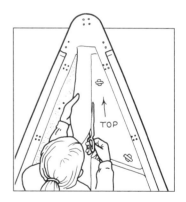

2 Do not cut patterns for only one side, expecting to reverse them for the opposite panels. Boats are almost never symmetrical, and hardware is certain to be in different locations. Cut a separate pattern for every panel. When all the patterns have been cut, tape them all in place and evaluate the overall effect before proceeding. Trace around each pattern with a pencil to outline the deck area to be coated with adhesive.

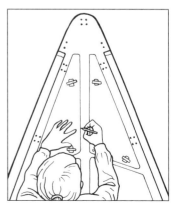

Cutting the overlay.

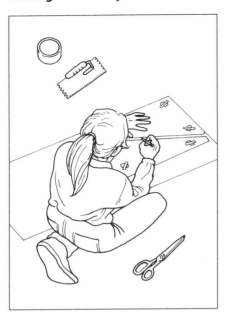

1 Place the patterns topside down on the back of the overlay material. Position all the patterns on your material to minimize the waste before making any cuts. Depending on the overlay you have chosen, it may be necessary to align the patterns; use the line you drew on each pattern for this purpose, aligning it parallel to the long edge of the sheet of material. Trace each pattern onto the overlay.

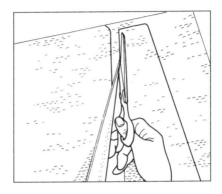

2 Cut out the pieces with tin snips or heavy scissors.

Applying the overlay.

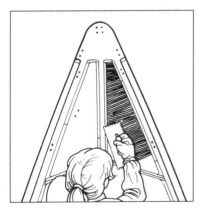

1 If the overlay manufacturer doesn't specify a different adhesive, glue the nonskid to the deck with thickened epoxy. Coat both the outlined deck area and the back of the nonskid with the adhesive, using a serrated trowel.

2 Position the nonskid on the deck and press it flat, beginning with pressure in the middle and working outward to all edges.

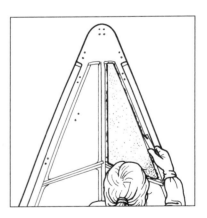

3 Pick up any squeeze-out with a putty knife, and clean away the residue with an acetone-dampened cloth. Continue applying each section in turn until all are installed.

CLEANING

Left untreated, good-quality teak would normally weather to an attractive ash gray, but the assault of modern-day air pollutants turns bare teak nearly black. Clean it with the mildest cleaner that does the job. Start with a 75/25 mixture of liquid detergent and chlorine bleach (no water), boosted with TSP. Apply this with a stiff brush, scrubbing lightly with the grain. Leave the mixture on the wood for several minutes to give the detergent time to suspend the dirt, and the bleach time to lighten the wood, then rinse thoroughly by flooding and brushing.

LIGHTENING

As good as chlorine is at bleaching cotton sweatsocks, it's not a very effective wood bleach. For that you need oxalic acid. You can get it by buying a commercial single-part teak cleaner—oxalic acid is the active ingredient in most—or for about one-tenth the price you can buy a can of Ajax household scouring powder. Whichever you select, brush the cleaner onto wet teak and give it time to work, then scrub the wood with Scotchbrite or bronze wool. (Never, ever, ever use steel wool aboard your boat—it will leave a trail of rust freckles that will be impossible to remove.) Oxalic acid dulls paint and fiberglass, so wet down surrounding surfaces before you start, and keep them free of the cleaner. Rinse the scrubbed wood thoroughly; brushing is essential.

For potential treatments for teak decks, see *Boat Refinishing* in this series.

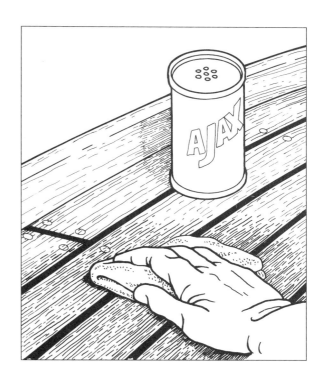

TWO-PART CLEANERS

Two-part teak cleaners are dramatically effective at restoring the color to soiled, stained, and neglected teak, but these formulations contain a strong acid—usually hydrochloric—and should only be used when all other cleaning methods have failed.

1 Wet the wood to be cleaned, then use a *nylon-bristle brush* to paint part 1 onto the wet wood, avoiding contact with adjoining surfaces. If you use a natural-bristle brush, the cleaner will dissolve the bristles; it is doing the same thing to your teak.

2 Scrub with the grain.

3 Part 2 neutralizes the acid in part 1 and usually has some cleaning properties. Paint sufficient part 2 onto the teak to get a uniform color change. Scrub lightly.

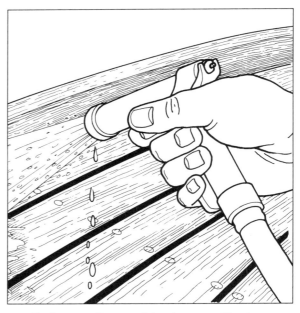

4 Flush away *all* traces of the cleaner and let the wood dry.

SURFACING

After a number of years, bare teak decks become rough and ridged. This unevenness traps dirt and harbors mildew, making the deck harder to clean and harder to keep clean. The solution is to resand the deck with a belt sander, using a 120-grit belt. Keep the sander moving at all times, and sand at about 15° to the grain.

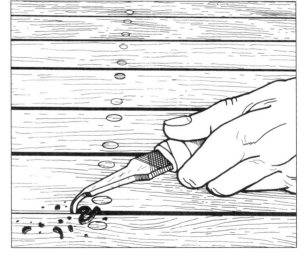

RECAULKING

The instructions that follow are for recaulking a section of a single seam, but the steps are the same for an entire deck.

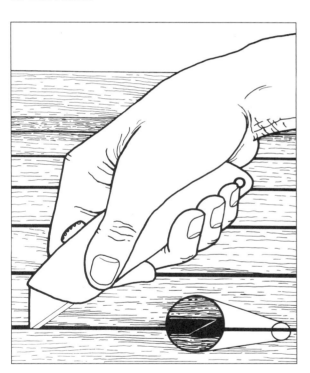

1 With a razor knife, cut the seam caulk at a diagonal a couple of inches beyond the bad section, then slice the section to be replaced free from the planks on either side, taking care not to nick the wood.

2 Dig out the old caulk. This is much easier with a rake made by heating the tail of a file and bending it about 90°. When every bit of the old caulk is off the planks, vacuum the scrapings out of the seam.

3 Use an acid brush or a Q-Tip to thoroughly prime both plank edges. Use the primer recommended for the caulk you are using. Two coats are generally required.

4 Mask the surface of the planks.

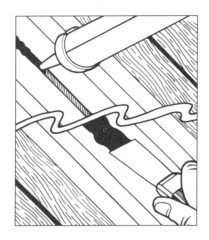

5 The "right" caulk for deck seams is two-part polysulfide. Mix the catalyst into the sealant per label instructions, taking care not to introduce bubbles, then fill an empty caulk tube with the mixture. (For limited repairs, a single-part polysulfide will also give good results and may be more convenient.) Cut the tip of the tube and fill the seam from the bottom. When the entire seam is slightly overfilled with sealant, compress it into the seam by dragging a putty knife over it firmly. The sealant will hump up slightly behind the knife, but it will shrink almost flush as it cures. Remove the masking carefully while the caulk is still tacky.

IDENTIFYING DECK CAULKING FAILURE

HOW DO YOU KNOW WHEN THE SEAM caulking on a teak deck has released its grip on the wood? The wood usually tells you. On a sunny day, scrub the deck, then keep it wet for half an hour or so before letting it dry. Areas along the seams that stay wet longer than the rest of the deck are suspect; spots that stay dark a lot longer definitely indicate caulk failure. Using the point of a knife, you will see that you can separate the caulk from the wood, and the edge of the plank will be wet. Repair all "flagged" seams before they result in bigger problems beneath the teak.

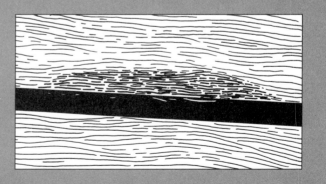

BUNG REPLACEMENT

The most common problem of teak decks is popped bungs. Years of scrubbing thins already-thin overlay planks until the grip of the bungs is insufficient to hold against flexing or expansion.

Just tapping a new bung in place will be a temporary repair at best. Deck overlay bungs require special procedures.

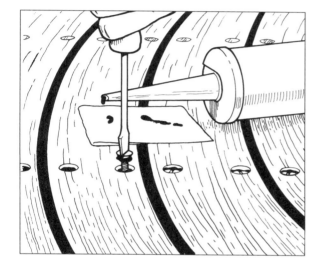

1 Remove and rebed the screw in polysulfide. Deck core problems often occur beneath teak overlay because the screws holding the overlay penetrate the top skin of the deck. *Always* rebed exposed screws. It is a good idea to check the core for sponginess with a piece of wire. (See "Core Problems.")

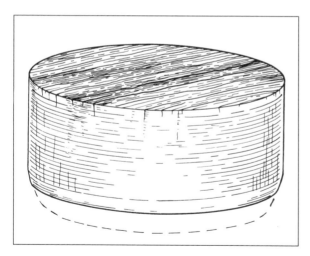

2 Reduce the plug bevel. In a shallow hole, you cannot afford the generous bevel found on most commercial teak plugs; sand the bottom of the plug to reduce it to the bare minimum.

3 Use a Q-Tip to wipe both the hole and the plug with acetone to remove surface oils. Wait 20 minutes to install the plug.

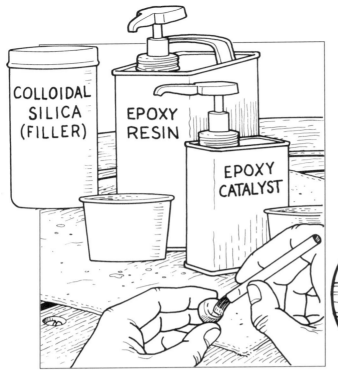

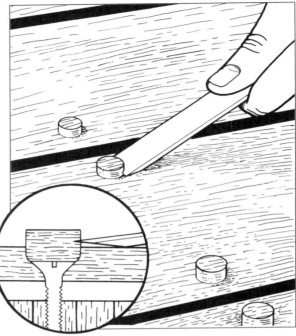

4 Mounting a plug permanently in a shallow hole requires the tenacious grip of epoxy glue. Paint the hole and sides of the bung with unthickened epoxy. Thicken the epoxy to catsup consistency with filler (colloidal silica) and coat the sides of the hole and plug. Tap the coated plug into the hole as far as it will go. Wipe up the excess glue.

5 After the epoxy is dry, place the point of a chisel—beveled side down—against the plug about 1/8 inch above the surface of the plank and tap the chisel with a mallet. The top of the plug will split away.

6 Working from the lowest edge of the trimmed plug, pare away the plug until it is nearly flush with the plank. Finish the job by block sanding the plug with 120-grit sand-paper.

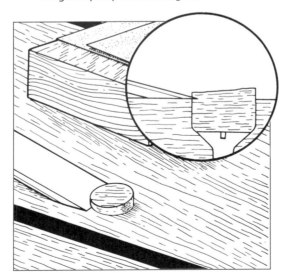

PLANK REPLACEMENT

Occasionally a teak plank splits or is otherwise damaged and requires replacement. More often teak overlay problems have an underlying cause—usually a wet core—and to effect repair the teak must be removed. Since the cause of the leakage often turns out to be the screw holes, some boat-owners elect not to replace the overlay, but most are unwilling to give up the beauty and footing of teak decks. A careful installation minimizes the risk to the core.

1 If you are replacing more than a single plank, number and crosshatch all the planks to be removed so you can put them back properly.

2 With bungs removed (carefully if you will be reusing the plank), extract the screws. Slice the plank free of the caulking all around, and pry the plank up from the bedding compound, using a block under your prying tool to protect the adjacent plank. If the plank is bedded in an organic compound, it should slowly pull free—like a gum-stuck heel. If it is bedded in polysulfide, you are likely to have to destroy the first plank to remove it. With side access, you should be able to separate the rest of the planks from the deck with a thin, sharpened putty knife. A length of steel leader wire connected to two lengths of dowel is sometimes effective in "cutting" deck planks free.

3 Scrape and sand away all old bedding compound.

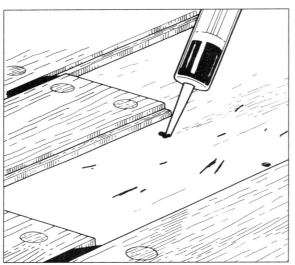

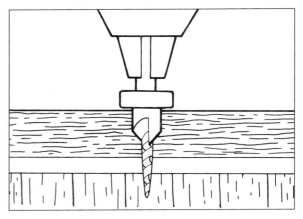

4 If the core is dry, protect it by injecting the hole full of epoxy. Give the epoxy a few minutes to saturate the edge of the core, then draw out the excess with a small brush or a stick. For greater security, drill each hole oversize and, after painting the sides with unthickened epoxy, fill each hole with epoxy thickened with colloidal silica. When the filler hardens, redrill the center for the screw.

5 If you are installing new planks, fill the old screw holes with epoxy putty, but don't redrill them. Wedge the new plank into position, then drill the plank and the deck. Counterbore the hole in the plank at least half the plank's thickness but not more than two-thirds. Epoxy the new holes in the deck in one of the two methods just detailed.

6 Wash the deck and the underside of the plank with acetone. For better adhesion, prime the teak. Coat the deck with black polysulfide (two-part preferred) and distribute it evenly with a saw-toothed spreader.

7 Hold or wedge the plank in position and screw it down. Select Philips-head screws and you will be much less likely to damage the edge of the bung hole with the screwdriver.

8 Install bungs (with epoxy) and trim them. Caulk the seams. Belt sand the deck fair.

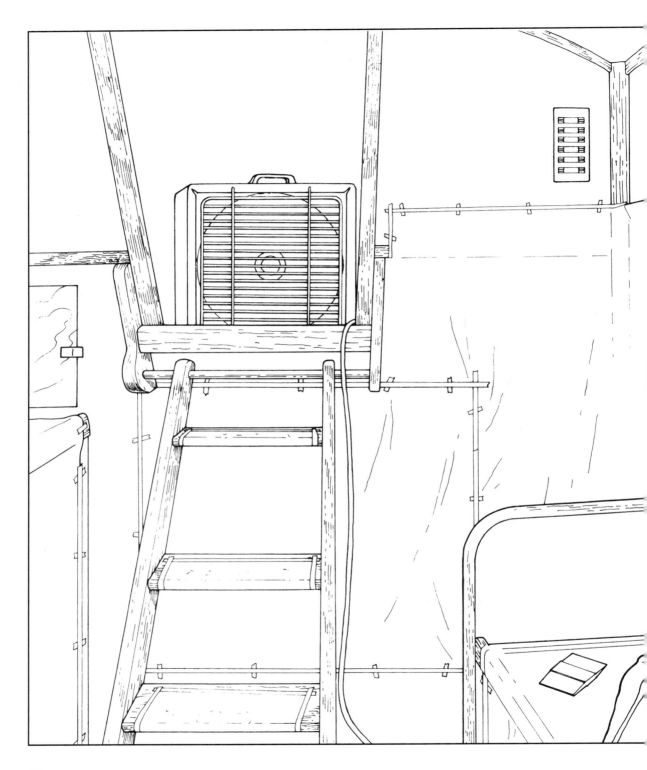

LAMINATE REPAIR

The hull and deck repairs described so far have either been cosmetic or leak related—problems that can be fixed with a proper topical application of one glop or another. But sometimes the problem is below the surface: the original laminate lacks the requisite stiffness; moisture has caused disintegration or delamination; or the glass fibers have been broken by impact. These problems require more extensive repair.

Fiberglass has become the predominant boatbuilding material because of its durability, but it is repairability that accounts for the near immortality of fiberglass boats. The most horrifying hole in a fiberglass hull is quickly healed with a bit of glass fabric, a can of resin, and equal parts skill and care. And the repair is less patch than graft—a new piece of skin indistinguishable from the old.

Fiberglass lay-up can hardly be simpler. It is nothing more than layers of glass fabric saturated with polyester (or epoxy) resin. With a paint brush, a cup of water, and a piece of old T- shirt, you can practice all the requisite skills for fiberglass lay-up.

Don't misunderstand: because of blocked access or complex shape, laminate repair cannot always be honestly characterized as easy, but such problems aren't what make most boatowners shy away from attempting a repair. It's the lay-up. Most boatowners imagine a self-applied laminate as only slightly more durable than a wet Band-Aid. That is a false concern. Follow a few simple rules— provided in this chapter—and your lay-up will be as good as or better than what you can expect a yard to do. And it will remain that way a decade down the road.

UNDERSTANDING POLYESTER RESIN

Polyester resin is the glue that binds glass fibers into the hard substance we call fiberglass. On the other side of the Atlantic, the same product is called GRP—glass-reinforced plastic. As usual, the British take more care with the language than we do; glass-reinforced plastic is *exactly* what it is.

Polyester resin, when catalyzed, hardens into plastic—not one of those tough plastics that deflects bullets or that you can use as a hinge for 100 years—but an amber-colored, rather brittle plastic that seems more like rock candy than boat-building material. But when polyester resin is combined with glass fibers, the sum is greater than the parts.

Polyester resins come in various formulations (see sidebar), although you can't always tell what kind a particular brand is from the label. Generally, you don't need to know. When polyester is appropriate for the repair (sometimes epoxy resin is a better choice), whatever laminating resin your supplier carries should prove satisfactory. Below-the-waterline repairs are the exception; avoid ortho resins if the repair will be continuously immersed.

LAMINATING VERSUS FINISHING

1 You do need to choose between laminating and finishing resin. Laminating resin is "air-inhibited," meaning that the resin will not fully cure while exposed to air. That may sound odd, but remember that polyester solidifies not by drying, like paint, but by a chemical reaction (called cross-linking) induced by adding a catalyst. Air interferes with this curing process.

For any job that requires the laminates to be applied in more than one operation, you need laminating resin. The fact that the surface remains tacky after the resin sets allows you to apply the subsequent laminates without any intermediate steps, and the new application will link chemically with the previous one to form a powerful chemical bond. For a tack-free surface on the final application, coat the resin with polyvinyl alcohol (PVA) or seal it with plastic wrap.

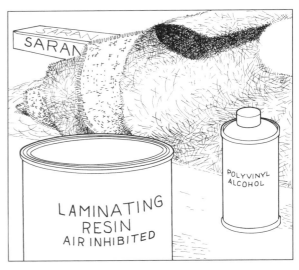

2 Finishing resin is identical to laminating resin, but with an additive that "floats" to the surface of the curing resin. This surfactant (once wax, but now usually a drying oil) seals the resin from the air, thus allowing the surface to fully cure to a tack-free, sandable state.

Use finishing resin for laminate jobs that can be done in a single operation. Finishing resin can also be used for the final layer of a multilayer lamination.

HOW MUCH CATALYST?

The catalyst for polyester resin is methyl ethyl ketone peroxide, or MEKP. Do not confuse MEKP with the common solvent MEK; they are *not* the same.

Polyester resin usually requires 1 to 2 percent of hardener by volume (follow the manufacturer's instructions). As a rule of thumb, four drops of hardener will catalyze 1 ounce of resin at 1 percent. Be certain to stir the catalyst in thoroughly or part of the resin will be undercured, weakening the lay-up.

You can adjust the cure time by adding more or less catalyst. Temperature, weather, and the thickness of the laminate all affect curing times. Some experimentation is generally required. The mix shouldn't kick (start to harden) in less than 30 minutes. Hardening in about two hours is probably ideal, but overnight is just as good unless the wait will hold you up. *Always err on the side of too little catalyst;* if you add too much, the resin will "cook," resulting in a weak lamination.

FIBERGLASS MATERIAL

Fiberglass material is exactly what it sounds like, a weave of glass fibers. For boat construction and repair, the glass comes in chopped-strand mat, roving, and cloth.

Chopped-strand mat.

Chopped-strand mat is made up of irregular lengths of glass strands glued together randomly. Generally speaking, *CSM* is the easiest fabric to shape, gives the best resin-to-glass ratio, yields the smoothest surface, is the most watertight, and is the least subject to delamination, but the short fibers do not provide the tensile strength of a woven material.

Mat is sold by the yard from a roll and comes in various weights designated in ounces per square *foot;* 1½-ounce mat is a good choice for general use.

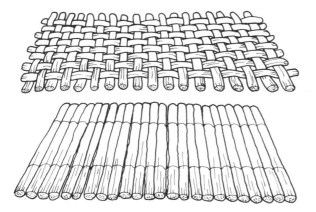

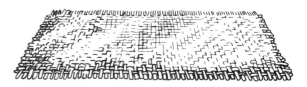

Fiberglass cloth.

Fiberglass cloth looks like shiny canvas but not woven as tightly. Cloth is stronger for its weight than roving, less prone to pulling and unraveling in the laminating process, and the finished product looks better. While manufacturers generally use alternating layers of mat and roving, mat and cloth is a better choice for most repair work.

Cloth is commonly available in weights from 4 to 20 ounces. For any boat over 15 feet, there will be little, if any, fiberglass work that you cannot do with 1½-ounce mat and 10-ounce cloth. If you have a choice, buy 38-inch width.

Roving.

Roving is parallel, flat bundles of continuous glass strands. In *unwoven roving* parallel bundles are cross-stitched together; *woven roving* assembles the bundles in two directions in a loose weave.

The straight, continuous strands in unwoven roving add excellent strength in the direction of the strands but little strength perpendicular to them. For hull and deck repairs, woven roving is usually a better choice because it provides full strength in two directions and good strength in all directions. (You can accomplish the same thing by rotating the orientation of alternating laminates of unwoven roving, but unless you need additional strength in a particular direction, using woven roving is simpler.)

Roving laminated to roving, either unwoven or woven, is unacceptably easy to peel apart. *Always* bind layers of roving together by using a layer of mat between each layer of roving.

For most repairs, select 18-ounce roving. That may sound heavy relative to 1½-ounce mat, but don't be confused. Weight designations for mat are per square foot, while for roving (and cloth) they are per square *yard*; 18-ounce roving weighs the same as 2-ounce mat.

OTHER MATERIALS

GLASS ISN'T THE ONLY MATERIAL THAT CAN be combined with resin. Increasingly, boatbuilders are using "exotic" materials to create composites with special characteristics—light weight, rigidity (or flexibility), impact resistance, tensile strength, or others. These materials include graphite (carbon fiber), Kevlar, polypropylene, xynole-polyester, Dynel, and ceramic. None of these is essential for the typical hull or deck repair. You should understand a material's strengths and weaknesses before you use it.

GRINDING IS ESSENTIAL

During the lay-up process, each application of resin links chemically with the previous application to form a solid structure—as though all the layers were saturated at once. The layer-cake look of fiberglass is deceiving; a better analogy is Jell-O salad—the fruit may be in layers, but the encapsulating Jell-O is solid. Chemical linking between resin layers occurs because each layer is applied before the previous one fully cures (the reason for using air-inhibited resin).

Unfortunately, no matter how strong the laminate-to-laminate bond, the initial bond of any *repair* is mechanical, not chemical. This need not weaken the repair as long as the surface is properly prepared. That means grinding.

1 Before grinding, always wash the area thoroughly with a dewaxing solvent. The original fiberglass will have traces of mold release on the outer surface and wax surfactant on the inner surface. If you fail to remove the wax first, grinding will drag it into the bottom of the surface scratches and weaken the bond.

2 Protect your eyes with goggles and your lungs with a good dust mask. A paper mask is inadequate for all but the smallest grinding task. Long sleeves will reduce skin irritation.

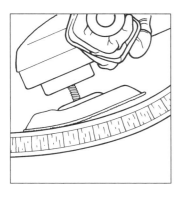

3 Outline the area of the bond and grind inside the outline with a disk sander loaded with a 36-grit disk. Tilt the sander so that only one side of the disk is touching the surface and the dust is thrown away from you.

4 Brush away the dust and wipe the area with an acetone-dampened rag. The surface should have a uniform dull look; if any areas remain glossy, make another pass over them with the sander.

PREPARATION

1 Dewax and grind the surface the lay-up will be applied to. Specific types of repairs—detailed later—require additional surface prep.

2 Protect all surrounding surfaces by masking. Waxing below the repair area will make unanticipated runs easier to remove.

3 Cut the fiberglass pieces to the correct size and lay them out in the order you will be applying them. As a rule, apply the smallest piece first, the largest piece last. Always start and finish with mat.

APPLYING THE INITIAL LAYER

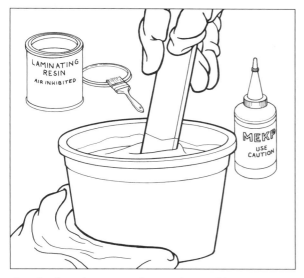

1 Catalyze the resin and mix it thoroughly.

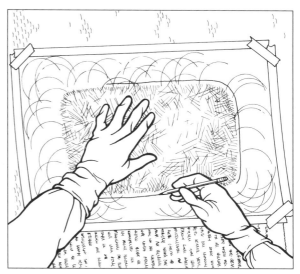

2 Hold the first layer of fiberglass in place and pencil a line around it. Use a throw-away brush to coat the outlined area with resin.

3 Apply the first layer of mat to the wetted surface. On a vertical surface, use masking tape to help hold the fabric in place. Use a squeegee to smooth the mat and press it into the resin.

4 With a brush or a roller, wet the mat with resin until it is uniformly transparent. White areas are dry spots and require additional resin. Brush or roll gently to avoid moving the fabric or introducing bubbles.

ADDITIONAL LAYERS

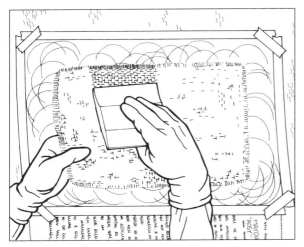

1 Apply the next layer—cloth or roving—on top of the saturated mat. Smooth it against the mat with a squeegee.

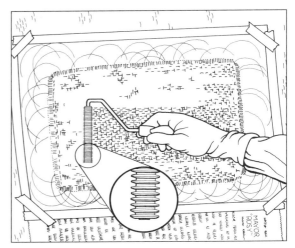

2 Wet out the cloth with resin. Use a squeegee or a grooved roller to compact the laminates and force any air bubbles to the surface. Remove excess resin with the squeegee.

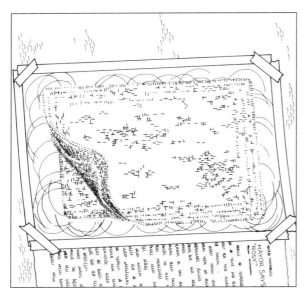

3 If the weather is cool or the repair area is small, you can apply up to two more layers without risk of the cure generating so much heat that it cooks the resin or warps the repair.

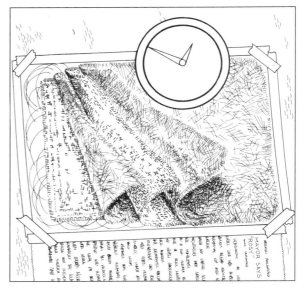

4 Allow the resin time to gel, then mix fresh resin and apply two (or more) additional layers, repeating this process until all the layers have been laminated.

FINISHING

1 For a smoother finish, roll an additional coat of resin over the final layer of material.

2 After the last coat of resin kicks, brush or spray on an unbroken coating of polyvinyl alcohol (PVA) to seal the surface so it will cure fully. This isn't necessary if you use finishing resin for the last coat.

WORKING OVERHEAD

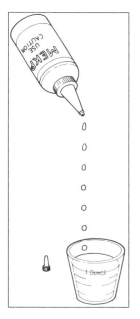

1 Alternative application techniques are required to laminate fiberglass overhead. Mix a small batch of resin, adding more catalyst than usual, and use it to wet the repair area.

2 For an overhead repair, always work with small pieces of fabric. Roll the first piece of mat onto a dowel or a cardboard tube, and wait for the resin to start to kick. When the surface feels tacky, carefully position the edge of the mat and unroll it, taking care to keep it smooth. The tacky resin will hold the mat in place while you saturate it with fresh resin. Use a roller or squeegee to distribute the resin.

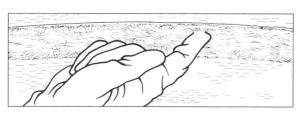

3 When the surface of the mat is tacky, you can unroll the next layer of fabric onto it, again taking care to keep it smooth. Saturate this layer with fresh resin. Repeat this step for each layer until all the laminates are in place.

Epoxy is almost always better than polyester resin for repair work because the mechanical bond it forms—the weakest link in any repair job—is stronger. Laminate made with epoxy is also superior— stronger and more durable—but because the cost of epoxy is more than twice that of polyester, manufacturers rarely use it for laminating. For repair work the additional expense is less significant, and the added strength is well worth the cost.

Do not use epoxy if the repair will be finished with gelcoat. While epoxy bonds tenaciously to polyester, the reverse is not true; the bond between polyester gelcoat and an underlying epoxy repair will not be strong. If the surface will be gelcoated, use polyester resin for the repair.

SELECTING EPOXY

Don't buy epoxy by the tube.

Select an epoxy formulated for boatbuilding. The two most common brands are West System (Gougeon Brothers) and System Three, but there are others. The main difference you are likely to notice between competing brands is the mix ratio, but metered pumps tend to make this difference of little consequence.

ADDITIVES

For saturating fiberglass laminates, use the epoxy as it comes—catalyzed, of course—but for bonding and filling, additives thicken the epoxy and give it specific characteristics. Three of these are especially useful for hull and deck repairs.

Fibers. Fibers added to epoxy will thicken it for filling and for bonding where there is a gap between the surfaces being bonded. You can snip glass cloth diagonally to generate short fibers for small putty needs, but for more than that buy packaged microfiber filler. Fiber fillers are easy to mix, provide good strength, and have excellent finish properties.

Microballoons. Microballoons are tiny hollow beads of plastic. Added to epoxy to produce a fairing compound, microballoons yield a putty that spreads and sands easily. Microballoons reduce the strength of the epoxy and should not be used for bonding or laminating. Also avoid using microballoons below the waterline because the resulting putty is porous and will absorb water.

Colloidal Silica. Silica is perhaps the most versatile of fillers. It provides better strength than microfibers and it doesn't affect the permeability of the cured epoxy. Silica-thickened putty cures with a rough texture and resists abrasion—including sanding.

MIXING

Metering pumps.

While polyester requires only a few drops of catalyst per ounce of resin to start the chemical reaction, the combination ratio for epoxy is much less one-sided. The resin-to-hardener ratio is typically at least five to one, but some formulations call for a two-to-one mix. Epoxy manufacturers typically have calibrated pumps available that will meter out the correct ratio—one pump of hardener to one pump of resin. Epoxy is very sensitive to mix ratio, so the purchase of metering pumps is strongly recommended. Stir the two parts together *thoroughly*, using a flat mixing stick to scrape the sides, bottom, and corners of the container.

Regulating cure time.

Unlike polyester, the cure time of epoxy cannot be adjusted by altering the amount of hardener. The specified proportion of hardener must always be used. However, epoxy manufacturers generally offer at least two hardeners—fast and slow—and they often have additional hardener formulations for special requirements, such as tropical use. Pot life varies with ambient temperature, but you will quickly learn how much time is available. Limit batch size to the amount of epoxy you can use in that amount of time. Epoxy cures faster in the pot, so the quicker you apply it, the longer you will have to work it.

Thickening.

Always add the thickening agent after the resin and hardener have been mixed. Stir in the thickener until the mixture reaches the desired consistency.

PRECAUTIONS

People in significant numbers develop a sensitivity to epoxy so that any exposure results in skin irritation and rash. Avoiding all skin contact is the safest course. Wear plastic gloves when working with epoxy. Goggles are recommended.

Avoid breathing the fumes of curing epoxy. Be sure you have good ventilation.

The heat generated by curing epoxy is sufficient to melt a plastic container and may even ignite into flames if you leave the mixture in the pot too long. If the mixture begins to heat up, put it outside—away from anything flammable—until it cools.

LAMINATING WITH EPOXY

1 Do *not* use chopped-strand mat when laminating with epoxy resin. The binder holding the strands together may react with the epoxy, affecting both the adhesion and the permeability of the epoxy. Epoxy is a strong enough adhesive to bind cloth to cloth without risk of delamination, and multiple layers of fiberglass cloth create an extremely strong laminate.

2 Prepare and wet the repair area as with polyester resin, but then thicken some epoxy to a catsup consistency with colloidal silica. Paint the repair area with the thickened epoxy. This serves much the same function as an initial layer of mat, filling voids and depressions and providing a good contact area for the initial layer of cloth.

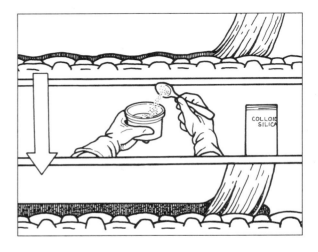

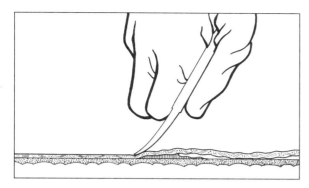

3 Put the initial layer of cloth in position and use the squeegee to smooth it. Wet it out thoroughly.

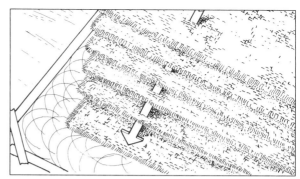

4 Place, wet, and smooth succeeding layers until you achieve the desired thickness. As long as the previous application has not reached a solid cure, which generally takes several hours, you can continue to add layers without any intermediate steps—cleaning, sanding, etc. Each application will chemically bond to the previous one. (If you let the epoxy reach solid cure, you must scrub then grind the surface before applying the next layer—so plan your repair accordingly.)

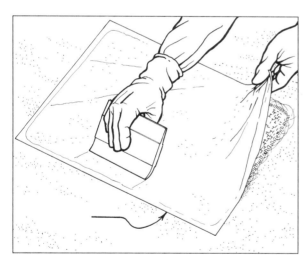

5 For smoothest results, cover the last layer with *peel ply*—a coated fabric epoxy will not adhere to—and use a squeegee to smooth and press the fabric. Scrape away excess epoxy.

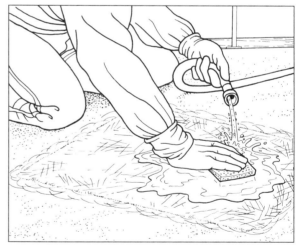

6 When the epoxy has cured thoroughly, remove the peel ply; the waxy amine blush that normally appears on the surface of cured epoxy will peel away on the fabric. If you haven't used peel ply, scrub the cured epoxy with a Scotchbrite pad and water before applying any coating (paint, etc.).

CORE PROBLEMS

Cored deck (or hull) is a sandwich construction of wood, plastic foam, or other material captured between two fiberglass skins. As long as the skins are attached to the core, this box-beam-like construction provides the desirable combination of stiffness and light weight; but if the bond between the skins and the core fails, the three components, none stiff individually, simply slide over each other as they flex—like leaf springs on an old car—and the stiffness is lost. Unfortunately the skin-to-core bond fails often.

This problem is made worse by moisture. Relaxed vigilance in maintaining the seal around deck-mounted hardware leads to water finding its way to the core. The core material, porous by design, absorbs the water like a sponge. The tenuous grip the resin had on dry core material is soon lost, not unlike a bandage releasing its grip under water.

It gets worse still. When water finds its way into the cavity between the two skins, it is captured as surely as if you had poured it into a capped jug. Once core gets wet, it is likely to stay wet until you take steps to dry it out.

The simplest of core problems is delamination, often signaled by cracking sounds underfoot when you walk on the deck. Sound also provides a clue to the more serious problem of wet core, but in this case the sound is a squish. Water may also squirt or weep from around hardware or a crack in the skin when you step on the wet area. Wet core should be attended to immediately. If left unattended until the deck feels spongy underfoot, the core is probably rotten and the repair job you face formidable.

The message here is simple: keep the deck well sealed and you won't need most of the information in this chapter. If it is already too late for this advice, then here are the instructions for dealing with core problems.

DELAMINATION

Sailboats rarely suffer from delamination except where core is involved. Between laminates of fiberglass, the bond is chemical (as long as the fabricator didn't delay too long between layers) and strong, but between the skins and the core the bond is mechanical and weak. Most often the outer skin separates from the core.

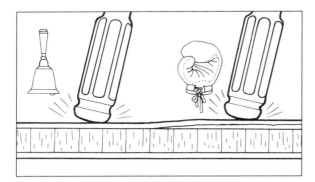

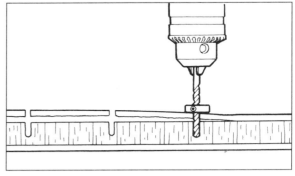

1 Map the area of delamination. Use a plastic mallet or the handle of a screwdriver to tap the area, listening for the telltale difference in sound. A void will have a dull, flat tone compared to the resonance of solid laminate.

2 Drill several small holes (³/₁₆") just inside the outline. The holes should penetrate the outer skin and the core, but not the inner skin. Check the core material pulled out by the drill to make sure the core is dry. Poke at the core through the holes with a piece of wire to make sure it is solid. If the core is wet or spongy, additional steps are required—see the next two sections.

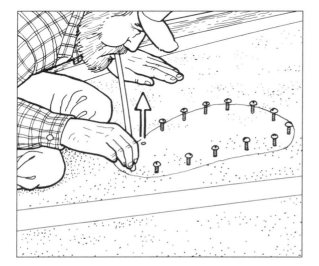

3 Identify the lowest hole in the pattern and mark it. With sheet metal screws, close all the other holes except one. Place the square-cut end of a piece of vinyl hose tightly over the marked hole and blow on the hose. Air should pass freely out the one open hole. Move one screw, closing the tested hole and opening another one, and check again for air flow. Check every hole. If any holes fail to pass air, drill a new hole 1 inch farther inside the outline and test it.

If any hole is more than 5 or 6 inches from another one, drill an intermediate hole to limit the between-hole distance to about 6 inches.

Remove the screws.

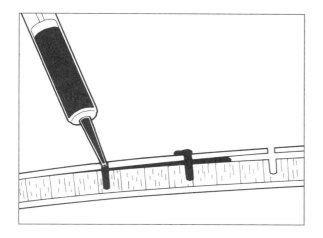

4 Mix an appropriate batch of epoxy and thicken it slightly—to catsup consistency—with colloidal silica. Cut the tip of an epoxy syringe (available from your epoxy supplier) at an angle and fill the syringe with the mix. Insert the tip tightly into the marked hole and inject the void with epoxy. As each hole begins to discharge epoxy, close it with tape. If you flow the epoxy in from the highest hole rather than injecting it from the lowest, you run the risk of trapping air resulting in an incomplete bond.

On a large void, you will need to inject a section at a time, using one of the outlet holes from the previous injection as the new fill hole. Keep flowing epoxy into the void until it flows out all the holes.

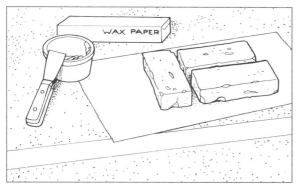

5 Weight (horizontal surface) or brace (vertical surface) the outer skin to compress it against the core. Take care not to deflect the skin out of shape. Use wax paper under the weights or braces, and pick up any excess epoxy that vents from the holes.

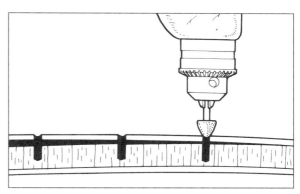

6 Repair the holes like any other surface damage. In this case, the underlying epoxy does not preclude the use of gelcoat to patch the holes as long as the epoxy is scrubbed clean of amine and the surface is dimpled and roughened with a drill-mounted grinding point.

SOLID LAMINATE DELAMINATION

DELAMINATION FOUND IN SOLID LAMINATE should be analyzed carefully. The cause is almost always excessive flex, which tears the bond between the laminates. Occasionally the lay-up schedule is too light—i.e., too few layers of glass—for the duty required of the laminate. In such a case, repair efforts must also include adding additional layers or perhaps other stiffening actions to prevent any recurrence.

More often the delamination is the result of an "incident," usually a collision with a solid object. Provided there is no indication of any substantive damage to the glass fabric, the repair detailed for delamination of the top skin from the core—the most common circumstance—can be used with equal success to treat delamination in solid lay-up.

WET CORE

If the drill bit brings out wet core material, it must be dried out before it can be rebonded to the skin.

SMALL AREA

Often the wet area is limited to the immediate proximity of the source of the moisture. Depending upon the core material and the extent of the saturation, one or all of the following drying methods may be applied.

Vacuum.
A shop vac will remove water from the cavity and extract some moisture from the core. Vacuum bagging using an air compressor and a vacuum generator will be more effective at drying the core. A refrigeration vacuum pump might also be adapted by threading a nipple into the skin.

Flushing.
Flooding the cavity with acetone can be effective. The acetone combines with water, carrying it away. Acetone left behind quickly evaporates, leaving dry core. Always keep in mind that acetone is extremely flammable.

Heat.
A hair dryer or a heat gun played over the wet area will effect some drying. (Be careful not to overheat the laminate; if it's too hot to touch, it's too hot.) The core must be sufficiently exposed for the heat-evaporated moisture to escape; otherwise the moisture simply rises to the underside of the skin and is reabsorbed by the core when the heat is removed. To expose the core, perforate the skin and the core with a pattern of holes drilled about every inch. Don't use a heat gun if you have flushed the core with acetone.

Dry air.
Simply leaving the core exposed will result in drying if the air is dry. Drill the pattern of holes in the wet core and leave the boat in a heated garage or other enclosed storage area for a few weeks, or tent the damaged area and leave a light bulb burning inside the tent to reduce the humidity.

LARGE AREA

If a large section of core is saturated, the only practical solution is to remove one of the skins to fully expose the core so it can be dried thoroughly.

Removing the inner skin.
Removing the inner skin is the preferable way of gaining access to saturated core because it leaves the glossy and the nonskid surfaces of the exterior hull or deck unmarked. Gaining access to the inner skin may require removal of furniture or an inner liner. If access will be very difficult, you are likely to be better off removing the outer skin instead.

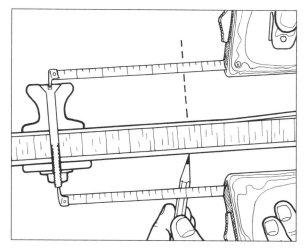

1 Outline the damaged area with straight lines and transfer this outline to the inner skin. This is easily done by making a paper pattern and measuring corner distances to some through-deck feature—a cleat-mounting bolt, for example.

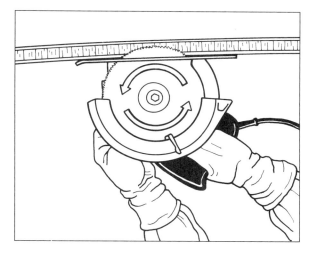

2 Drill an exploratory hole to determine how far the underside of the outer skin is from the surface of the inner skin. Fit a circular saw with a carbide-tipped plywood bit and set the cutting depth to slightly less to allow for some variation in this dimension. Cut around the outline.

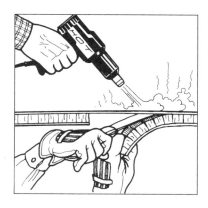

3 Finish the cut through the core with a razor knife. If the top-skin bond is completely broken, the cutout will drop out. If not, find a loose corner and pry down, then use a sharpened flexible putty knife as a chisel to free the rest of the core. Heat applied to the outer skin may help.

Removing the outer skin.

If access to the inner skin is difficult, removing the outer skin may be the better way to expose a saturated core.

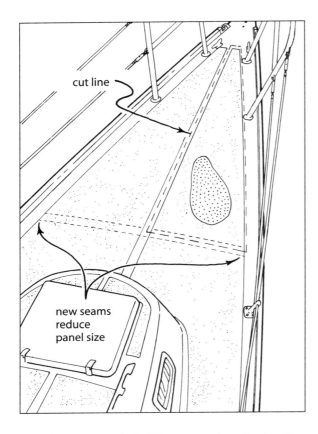

cut line

new seams reduce panel size

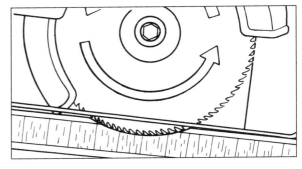

2 Set the cut depth to the *thickness of the skin only* and cut around the outline.

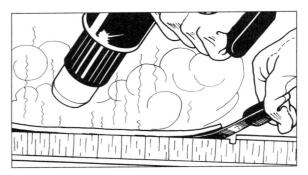

1 Plan your cut wisely. When you replace the skin, the cut will be much easier to hide if it is in a smooth section of the deck rather than across the nonskid. On the other hand, if you are going to cover the repair with a nonskid overlay, you will want to confine the cuts to the nonskid area. If the wet core is only on one side of the foredeck, adding a smooth seam to the nonskid on the centerline (when you reinstall the skin) will allow you to cut away a smaller section of deck.

3 If the skin is still partially bonded, pry up a free corner and use a sharpened flexible putty knife as a chisel to free it completely. Heat applied to the skin may help. Unless the skin is badly damaged, save it; you will reinstall it when the core is dry.

DAMAGED CORE

Once you have exposed the core, you may discover that it has already begun to deteriorate. In that case, replacement is the only sensible course.

1 Chisel the damaged core from the inner skin. Or . . .

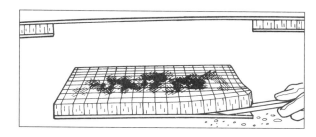

2 . . . Use a utility knife (for balsa or foam) or a saw (for plywood) to cut around the damaged area, taking care not to cut the inner skin. Remove the core with a chisel, shaving the inner skin completely clean.

3 Make a paper pattern of the removed section and use it to cut a new piece of core. It is important to use core material the same as the original, both in type and thickness, if possible. Balsa and foam are available from specialty suppliers. If the core is plywood, use only marine-grade plywood for the repair. Sand the surface of the plywood and clean it with a solvent-dampened rag.

Dry-fit the core into the cavity, trimming as necessary. When you are satisfied with the fit, grind the inner surface of the skin.

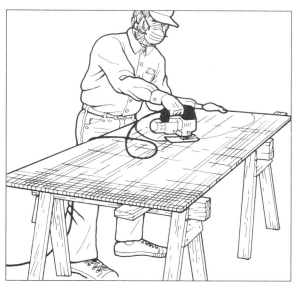

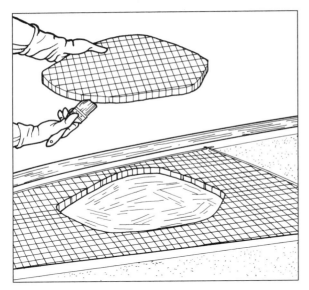

4 Wet out the surface of the skin, the appropriate surface of the new core, and all core edges (old and new).

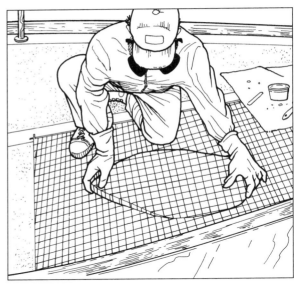

5 Thicken the epoxy to mayonnaise consistency with colloidal silica, and coat all the bonding surfaces generously. Install the new core.

6 Brace the core in position. Or . . .

7 . . . Weight the core in position. Protect the weights (or braces) with plastic sheeting or wax paper. A sand-filled garbage bag makes an excellent weight for this purpose because it conforms to the shape of the deck. The core should be solidly bedded in the thickened epoxy, and epoxy should squeeze out the cut line all the way around the new section. Remove the squeeze-out, then let the epoxy cure fully before removing the weights.

REINSTALLING THE SKIN

If the old skin was in poor condition, you will have to lay up a new one with glass cloth and epoxy, but most of the time the old skin can be simply put back in place. Whether you have dried the old core or replaced it, reinstallation of the removed skin is the same.

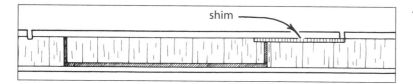
shim

1 Dry fit the removed section. Grind or shim until the skin section realigns properly with the surrounding surfaces.

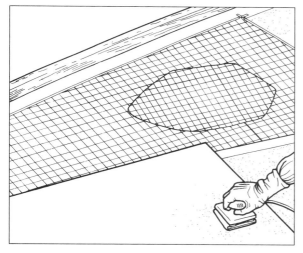

2 Sand the surface of the core and the underside of the skin. Clean with a solvent-dampened rag.

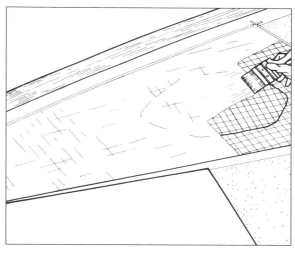

3 Wet out the sanded surfaces with epoxy.

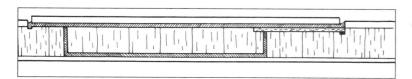

4 Thicken the epoxy to mayonnaise consistency with colloidal silica and coat both surfaces. Be sure to apply enough epoxy putty to solidly fill the space between the skin and the core.

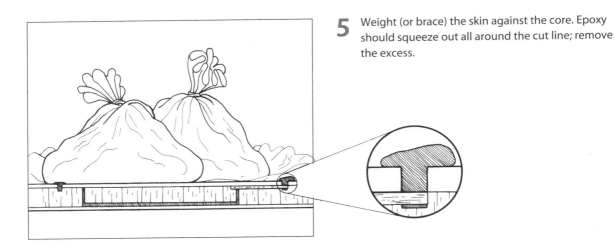

5 Weight (or brace) the skin against the core. Epoxy should squeeze out all around the cut line; remove the excess.

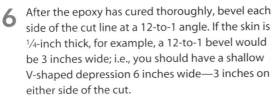

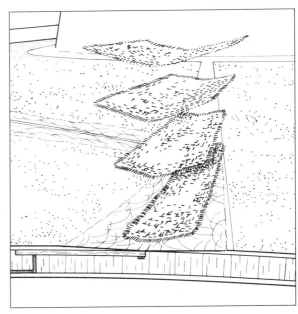

6 After the epoxy has cured thoroughly, bevel each side of the cut line at a 12-to-1 angle. If the skin is ¼-inch thick, for example, a 12-to-1 bevel would be 3 inches wide; i.e., you should have a shallow V-shaped depression 6 inches wide—3 inches on either side of the cut.

7 Cut fiberglass cloth into narrow strips (tape) and laminate them into the depression with epoxy resin. Each strip should be about 1 inch wider than the previous one. Remember not to use mat with epoxy. Sand the cured surface and paint it or cover it with nonskid overlay.

STRENGTHENING

Fiberglass boats sometimes flex alarmingly under pressure or exhibit a pattern of surface cracks around hardware, in corners, or at other stress points. These can both be signs of excessive weakness.

Fiberglass laminate is easily strengthened by adding additional layers. Strengthening layers are most often added to the inner surface of the hull or deck.

1 Outline the area to be strengthened. Dewax and grind the surface thoroughly to prepare it.

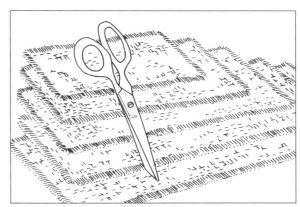

2 Cut multiple layers of fiberglass cloth, the first to the size of the outline, then each a little smaller than the previous. If you are reinforcing a large area, keep the cloth pieces small enough to handle—generally not much larger than 1 square yard if you're working alone. Overlap joints on an inside surface; butt them if you are adding laminates to the outside surface.

3 Follow the laminating procedures detailed in "Laminate Repair." Use epoxy resin. Epoxy is stronger than polyester resin, and strength is what you're after. Epoxy also binds the new laminates to the old more securely.

Glass-reinforced plastic is by nature flexible, and excessive flexing more often indicates inadequate stiffness rather than inadequate strength. Adding layers does stiffen the area as well, but when strengthening is not required, stiffening is usually better accomplished with a sandwich construction.

Doubling the thickness makes a panel eight times as stiff. This is the reason the manufacturer put core in the deck. You can add a layer of core and a thin inner skin to a panel to stiffen it, but often a reinforcing member or two will do the same job with less work and less weight.

Called "hat-shaped stiffeners," reinforcing members are rib- or stringer-like fiberglass constructions formed over a strip of wood or other material. Judiciously spaced, hat-shaped stiffeners are quite effective.

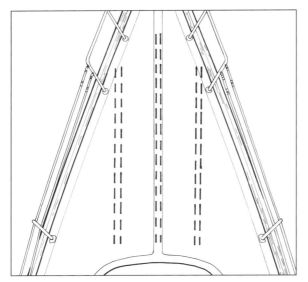

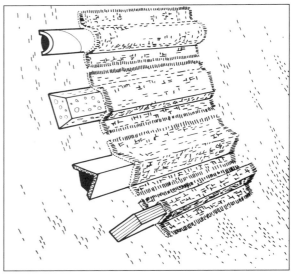

1 Decide where the stiffeners will go. Generally they should run parallel to the long dimension of the area you are stiffening. The number of stiffeners required will depend on the flexibility of the panel; add them one at a time until you are satisfied with the result.

2 Select a core material. Wood adds some strength, but most of the stiffness comes from the box construction, so an easier-to-form core material such as foam, hose, or cardboard may be a better choice. If you use wood, taper the ends to avoid creating a hard spot where the stiffener ends.

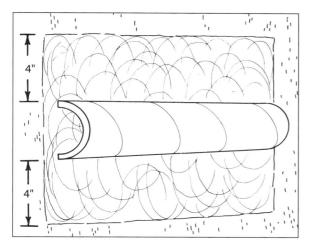

3 Dewax the panel where the first stiffener will be attached. Grind an area about 8 inches wider than the width of the core material.

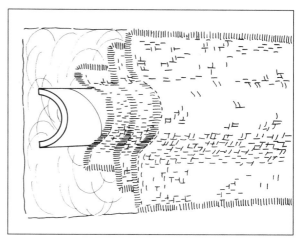

4 Cut a strip of 10-ounce cloth wide enough to extend out 2 inches on either side of the core. Cut a second strip 2 inches wider and a third 2 inches wider still. The recommended height-to-skin-thickness ratio is 30 to 1, so three layers of 10-ounce cloth (which yield a cured thickness of about 0.120 inch) are generally sufficient for stiffeners up to about 3½ inches high. If you want extra layers for added assurance, cut each strip 2 inches wider than the previous one. Be sure to grind a wider area.

5 Tack the core material in place with hot glue or quick-set epoxy. Since the glass cloth won't form into a sharp corner, it is good practice to put an epoxy-putty fillet along the edges of the core.

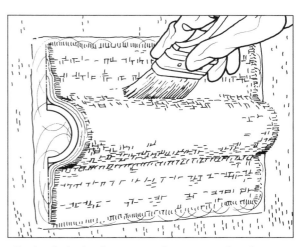

6 Apply the laminates over the core, wetting them out with epoxy resin. (Polyester resin is inferior for this use for reasons already mentioned, but if you use polyester, substitute mat for the odd-numbered layers.)

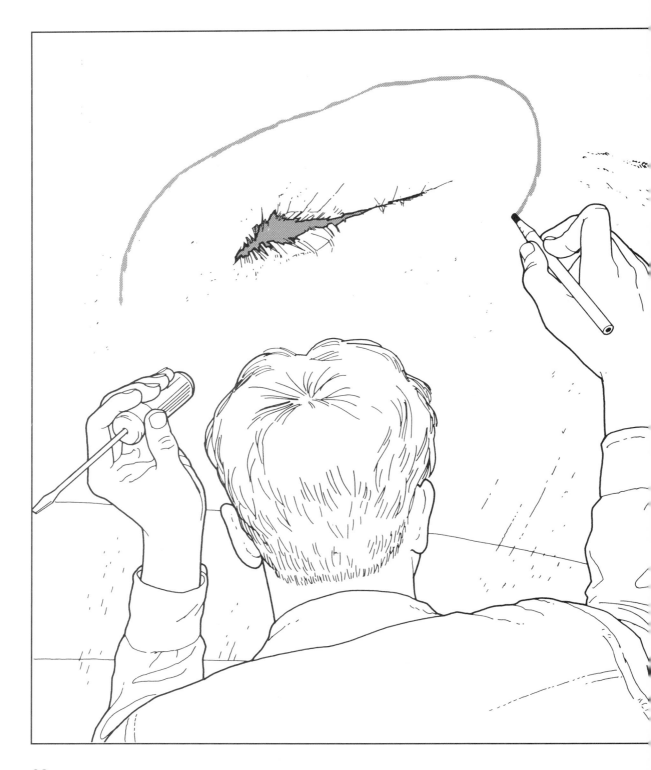

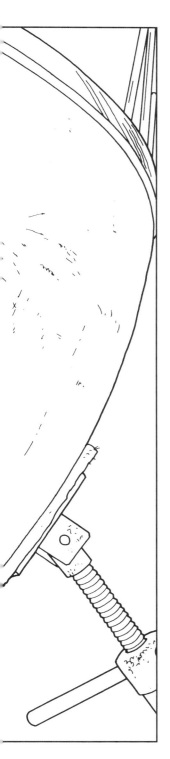

HULL REPAIRS

Fiberglass hulls are incredibly durable. They don't rot like wood or corrode like steel. They don't dry out and open seams. They don't get worms. In fact, after 35 years of production-line manufacture, the only insidious problem to surface in these boats is blistering, and even that affects only a small percentage of hulls.

If you keep the hull away from solid objects such as rocks, docks, and other boats, required repairs are likely to be limited to restoring the surface gloss. Unfortunately avoiding the occasional "kiss" sounds simpler than it really is. Boats aren't always where you think they are. Underwater obstructions aren't always charted. And rare is the skipper who hasn't watched in disbelief as his boat—a moment before in complete control—suddenly rushes sideways for the dock. Even impeccable navigation and iron-fisted control doesn't prevent another captain's lapse from leaving you with broken glass.

Because the hull is relatively featureless compared with the deck, hull repairs are generally less complicated. But they are more visible; a poor topsides repair can stand out like a shirt-pocket ink stain. You must take the time to fair the patch and match the color for the repair to be invisible.

While it is true that the hull is the part of the boat that "keeps the ocean out," there is no reason to approach hull repair with any greater trepidation than repairing the deck. Prepare the surface properly, which means little more than cleaning and grinding—always grinding—and follow the other steps carefully, and your patch will be just as strong as the surrounding hull, maybe stronger.

GOUGES

In "Restoring the Gloss," we looked at scratch repair, but sometimes an encounter with a sharp, solid object gouges into the underlying laminate. In this instance, the laminate must be repaired before the gelcoat is restored. How you make the repair depends on the extent of the damage.

REPAIRING A SHALLOW GOUGE

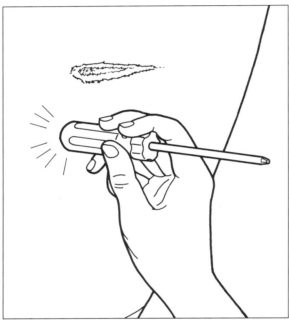

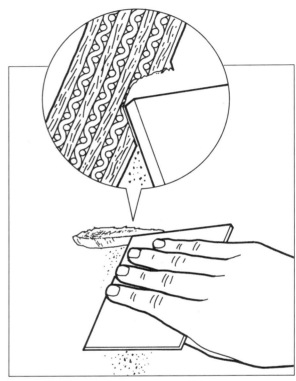

1 Sound the area for delamination. The impact that caused the gouge may have torn the underlying laminates. If the tapping sounds dead rather than resonant, repair the damage as if it were a deep gouge (see below).

2 Use the edge of a cabinet scraper to open the damage, putting a smooth chamfer on each side of the gouge.

3 Catalyze a small quantity of polyester resin and thicken it with chopped glass. Wipe the V with styrene to reactivate the surface of the cured resin, then fill the V to the *bottom* of the gelcoat layer with the thickened resin.

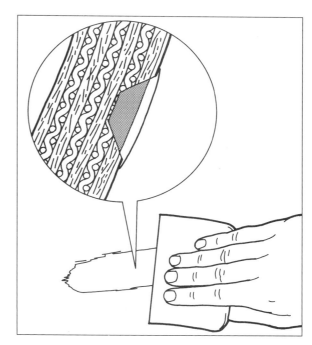

4 When the resin kicks, fill the remaining depression with color-matched gelcoat paste, letting it bulge slightly above the surface. When the gelcoat begins to gel, seal its surface with plastic or a coat of PVA.

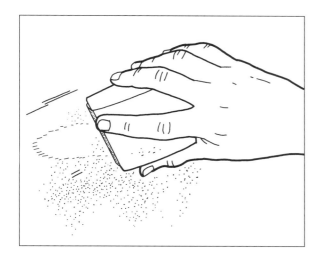

5 Allow the gelcoat to cure fully, then fair and polish the repair.

DEEP GOUGES

1 If the damage extends through more than the upper two or three laminates and the gouge is more than a couple of inches long, restoring hull integrity requires replacing the damaged fiberglass. Grind the damaged area into a depression with a 12-to-1 chamfer all around. Grind through all broken layers, using the first undamaged laminate as the bottom of the depression.

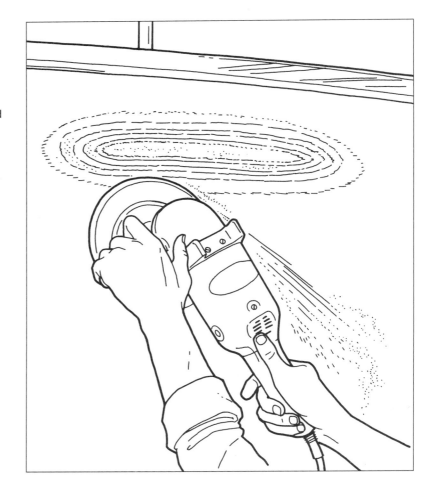

2 Cut alternating layers of mat and cloth, beginning with mat and making each layer larger than the previous one.

3 Wipe the depression with styrene, then coat it with polyester resin. Place the first layer of mat in the depression and wet it out. Position and wet out succeeding layers until the repair is even with the bottom of the gelcoat layer.

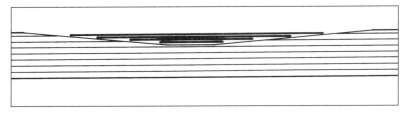

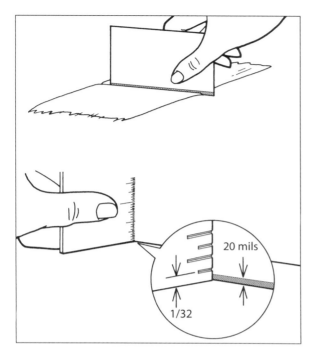

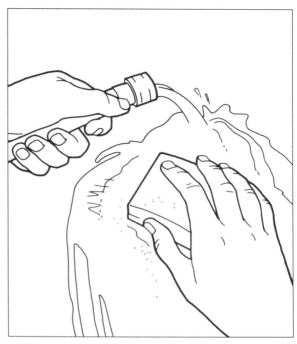

4 When the resin kicks, spray on several coats of color-matched gelcoat (see *Sailboat Refinishing* in this book series), or brush on at least 20 mils of gelcoat paste. Check the thickness by slicing the gelcoat with the edge of a piece of paper and comparing the height of the gelcoat on the paper to a ¹⁄₃₂-inch mark on a ruler, which is about 30 mils.

5 Allow the gelcoat to cure fully, then fair and polish the repair.

WHEN TO USE EPOXY

IF YOU PLAN TO PAINT THE REPAIR RATHER THAN finish it with gelcoat, repair the gouge with epoxy for a better bond. For a shallow gouge, wet the V first with unthickened epoxy, then fill it to the surface with the epoxy thickened with colloidal silica to peanut butter consistency. Try to get the surface as fair as possible before the epoxy kicks because the silica-thickened epoxy is hard to sand, especially without damaging the much softer gelcoat that surrounds it.

For deep gouges, use epoxy and 10-ounce cloth—no mat. Wet out the depression, then coat it with epoxy thickened with colloidal silica to catsup consistency. Apply the layers of cloth and saturate them with epoxy resin. Give the last layer two or three extra coats of resin to completely hide the weave of the cloth. Coating the surface with peel ply is recommended (see "Laminate Repair"). Fairing is likely to be required, but wait until the laminates have fully cured, then thicken some mixed epoxy to peanut butter consistency—use microballoons if the repair is above the waterline, colloidal silica if it is below—and fill all voids and depressions. Sand and paint.

BLISTERS

Fiberglass blisters occur because water passes through the gelcoat. Water-soluble chemicals inside the laminate exert an osmotic pull on water outside, and some water molecules find a way through. As more water is attracted into the enclosed space, internal pressure builds. The water molecules aren't squirted back out the way they came in because they have combined with the attracting chemicals into a solution with a larger molecular structure. Instead, the pressure pushes the covering laminate into a dome—a blister.

There has been a great deal of hysteria about blisters, but the reality is that the number of boats that develop *serious* blister problems is extremely small. An occasional blister or two is *not* a serious problem, any more than is an occasional gouge in the hull. Some boats seem to exhibit a greater propensity to blister, presumably due to the chemical components used and/or the lay-up schedule, but all boats are at some risk. Surveys suggest that about one boat in four develops blisters. Maybe the factory was training a new fiberglass crew the day your boat was built, or the humidity was abnormally high. Or maybe the factory did everything right but somewhere along the way someone sanded the gelcoat excessively, or even sandblasted it to prepare it for bottom paint. It is pointless to speculate. If your boat develops blisters, deal with them; if it doesn't, forget about it.

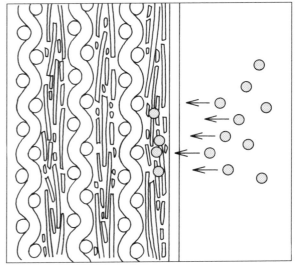

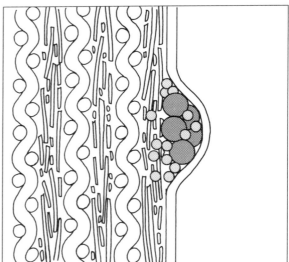

MINOR BLISTERING

1 Open the blister. Wear eye protection; internal pressure can be double that of a champagne bottle, and the fluid that blasts out when you pop the dome is acid. Use a 36-grit disk to grind the blister into a shallow depression.

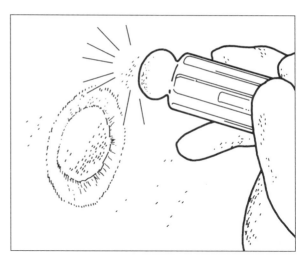

2 Sound around the blister to make sure there isn't any additional delamination.

3 Flush the open blister with water, then scrub it with a TSP solution. Rinse thoroughly.

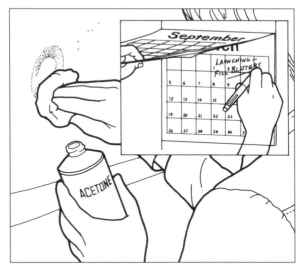

4 Allow the blister to dry for as long as practical. If you dry-store your boat for the winter, open blisters at haulout but don't fill them until launch time. Just before filling, wipe out each blister depression with a rag *dampened* with acetone.

5 Epoxy is the resin to use for blister repairs. It is less permeable than polyester and it forms a much stronger bond. Wet out the cavity with epoxy.

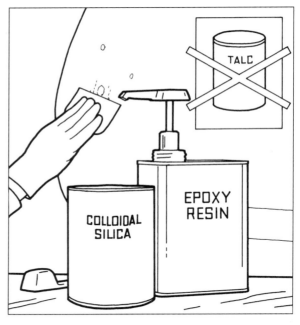

6 For small blisters, thicken epoxy to peanut butter consistency with colloidal silica and fill the cavity, using a squeegee to compress and fair the filler. Silica-thickened epoxy is hard to sand, so fair it well before it kicks. *Never* use microballons or any other hollow or absorbent (talc, for example) fairing compound to fill blisters.

7 Before the repair reaches full cure, paint it with at least two coats of unthickened epoxy.

BOAT POX

Boat pox is a much more serious condition, related to the occasional blister like acne to the occasional pimple. If the bottom of your boat is covered with blisters, filling them won't cure the problem. Pox is a systemic condition and requires remedial action.

1 Examine the bottom as soon as your boat is lifted. Out of the water, blisters can shrink and even disappear altogether. If the bottom is covered with hundreds of blisters, your boat has pox, and the condition will only worsen unless you take the cure.

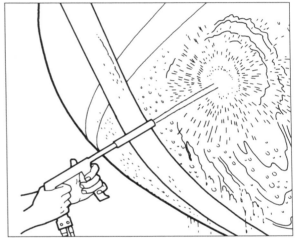

2 Scrub the hull to remove growth, oil, and all other contaminants.

3 Open a few blisters to determine their location. Usually blisters occur between the gelcoat and the first layer of laminate. If they are deeper, see the sidebar.

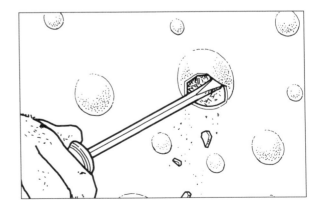

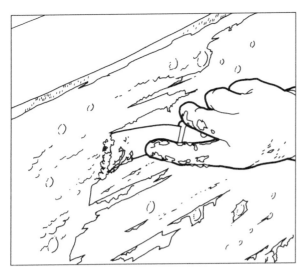

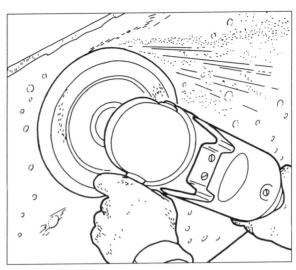

4 Unless you are having the hull machine peeled, chemically strip the bottom of all paint. Curing boat pox requires the removal of *all* the gelcoat below the waterline; but if you do not remove the paint first, gelcoat stripping will contaminate the underlying laminate with paint particles, weakening the bond of the barrier coat.

5 Grind away all the gelcoat below the waterline. You cannot just grind open the blisters because a hull with pox is saturated throughout and will not dry out unless the gelcoat is removed. Despite the time required, the best way to strip the gelcoat is with a lightweight disk sander and 24-grit disk on a foam pad. Boatyards prefer to sandblast the hull to remove the gelcoat, but this harsh treatment damages the underlying laminate. If you give in to the expediency of sandblasting, be sure the pressure is low and the sand is directed at the hull at a shallow angle—less than 30 degrees.

Some yards now have peelers that work like a power plane. Set to the appropriate thickness, they shave the gelcoat from the hull in a single pass without any damage to the laminate. If you have the hull peeled, run a 50-grit disk over the peeled surface to remove any ridges and to provide tooth for the barrier coat.

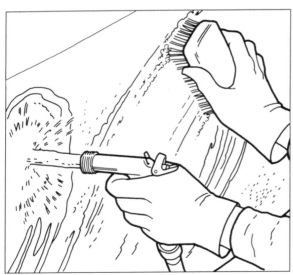

6 Wash the stripped surface with a stiff brush. It is imperative that all loose bits of gelcoat, paint, and grit are removed. Inspecting the surface with a magnifying glass is not overkill.

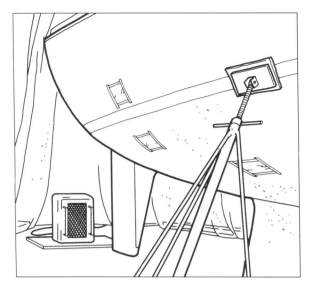

7 With the gelcoat removed and the laminate clean, allow the hull to dry out. This will take at least two weeks in hot temperatures and as long as six months in the cold. Tenting the hull and running fans or a heater will shorten the time.

Keep track of the drying process by taping 6-inch squares of plastic cut from heavy freezer bags to a dozen or more places on the hull—2 or 3 above the waterline. Seal the plastic all around with electrician's tape. Sun on the plastic will cause moisture in the hull to condense on the plastic. Open the plastic and wipe it and the hull dry every few days, then seal it back in place. When condensation ceases to form in any of the test panels on sunny days, the hull is sufficiently dry to reseal.

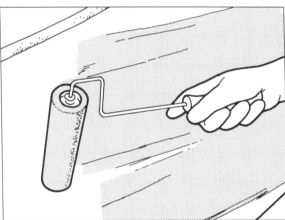

8 Select an epoxy barrier coat product such as InterProtect or West System epoxy and apply it to the recommended thickness—usually 20 mils— according to the manufacturer's instructions.

DEEPER BLISTERS

IT IS FAIRLY COMMON FOR BLISTERS TO OCCUR beneath the first laminate instead of in the laminate/gelcoat junction because of the all-too-common practice of getting the first layer applied to the wet gelcoat, then waiting a day to continue the lay-up. This makes the bond between the first and second laminate weak and susceptible to blistering. Blisters deeper in the laminate are thankfully much rarer, occurring only after less serious blisters have been long ignored.

If you discover that the blisters on your hull are beneath the initial layer of laminate, remove this layer along with the gelcoat. When the hull is dry, apply a replacement layer of 10-ounce cloth to the hull, working with manageable pieces and butt-joining (not overlapping) the sections. Wet each section with epoxy resin, then apply a layer of silica-thickened epoxy before positioning the cloth. Wet out the cloth and seal it with peel-ply and a ply of plastic. Sand and fair the entire surface before applying the barrier coating.

IMPACT DAMAGE

ew things are more disheartening to the boat-owner than staring at the fuzzy edge of broken fiberglass. Most don't realize that the repair of impact damage is usually only a step or two more complicated than filling a gouge or a blister. You don't believe it? If you cut out the damaged area and bevel the edges, then close one side of the hole by laying up a single ply of fiberglass over it,

how is the resulting depression different from a deep gouge except in size? It isn't.

Sometimes, of course, damage is so extensive that a significant section of the hull or deck (or both) has to be rebuilt, but even then the lay-up process is the same. The only difficulty comes in getting the new laminates to take on the right shape.

CUTTING AWAY THE DAMAGED LAMINATE

1 Impact damage almost always has some associated delamination. Tap the area to determine the extent of the damage and map it. Enclose the marked area in a circular or oval trace.

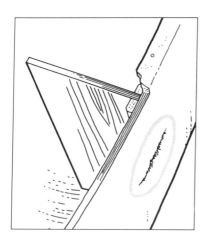

2 Go inside the boat and determine if anything will be in the way when you cut out the damaged section. Sometimes it is best to cut out an offending member; in other cases you may want to leave it intact, cutting the damaged skin free. If you don't have inside access, use a 3- or 4-inch hole saw to remove a circular plug from the hull so you can look and feel inside before making the full cutout.

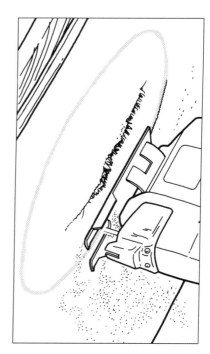

3 Saw around the oval trace. Never try to save damaged fiberglass; always cut it out and replace it with new laminate. You can make the cut with a circular saw fitted with a carbide blade or a cut-off wheel, or with a saber saw fitted with a blade for cutting fiberglass.

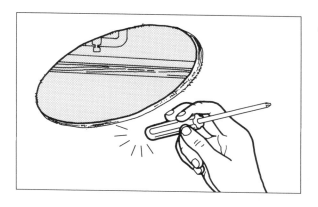

4 Check all the edges to make sure the laminate is solid, and tap again around the hole. Enlarge it if you find any additional delamination.

WORKING FROM THE INSIDE

There are two reasons you should make hull repairs from the inside whenever possible, especially when the damaged area is small. First, you are going to bevel the edge of the hole with a 12-to-1 chamfer. If you repair a 3-inch diameter hole through a ½-inch-thick hull from the outside, you end up with about 15 inches (diameter) of surface damage to refinish, but if you repair it from the inside, you have only a 3-inch hole to refinish.

The second reason is that if you back the hole on the outside with a polished surface, you can in effect create a mold that allows you to lay up the repair the same way the boat was built—gelcoat first—and very little finish work will be required.

1 Dewax the interior surface of the skin around the hole.

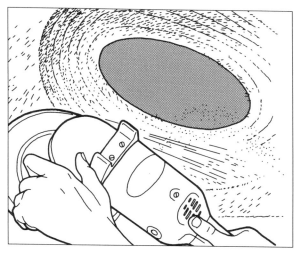

2 Grind the edge of the hole into a 12-to-1 bevel. Also grind a rectangular area of the inner surface a few inches beyond the hole to accommodate a finishing layer of cloth.

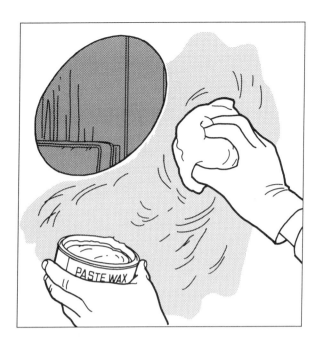

3 Give the exterior surface of the skin around the hole a heavy coat of paste wax, taking care not to get any wax on the edge or inside the hole. The purpose of the wax is to prevent any resin runs from adhering to the exterior surface; paint the wax with PVA to be sure. Mask the area below the hole.

4 Cut a scrap of smooth plastic laminate (Formica) or thin clear acrylic (Plexiglas) a foot larger than the hole. Wax this backer and paint it with PVA, and screw or tape it to the outer surface. If the hull is flat or curving in only one direction in the damage area, the backer will assume the correct curve—check from inside to make sure it seats against the skin all around the hole. If the hull is spherical, i.e., curving in two directions, a sheet backer usually won't work. Where the compound curvature is slight, acrylic screwed to the hull will bend into the correct shape if warmed with a heat gun. Otherwise you will need to make a backer following the instructions in "Taking Off a Mold" below.

5 Cut the fiberglass material to fit the hole. Unless you have reason to follow a different schedule, begin with two layers of 1½- or 2-ounce mat, then alternate mat and 10-ounce cloth. The number of laminates will be determined by the thickness of the hull; you will need roughly one layer for every $\frac{1}{32}$ inch. Cut the first layer of mat 1 inch larger than the hole, overlapping the bevel by ½ inch all around. Subsequent pieces should be ½ inch larger all around than the previous one.

6 From inside, spray or brush 20 mils of color-matched gelcoat onto the waxed backer. Check the gelcoat thickness with a toothpick—$\frac{1}{32}$ inch is about 30 mils.

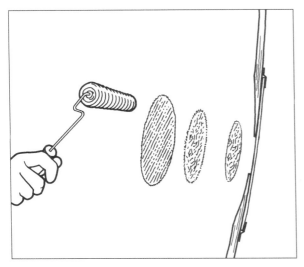

7 When the gelcoat kicks, wet it with polyester resin and lay up the first two layers of mat and one layer of cloth, compressing them against the gelcoat and working out all voids and bubbles with a resin roller and/or a squeegee.

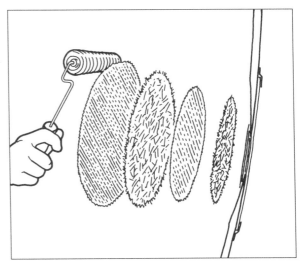

8 Let the first three plies kick, then lay up four additional plies. Never lay up more than four plies at a time or the generated heat may "cook" and weaken the resin. Continue the lay-up four plies at a time until the repair is flush with the interior surface.

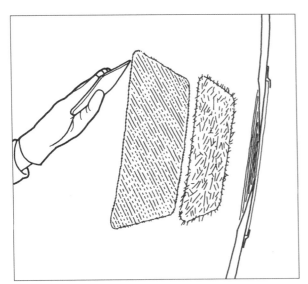

9 For a finished look, cut a rectangular piece of mat and one slightly larger of cloth and apply these over the patch, smoothing them with a squeegee. Seal this top layer with plastic or PVA to allow a full cure.

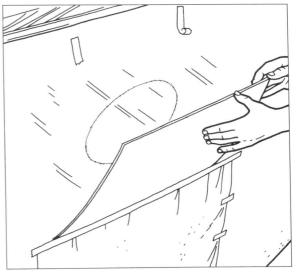

10 Remove the backer from the exterior surface. Fill any imperfections in the new gelcoat with gelcoat paste and allow it to cure fully. Clean the area around the patch, then sand—if necessary—and polish the repair area.

TAKING OFF A MOLD

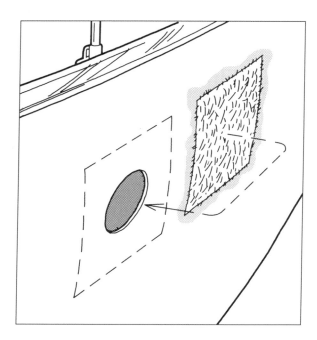

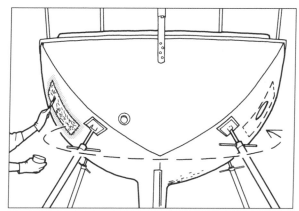

1 At its simplest, taking off a mold involves no more than waxing and release-coating (with PVA) a section of hull with approximately the same contour as the damaged area, usually adjacent to the damage. Coat the waxed area with resin, then lay up two plies of 2-ounce mat. When the lay-up cures, peel it from the hull, coat it with releasing agent, and tape in place over the hole to provide a contoured backer/mold for your repair.

2 For more extensive damage or damage in an area where the shape of the hull is rapidly changing, try taking a mold from the same spot on the other side of the hull. Reversed top to bottom, the molded piece should give you a close approximation of the correct contour. Transfer it while it is still "green," that is before full cure, and it should conform perfectly to the damaged side.

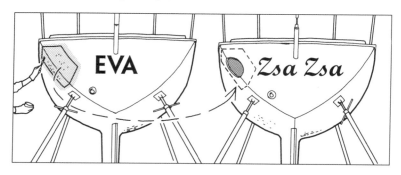

3 When damage is extensive or to an area of the hull with a feature, such as the intersection of the quarter and the transom, taking a mold from the opposite side may not give satisfactory results. If you can locate a sistership with a willing owner, you can lay up a perfect mold. If the mold is sizable, stiffen it with a few hat-shaped stiffeners. Be sure you know how to thoroughly coat a hull to assure release—meaning try it on your own hull first—before you paint someone else's hull with resin.

OUTSIDE REPAIR

If you are going to paint the repair rather than trying to match gelcoat, make the repair from the outside. Working outside is somewhat easier and a lot more comfortable—you're not engulfed in resin fumes or wedged into some impossible corner. If you're not going to finish the repair with gelcoat, you should also use epoxy for its superior bonding strength.

1 Dewax around the hole and grind the edge into a 12-to-1 bevel.

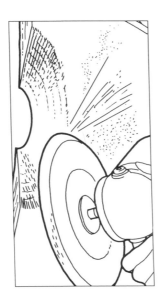

2 Wax the exterior surface of the skin around the hole, taking care not to get any wax on the bevel. Mask the area below the hole.

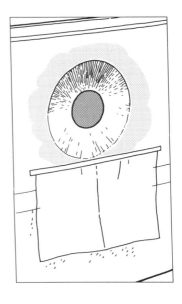

3 Wax and release-coat a scrape of smooth plastic laminate or thin acrylic and *brace* it tightly against the hole from inside. If the hull is flat or curving in only one direction in the damage area, the backer will assume the correct curve. If the hull is spherical, lay up a backer using an adjacent section of the hull as the mold.

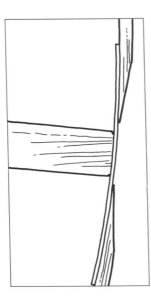

4 If you are using epoxy resin for the repair, cut all your repair pieces from 10-ounce cloth. Mix a batch of epoxy and wet out the first layer of cloth. Use a squeegee to smooth the cloth into position and to

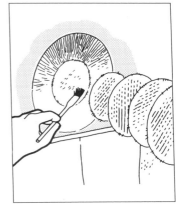

remove trapped air. Apply the second layer and wet it out. Squeegee. Repeat this process a layer at a time, mixing fresh epoxy as needed, until the repair is slightly below the surface.

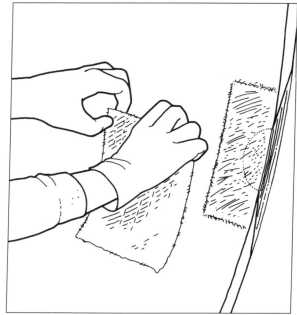

5 Give the final layer a coat or two of epoxy, then cover the surface with peel ply and a layer of plastic and smooth it with the squeegee. Wipe up any squeeze-out.

6 Remove the backer from the interior surface and grind the repair and a rectangular area around it. Cut a piece of fiberglass cloth to the size of the ground rectangle and laminate it in place to give the interior of the repair a finished look.

7 Fair the repair with epoxy thickened with micro-balloons (below the waterline, use colloidal silica) and paint.

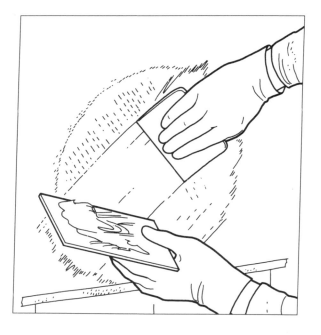

NO INSIDE ACCESS

Often there is a liner or a tank or some other interior obstruction in the damage area that denies you access to the inside of the hull. Except for the need to take extra care when cutting out the damaged section and the effect working from the outside has on the size of the repair requiring finish work, lack of inside access isn't a very big problem.

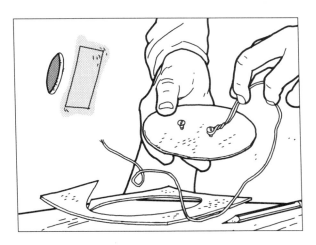

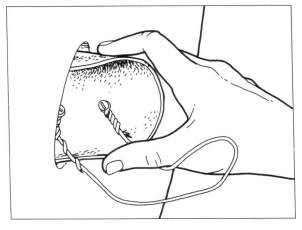

1 Using epoxy resin and two layers of 10-ounce cloth, lay up a backer piece on a waxed and release-coated section of the hull near the damage.

2 When the backer is thoroughly cured, peel it up and trim it to 1 inch larger than the hole. Screw two sheet-metal screws into the backer and twist the ends of a length of wire to the screws to provide a loop handle.

3 Dewax the inside of the skin around the hole by reaching through the hole with a solvent-soaked rag. Sand the dewaxed area with 50-grit paper, again by reaching through the hole. Scrub (with water) and sand the handled surface of the backer.

4 Bend the backer slightly and push it through the hole, holding on to the wire loop. No bending will be necessary if the hole is oval shaped.

5 Thicken a small amount of epoxy to peanut butter consistency with colloidal silica, and butter it onto the perimeter of the backer. Center it and pull it tightly against the opening. Hold it in position with a string tied from the loop to some fixed object. Wipe up all the epoxy that squeezes out and let the bond cure completely.

6 Remove the screws and proceed with the repair as described above.

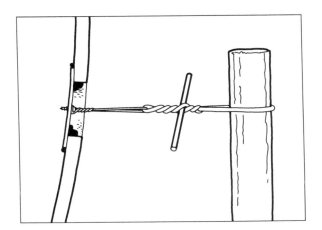

CORED HULL

Impact damage to a cored hull requires repair to the three components—inner skin, core, and outer skin—separately.

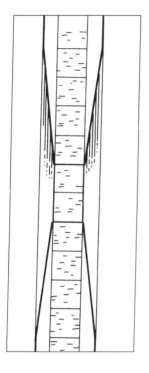

With inside access. Cut out the damage, then grind a 12-to-1 bevel on both skins. Bond a new section of core into the hole, then lay up new skins on either side.

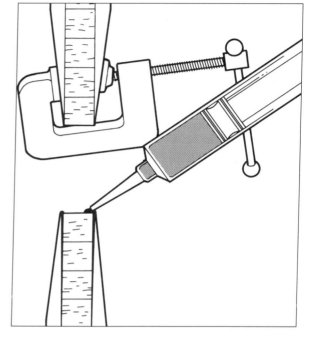

Without inside access.
Outline the damage and cut away the outer skin and the core with a router, taking care not to cut the inner skin. With the inner skin now exposed, cut out the damage to it. Bevel both the inner and outer skins, then lay up a new section of inner skin as previously detailed. Install new core on top of the new inner skin and lay up a new outer skin.

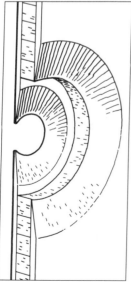

Core delamination.
The flexing associated with an impact is likely to result in delamination beyond the area of skin damage. With edge access, inject the separation with epoxy and clamp it until it cures, then complete the repair.

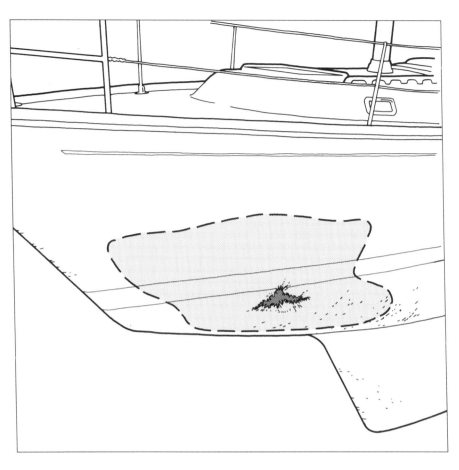

Wet core.

Impact damage below the waterline of a cored hull usually results in wet core. With inside access, determine the extent of the saturation by removing progressively larger sections of the inner skin. Let the core dry or replace it, then complete the repair normally.

Without interior access, it is the outer skin you will have to remove. If the damage to the outer skin is small but the area of saturation large—you should be able to tell by tapping the hull—you may do better to remove the large section of skin intact. While the core dries (if it hasn't been wet long, you will only need to replace the damaged portion), you can repair the damage to the panel of skin. Reglue the repaired skin to the core, then grind a 12-to-1 bevel around the cut line and bond the piece back in place.

KEEL AND RUDDER DAMAGE

Unless you are an extremely timid navigator, eventually you find yourself coming to a rude stop as you unexpectedly run out of water. In such events, it is generally the keel or the rudder that takes the brunt of the impact. When the bottom is ooze or sand, little if any harm is done, but fetching up against solid granite or jagged coral is likely to result in serious trauma.

Scrape and gouge repairs to a fiberglass keel and rudder are no different than on other parts of the hull, except that damage to the keel or rudder should make you suspicious. Because of the lever effect, even a modest impact near the bottom of an underwater appendage applies off-the-scale stresses at the attachment points; fin keels can break the hull like the pull ring on a pop-top can, and rudders can tear free of their stocks. Even an encapsulated full keel, less at risk at the bilge, may deform from the inertia and split, letting seawater into the ballast cavity. All repairs to keel and rudder should include a thorough inspection for collateral damage.

WEEPING KEEL

A grounded boat lifted and dropped by even modest wave action hits the bottom with the force of a pile driver. If even a fist-size rock is under the hammering keel, the bottom laminate may be distorted enough to split. Such damage is likely to go unnoticed unless you look for it, or until the water intrusion causes iron ballast to swell and deform the keel.

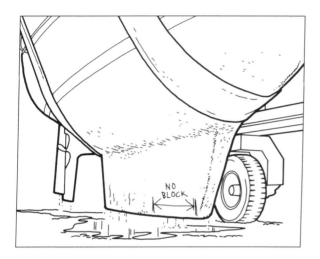

1 Examine the bottom of the keel while the boat is in the hoist and mark where you want yard workers to place the blocks so your access to any suspect areas won't be obstructed.

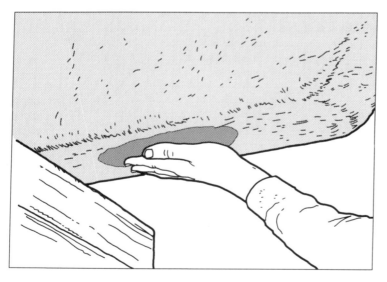

2 Scrape, scrub, and rinse the keel, and let it dry completely. Now examine it with eye and finger for any signs of lingering wetness. Weeping that persists for more than a day indicates a crack in the keel laminate.

3 Grind away the bottom paint in the wet area. The crack should become visible.

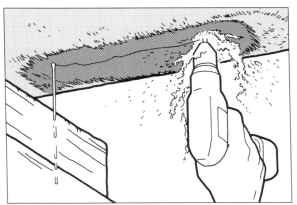

4 Find the ends of the crack and use a battery- or hand-powered drill to drill a ³⁄₁₆-inch hole at each end. These holes mark the break, relieve stress that might lengthen the crack, and allow the cavity to drain. Don't use a plugged-in drill because water may pour out when you puncture the skin.

5 Grind a V along the crack line and beyond each end. The bottom of the V should reach the last layer of laminate, and the sides should have a 12-to-1 bevel.

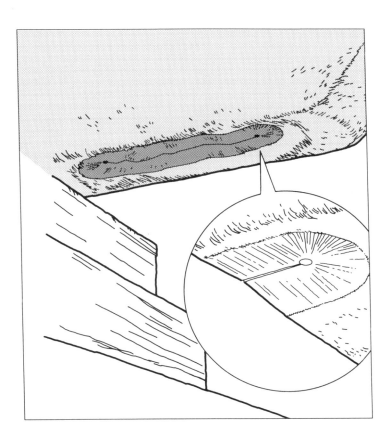

6 Laminate the repair using epoxy resin for a better bond and better resilience to future groundings.

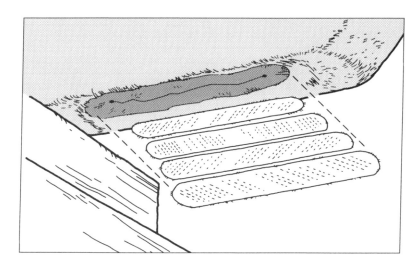

7 If a cracked keel is a recurring problem, grind the entire bottom of the keel and a few inches up the side, then add several additional laminates of glass cloth. You can also strengthen the keel by adding additional laminates from inside if you have interior access.

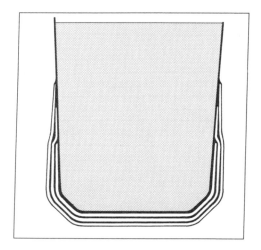

INTEGRAL FUEL TANKS

SOME OF THE SPACE INSIDE A MOLDED KEEL is often used for tankage. This is fine as long as there is actually a tank, but occasionally, to save money, designers or builders have opted for what is known as an integral tank—the designated space in the keel is simply partitioned off from the ballast area and given a fiberglass top. In such cases, a crack in the keel in the tank area may weep fuel.

Permanent repairs will not be possible without emptying the tank and removing all fuel residue from the fiberglass in the area around the crack. The best way of dealing with an integral tank is to replace it with a proper tank; the type will depend on the liquid it is to hold. Cut and grind away the top of the integral tank and steam clean the cavity, then take the dimensions to your tank builder. Meanwhile you can repair the crack in the skin.

KEEL/CENTERBOARD PIVOT PROBLEMS

In a keel/centerboard boat, the pivot pin usually passes through the stub keel, sealed on each end by a plug of mish-mash (a putty made from resin and chopped fiberglass). Leaks around the pivot only rarely get inside the boat because the space in the stub keel, presumably filled with ballast, is usually sealed off from the interior of the hull. A leaking pivot pin is nevertheless a serious problem, especially if the ballast is iron or steel. Rusting steel can expand with enough force to split the confining fiberglass. Even if the ballast is lead, water inside the ballast cavity can freeze during winter storage and rupture the hull.

How do you know if the pivot pin is leaking? If the area around the pin stays wet for days after the boat is hauled (check *inside* the trunk), water is weeping out of the keel and the pin is leaking.

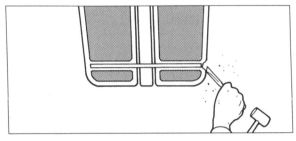

1 Drill and chip out the putty covering both ends of the pivot pin. Take the weight off the pin by supporting the centerboard, and tap out of the keel stub. Remove the centerboard or shift it out of the way.

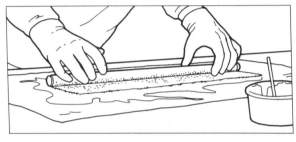

2 Because of hollows in the keel stub, drilling the hole oversize and filling it with epoxy is often impractical. Wax the pin *heavily*. Wet out a piece of light cloth with epoxy and roll it onto the pin to form a fiberglass tube. When the epoxy cures, slip the pin out of the tube.

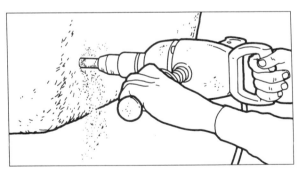

3 Enlarge the hole in the keel stub to the outside diameter of the tube. This can be the most difficult part of the job if the hole is large and the ballast is iron.

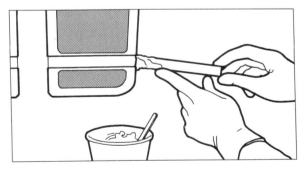

4 Cut the tube into two shorter pieces and epoxy them into the keel. Use epoxy thickened with colloidal silica, and take great care to get a good seal at each end of the tubes where they pass through the fiberglass. Reinstall the pin and seal each end with epoxy putty.

HULL DAMAGE AROUND FINS AND SKEGS

A bolted-on fin keel or skeg functions like a pry-bar when it hits something, deflecting the hull laminate into an S-shape as the front of the fin tries to tear away from the hull and the back tries to drive into the interior.

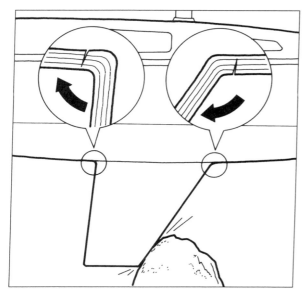

1 Check the hull, both inside and outside, all around the base of the fin. Expect ruptures to be in the outside laminates forward of the fin and in the inside laminates aft. Splits beside the fin can be inside or outside depending on how the fin was stressed.

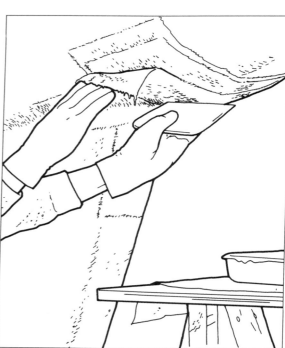

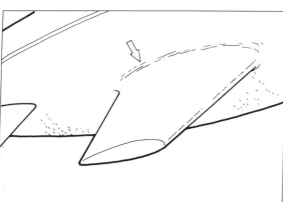

2 Don't confuse gelcoat cracks with laminate damage. Gelcoat often cracks around the fin or skeg because it is more brittle than the underlying laminate. Repair the gelcoat cracks to protect the laminate from water intrusion.

3 Hull damage around a fin almost always signals inadequate laminate strength. After you grind open and repair all damaged laminate, add additional layers of fiberglass to strengthen the hull in the affected area.

DAMAGED RUDDER

In order to neither float nor sink when the boat heels, a sailboat rudder needs to have nearly neutral buoyancy. That means fiberglass rudders are usually foam filled or hollow. Unfortunately, too many are also so lightly built that the slightest brush with anything more solid than a jellyfish breaks the skin. Weeping rudders are even more common than weeping keels, and water inside your rudder is almost certainly doing damage.

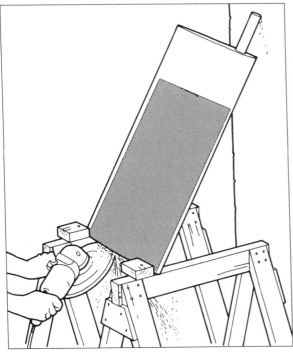

1 If you have a weep anywhere on your rudder, the entire rudder may be full of water, but it will only drain out to the level of the crack. It is almost always a good idea to grind a hole in the lowest part of the skin to allow complete drainage.

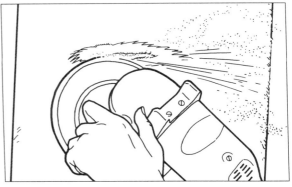

2 Grind the damaged area open to expose enough of the core to determine what it is and whether it is saturated.

3 If the core is saturated, you should have some success in drying it by removing the rudder and drilling a pattern of $^3/_{16}$-inch holes all the way through it on about 6-inch centers. Put the rudder on sawhorses in an enclosed area and place a small electric heater beneath it. After a few days drill a couple of test holes to see if all the core is drying. If closer spaced holes seem necessary, removing one of the skins may be easier.

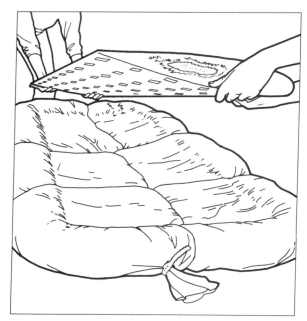

4 When the core is dry, tape the holes closed on one side. "Nestle" that side of the rudder into sand-filled garbage bags, then put it back on the horses, supported as close to the ends as possible to keep from compressing the skin against the core.

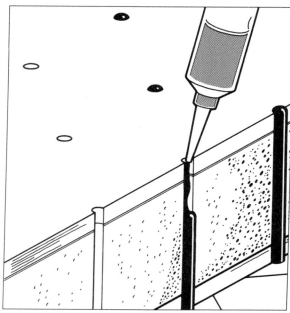

5 Mix epoxy to catsup consistency with colloidal silica and inject each hole until it refuses to accept more or until epoxy appears in the adjacent holes. Fill every hole, working quickly to fill all the voids before the epoxy begins to set. A slow hardener is helpful.

6 Position the rudder in the sandbag "nest" and weight the top surface with sandbags to compress the skins against the core.

7 When the epoxy has fully cured, regrind the damaged area and the drain hole and repair both with lay-ups of fiberglass cloth and epoxy.

BLADE/SHAFT MOVEMENT

Resin doesn't attach to metal very well, so to keep a rudder from spinning on the stock, it is usually built around a metal framework welded to the stock. Done well, this provides a strong and trouble-free assembly, but too often the framework is little more than two or three metal straps or rods spot-welded to the stock. When (not if) these welds break, the rudder swings freely.

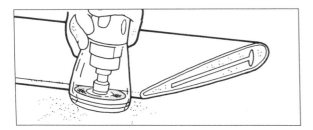

1 Fiberglass rudders are almost always built in two halves and glued together over the stock assembly and the core. Split the seam with a circular saw or rotary tool (dremel) and separate the two halves. Making the cut to one side will make reassembly easier.

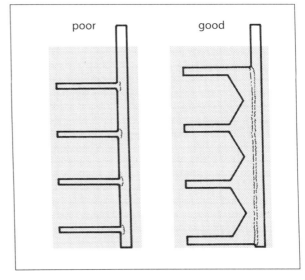

poor   good

2 Remove the broken framework and have a new one constructed that attaches to the stock for almost the full length of the rudder.

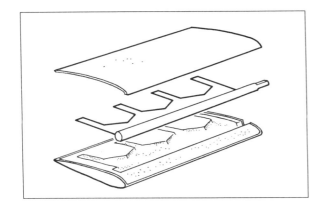

3 Relieve the foam core to fit the new framework and join the two halves with epoxy thickened to mayonnaise consistency with colloidal silica. Clamp or weight the halves together until the epoxy cures fully, then grind a bevel on the cut line and laminate the two halves together.

Grounding an external lead keel is less likely to damage the hull, but the keel may be deformed by the impact.

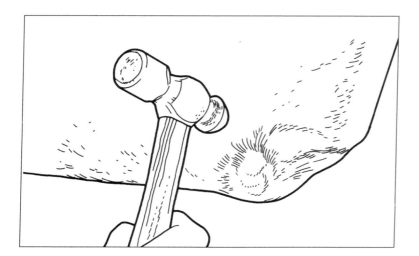

1 Use the round end of a ball-peen hammer to reshape the lead. You are trying to "push" the displaced lead back in shape; be careful not to shear the bulge.

2 Remove high spots with a body plane or a sharp block plane. Lubricate the lead with petroleum jelly before you plane it.

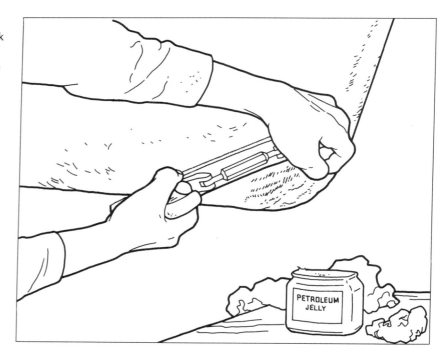

PETROLEUM JELLY

3 Fill low spots with epoxy and microballons. First remove the petroleum jelly and any other contaminants with acetone, and wire brush the lead. Wet the lead with epoxy, then wire brush it again to expose fresh lead to the epoxy without exposing it to the air. Thicken your epoxy to peanut butter consistency with microballoons (there is no blister risk here) and flush-fill the depressions.

4 Sand the entire area after the filler cures and coat it with unthickened epoxy, sanding the wet surface to expose fresh lead to the epoxy. Give the repair two additional coats of epoxy, then sand and paint.

INDEX

Acrylic vs. polycarbonate, 22
Additives, epoxy, 74
Air drying, wet core, 82
Alligatoring, 47–50

Backing plates, 43
Bedding
    deadlights, 18–20
    deck hardware, 10–13
    hull-to-deck joints, 25–26
    portholes/portlights, 18–20
Blade/shaft movement, 123
Blisters, 98–103
    boat pox, 101–103
    deep, 103
    minor, 99–100
Boat pox, 101–103
Boatlife, LifeSeal, 10
Buffing, 32–33
Bung replacement, 59–60

Catalysts, for polyester resin, 67
Catalyzing gelcoat resin, 37
Caulking teak decks, 57–58
Centerboard pivot problems, 119
Centerboard trunks, 27
Chainplates, sealing, 15–16
Chopped-strand mat (CSM), 67
Cleaners, two-part, 56
Cleaning teak decks, 55
Cloth, fiberglass, 68
Colloidal silica, 74
Colors, matching, 36
Compounds and compounding, 33
Core, 79–91
    damaged, 85–86

delamination, 112
    inner skin removal, 83
    outer skin removal, 84
    skin reinstallation, 87–88
    skin stiffening, 90–91
    strengthening, 89
    wet, 82–84, 113
Cored decks, 79
    preparing for deck hardware, 14
Cored hulls, 79, 112–113
    repairs with inside access, 112
    repairs without inside access, 112
Cracks, repairing, 44
Crazing, 47–50
    localized, 47–48
    widespread, 49–50
CSM (chopped-strand mat), 67

Deadlights
    frames, 23
    rebedding, 18–20
Deck hardware
    bedding, 10–13
    preparing for, 14
    rebedding, 14
Decks, 41–63
    caulking failure, 58
    crazing (alligatoring), 47–50
    leaks, 5
    repairing cracks, 44
    stress cracks, 42–44
    voids, 45–46
Decks, cored, 79
    preparing for hardware, 14
Decks, fiberglass, 4–5
Decks, nonskid, renewing, 51–54

Decks, teak, 55–63
    bung replacement, 59–60
    cleaning, 55
    lightening, 55
    plank replacement, 61–63
    recaulking, 57–58
    surfacing, 57
Delamination, 80–81. See also Laminate
    core, 112
    solid laminate, 81
Dry air, to dry wet core, 82
Drying wet core
    large area, 83–84
    small area, 82

Ear plugs, 35
Epoxy
    additives, 74
    for gouges, 97
    for laminate repair, 74–77
    laminating with, 76–77
    metering pumps, 75
    mixing, 75
    precautions, 76
    regulating cure time, 75
    selecting, 74
    thickening, 75
    when to use, 74–77, 97
External ballast, reshaping, 124–125

Fiberglass, 4–5
Fiberglass layup
    additional layers, 72
    basics, 70–73
    finishing, 73
    for hull-to-deck joints, 27

initial layer, 71
    preparation for, 70
    working overhead, 73
Fiberglass material, 67–68
    chopped-strand mat, 67
    cloth, 68
    roving, 68
Fiberglassing. See Fiberglass layup
Fibers, epoxy additive, 74
Finishing resin vs. laminating, 66
Fins and skegs, hull damage around, 120
Fittings, through-hull, 27–28
Flushing wet core, 82
Frames, portlight, 23
Fuel tanks, integral, 118

Gelcoat, 31
    buffing, 32–33
    choices, 36
    hardening, 37
    polishing, 38–39
    restoring, 31–39
    sanding, 34–35, 38–39
    for scratch repair, 36
    spreading, 37
Gelcoat resin, catalyzing, 37
Gloss, restoring, 31–39
Gouges, 94–97
    deep, 96–97
    shallow, 94–95
Grinding, 69
Grit, encapsulated, 51–52

Hard spots, 42, 43
Hardware, deck
    bedding, 10–13
    preparing cored deck for, 14
    rebedding, 10–13
Hat-shaped stiffeners, 90
Headliners, 11
Heat, to dry wet core, 82
Hulls, 93–113
    blisters, 98–103
    cored, 79, 112–113
    damage around fins and skegs, 120
    gouges, 94–97
    impact damage, 104–113
    outside repairs, 109–110
    repairs from inside, 105–107
    repairs with no inside access, 111
Hulls, fiberglass, 4–5
Hull-to-deck joints, 24–26
    evaluating, 25
    fiberglassing, 27
    rebedding, 25–26

Impact damage, 104–113
Installing skin, 87–88
Isophthalic resin, 67

Joints
    hull-to-deck, 24–26
    through-hull fittings, 27–28

Keel/centerboard pivot problems, 119
Keels, 115
    reshaping, 124–125
    weeping, 116–118

Laminate, 65–77
    cutting away, 104–105
    delamination, 80–81, 112
    grinding, 69
    polyester resin, 66–68
Laminate, solid, delamination in, 81
Laminating
    with epoxy, 76–77
    laminating resin, 66
Layup, fiberglass
    additional layers, 72
    basics, 70–73
    finishing, 73
    for hull-to-deck joints, 27
    initial layer, 71
    preparation for, 70
    working overhead, 73
Lead keels, reshaping, 124–125
Leaks, 5, 7–29
    centerboard trunks, 27
    chainplates, 15–16
    deadlights, 18–20
    deck hardware, 14
    hull-to-deck joints, 24–26
    hybrid sealants for, 10
    mast boots, 24
    portholes, 17
    portlights, 21–22
    pressurizing to find, 29–30
    sealants for, 8–9
    through-hull fittings, 27–28
LifeSeal (Boatlife), 10
Lightening teak decks, 55

Mast boots, 24
Materials
    exotic, 68
    fiberglass, 67–68
MEKP (methyl ethyl ketone peroxide), 37, 67
Metering pumps, 75
Methyl ethyl ketone peroxide (MEKP), 37, 67

Microballoons, 74
Mold, taking off, 108

Nonskid, renewing, 51–54

Orthophthalic resin, 67
Overlay, rubberized, 53–54

Paint, matching colors, 36
Palm sanders, 35
Paste, gelcoat, 37
Plank replacement, teak, 61–63
Plastics, acrylic vs polycarbonate, 22
Polishing gelcoat repairs, 38–39
Polycarbonate vs. acrylic, 22
Polyester resin, 66–68
    catalyst for, 67
    isophthalic, 67
    orthophthalic, 67
    vinylester, 67
Polysulfide, 8, 9
Polyurethane, 8, 9
Portholes/portlights
    acrylic vs. polycarbonate, 22
    frames, 23
    rebedding, 18–20
    replacing, 21–22
    sealing, 17
Pox, boat, 101–103
Precautions for epoxy, 76
Pressurizing to find leaks, 29–30

Rebedding
    deadlights, 18–20
    deck hardware, 10–13, 14
    hull-to-deck joints, 25–26
    portholes/portlights, 18–20
Recaulking teak decks, 57–58
Reinstalling skin, 87–88
Reshaping external ballast, 124–125
Resins
    gelcoat, 31
    isophthalic, 67
    orthophthalic, 67
    polyester, 66–68
    vinylester, 67
Roving, 68
Rubberized overlay, 53–54
    applying, 54
    cutting, 54
    cutting patterns, 53
    preparing surface for, 53
Rubbing compounds, 33
Rudders, 115
    blade/shaft movement, 123
    damaged, 121–122

Sanders, 34
Sanding, 34–35
    gelcoat repairs, 38–39
    wet, 35
Scratch repairs, 36–39
    covering, 38
    gelcoat choices, 36
    to white boats, 36
Sealants, 8–9
    hybrid, 10
    polysulfide, 8, 9
    polyurethane, 8, 9
    silicone, 8–9
Shaping external ballast, 124–125
Silica, colloidal, 74
Silicone, 8–9
Skegs, hull damage around, 120
Skin
    reinstalling, 87–88
    stiffening, 90–91
Solid laminate, delamination in, 81
Stiffeners, hat-shaped, 90
Stiffening, 43
Strengthening, 89
Stress cracks, 42–44
    recognizing, 42
Surfacing teak decks, 57

Tanks, integral fuel, 118
Teak, plank replacement, 61–63
Teak cleaners, two-part, 56
Teak decks, 55–63
    bung replacement, 59–60
    caulking failure, 58
    cleaning, 55
    lightening, 55
    plank replacement, 61–63
    recaulking, 57–58
    surfacing, 57
Through-hull fittings, 27–28

Vacuuming wet core, 82
Vinylester, 67
Voids, 45–46

Weeping keel, 116–118
Wet core, 82–84, 113
    large area drying, 83–84
    small area drying, 82
Wet sanding, 35
White boats, scratch repairs to, 36
Woven roving, 68

International Marine/
Ragged Mountain Press

A Division of The McGraw·Hill Companies

Published by International Marine

10   9   8   7   6   5   4   3   2

Library of Congress Cataloging-in-Publication Data
Casey, Don.
      Sailboat hull and deck repair / Don Casey.
             p. cm. — (The International Marine sailboat library)
Includes index.
ISBN 0-07-013369-7
1. Sailboats—Maintenance and Repair—Amateurs' manuals. 2. Hulls (Naval architecture)—
Maintenance and repair—Amateurs' manuals. 3. Decks (Naval architecture)—Maintenance and
repair—Amateurs' manuals. I. Title. II. Series
VM351.C329 1995   95-31110
623.8'458'0288—dc20   CIP

Questions regarding the content of this book should be addressed to:

International Marine
P.O. Box 220
Camden, ME 04843

Questions regarding the ordering of this book should be addressed to:

The McGraw-Hill Companies
Customer Service Department
P.O. Box 547
Blacklick, OH 43004
Retail customers: 1-800-262-4729
Bookstores: 1-800-722-4726

Sailboat Hull and Deck Repair is printed on 60-pound Renew Opaque Vellum, an acid-free paper
that contains 50 percent recycled waste paper (preconsumer) and 10 percent postconsumer
waste paper.

Illustrations in Chapters 1, 2, 6, and 7 by Jim Sollers
Illustrations in Chapters 3, 4, and 5 by Rob Groves
Printed by R.R. Donnelley, Crawfordsville, IN
Design by Ann Aspell
Production by Molly Mulhern, Kate Mueller, Ann Aspell, and Mary Ann Hensel

Front cover photo by William Thuss
Polisher used on front cover courtesy Rockland Welding, Rockland, Maine

DON CASEY credits the around-the-world-voyage of Robin Lee Graham, featured in National Geographic in the late sixties, with opening his eyes to the world beyond the shoreline. After graduating from the University of Texas he moved to south Florida, where he began to spend virtually all his leisure time messing about in boats.

In 1983 he abandoned a career in banking to devote more time to cruising and writing. His work combining these two passions soon began to appear in many popular sailing and boating magazines. In 1986 he co-authored *Sensible Cruising: The Thoreau Approach*, an immediate best-seller and the book responsible for pushing many would-be cruisers over the horizon. He is also author of *This Old Boat,* a universally praised guide that has led thousands of boatowners through the process of turning a rundown production boat into a first-class yacht, and of *Sailboat Refinishing*, part of the International Marine Sailboat Library. He continues to evaluate old and new products and methods, often trying them on his own 25-year-old, much-modified, Allied Seawind.

When not writing or off cruising, he can be found sailing on Florida's Biscayne Bay.

THE INTERNATIONAL MARINE SAILBOAT LIBRARY

Sailboat Hull and Deck Repair has company. Check your local bookstore
or call us to order the following:

Sailboat Refinishing
by Don Casey
Hardcover, 144 pages, 350 illustrations, $21.95. Item No. 013225-9

Canvaswork and Sail Repair
by Don Casey
Hardcover, 144 pages, 350 illustrations, $21.95. Item No. 013391-3

The Sailor's Assistant: Reference Data for Maintenance, Repair, and Cruising
by John Vigor
Hardcover, 176 pages, 350 illustrations, $21.95. Item No. 067476-0

Inspecting the Aging Sailboat
by Don Casey
Hardcover, 144 pages, 350 illustrations, $21.95. Item No. 013394-8

Subjects to be covered in future volumes, all with *Sailboat Hull and Deck Repair*'s
step-by-step illustrated approach, include:

➤ *Sailboat Electrical Systems and Wiring*
➤ *Sailboat Diesel Engines*
➤ *Cruising Sailboat Design, Strength, and Performance*
and others